Solid State
Microbatteries

NATO ASI Series

Advanced Science Institutes Series

A series presenting the results of activities sponsored by the NATO Science Committee, which aims at the dissemination of advanced scientific and technological knowledge, with a view to strengthening links between scientific communities.

The series is published by an international board of publishers in conjunction with the NATO Scientific Affairs Division

A	**Life Sciences**	Plenum Publishing Corporation
B	**Physics**	New York and London
C	**Mathematical**	Kluwer Academic Publishers
	and Physical Sciences	Dordrecht, Boston, and London
D	**Behavioral and Social Sciences**	
E	**Applied Sciences**	
F	**Computer and Systems Sciences**	Springer-Verlag
G	**Ecological Sciences**	Berlin, Heidelberg, New York, London,
H	**Cell Biology**	Paris, and Tokyo

Recent Volumes in this Series

Series B: Physics

Solid State Microbatteries

Edited by

James R. Akridge

Eveready Battery Co., Inc.
Westlake, Ohio

and

Minko Balkanski

Université Pierre et Marie Curie
Paris, France

Plenum Press
New York and London
Published in cooperation with NATO Scientific Affairs Division

Proceedings of the NATO Advanced Study Institute and
International School of Materials Science and Technology
Fifteenth Course on
Solid State Microbatteries,
held July 3-15, 1988,
in Erice, Trapani, Sicily, Italy

Library of Congress Cataloging-in-Publication Data

International School of Materials Science and Technology (1988 :
   Erice, Italy)
    Solid state microbatteries / edited by James R. Akridge and Minko
 Balkanski.
       p.   cm. -- (NATO ASI series. Series B, Physics ; vol. 217)
    "Published in cooperation with NATO Scientific Affairs Division."
    "Proceedings of the NATO Advanced Study Institute and
 International School of Materials Science and Technology fifteenth
 course on solid state microbatteries, held July 3-15, 1988, in
 Erice, Trapani, Sicily, Italy"--T.p. verso.
    Includes bibliographical references.
    ISBN 0-306-43505-5
    1. Solid state batteries--Congresses. 2. Microelectronics-
 -Congresses. I. Akridge, J. R. II. Balkanski, Minko, 1927-
 III. North Atlantic Treaty Organization. Scientific Affairs
 Division. IV. Title. V. Series: NATO ASI series. Series B,
 Physics ; v. 217.
 TK2942.I57 1988
 621.31'242--dc20   90-6769
                                                                  CIP

© 1990 Plenum Press, New York
A Division of Plenum Publishing Corporation
233 Spring Street, New York, N.Y. 10013

Printed in the United States of America

SPECIAL PROGRAM ON CONDENSED SYSTEMS OF LOW DIMENSIONALITY

This book contains the proceedings of a NATO Advanced Research Workshop held within the program of activities of the NATO Special Program on Condensed Systems of Low Dimensionality, running from 1983 to 1988 as part of the activities of the NATO Science Committee.

Other books previously published as a result of the activities of the Special Program are:

SPECIAL PROGRAM ON CONDENSED SYSTEMS OF LOW DIMENSIONALITY

PREFACE

This Advanced Study Institute on the topic of **SOLID STATE MICROBATTERIES** is the third and final institute on the general theme of a field of study now termed "SOLID STATE IONICS". The institute was held in Erice, Sicily, Italy, 3 - 15 July 1988. The objective was to assemble in one location individuals from industry and academia expert in the fields of microelectronics and solid state ionics to determine the feasibility of merging a solid state microbattery with microelectronic memory. Solid electrolytes are in principle amenable to vapor deposition, RF or DC sputtering, and other techniques used to fabricate microelectronic components. A solid state microbattery mated on the same chip carrier as the chip can provide 'on board' memory backup power. A solid state microbattery assembled from properly selected anode/solid electrolyte/cathode materials could have environmental endurance properties equal or superior to semiconductor memory chips.

Lectures covering microelectronics, present state-of-art solid state batteries, new solid electrolyte cathode materials, theoretical and practical techniques for fabrication of new solid electrolytes, and analytical techniques for study of solid electrolytes were covered. Several areas where effort is required for further understanding of materials in pure form and their interactions with other materials at interfacial contact points were identified. Cathode materials for solid state batteries is one particular research area which requires attention. Another is a microscopic model of conduction in vitreous solid electrolytes to enhance the thermodynamic macroscopic Weak Electrolyte Theory (WET).

NATO funding was augmented by:

United States National Science Foundation
United States Army Research, Standardization, and Development
Group

The success of solid stated batteries does not rest on the
feasibility of fabrication of batteries--that has been clearly
demonstrated. The success or failure will depend upon
commercialization of the technology. As of this date a good
application for solid state batteries (micro- or macro-) has not been
discovered. International conferences on solid state ionics should
include as part of the program application areas for the technology.
The Solid State Microbatteries institute served to address practical
applications of solid state technology. More attention should be
given to application areas for solid electrolytes. The applications
will serve as a driving force for technical improvements and
fundamental understanding.

The editors are indebted to Cheryl L. Yanico who coordinated the
retyping of most of the manuscripts, redrafting of 55 figures,
correction of spelling and english syntax, and cutting, pasting, and
assembly of the manuscript into final camera ready format, Steven D.
Jones, Susan G. Humphrey, Roman E. Cehelsky for the book index and
Helen M. Friend who also corrected portions of the text.

James R. Akridge

Minko Balkanski

CONTENTS

OVERVIEW OF AQUEOUS AND NONAQUEOUS BATTERIES AND SEMICONDUCTORS

Forrest A. Trumbore

AT&T Bell Laboratories
600 Mountain Avenue
Murray Hill, New Jersey 07974

INTRODUCTION

The purpose of this "overview" of batteries and semiconductors is to convey the flavor of past accomplishments and future challenges in both fields, with hopefully some relevance to microbatteries and their potential applications. We consider first recent developments in the century-old lead-acid technology in the form of a round cell for standby power. More energetic, nonaqueous systems such as the high temperature sodium-sulfur and lithium-sulfur and the ambient temperature lithium rechargeable systems must yet prove themselves as viable commercial products. We consider here the rechargeable lithium-niobium triselenide system to illustrate some of the future challenges in the battery field. VLSI silicon technology is the obvious choice to demonstrate recent progress in semiconductors. We conclude with a discussion of heteroepitaxy and some emerging semiconductor devices and technologies.

BATTERIES

A battery consists of many or all of the following components: anode(s), cathode(s), solvent(s), electrolyte(s), separators(s), current collectors(s), conducting diluents, binders, contacts (welds), seals/feedthroughs, container(s) (can, jar, paper, other encapsulant), labels (the text very important in a litigious society!), insulators, vents, fuses, diodes and possibly electronic circuits involving microprocessors. Compatibility among these diverse components is key to the success of a given battery. Ideally, we want just the cell reaction to occur, and that only on demand. Actually, this goal is not met in the rechargeable (secondary) batteries in widest use; witness the high self-discharge rates of lead-acid and nickel-cadmium batteries. (A friend complained of being unable to start her new Jaguar after a 6 weeks holiday in Hawaii, a perfect example of self-discharge, even among the affluent!) Another key to success is often pure luck, e.g., in the form of passivating films or high overvoltages which allow some batteries to function, even though thermodynamically unstable.

Compatibility and luck may suffice for a primary battery, but the development and manufacture of a new secondary battery can be an awesome scientific and technological challenge. During the tens or even tens of thousands of charge-discharge cycles, the electrodes may change shape, unwanted reaction products may form, shedding of active material is common, the electrolyte solution may undergo decomposition, gases may be generated, separators may be breached with resulting internal short circuits, corrosion may occur, etc. We shall see examples of all of these below.

Solid State Microbatteries
Edited by J. R. Akridge and M. Balkanski
Plenum Press, New York, 1990

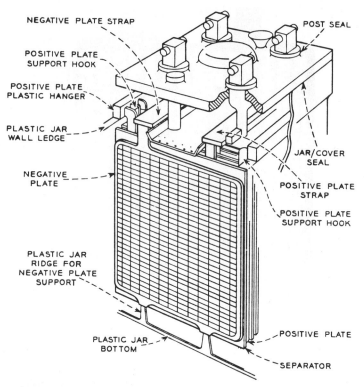

NEGATIVE PLATE STRAP

POSITIVE PLATE
SUPPORT HOOK

POSITIVE PLATE
PLASTIC HANGER

PLASTIC JAR
WALL LEDGE

NEGATIVE
PLATE

PLASTIC JAR
RIDGE FOR
NEGATIVE PLATE
SUPPORT

PLASTIC JAR
BOTTOM

POST SEAL

JAR/COVER
SEAL

POSITIVE PLATE
STRAP

POSITIVE PLATE
SUPPORT HOOK

POSITIVE PLATE

SEPARATOR

Fig. 1   Diagram of a rectangular lead-acid battery (after Ref. 5, reprinted by special permission. Copyright © 1970 AT&T).

Rounding of the lead–acid battery

The modern version of the lead-acid battery dates from 1881 [1] with the introduction of pastes of lead oxides, sulfuric acid and water supported on conducting lead alloy grids, which also serve as current collectors (Fig. 1). The pastes are converted into the positive and negative electrode active materials PbO_2 and "spongy" lead, respectively. The voltage of the lead-acid battery can be calculated from Table 1, which lists the reduction potentials of a number of half-cell reactions. The open circuit voltage (OCV) of an electrochemical cell is essentially

$$V_{cell} = V_+ - V_- \qquad (1)$$

TABLE 1 – Some Standard Reduction Potentials

Half Cell reaction	E° (volts)
(a) $Li^+ + e^- = Li$	−3.04
(b) $3LI^+ + NbSe_3 + 3e^- = Li_3NbSe_3$	−0.5 to −2.0
(c) $Cd(OH)_2 + 2e^- = Cd + 2OH^-$	−0.81
(d) $PbSO_4 + 2e^- = Pb + SO_4^=$	−0.36
(e) $2H^+ + 2e^- = H_2$	0
(f) $O_2 + 2H_2O + 4e^- = 4OH^-$	0.40
(g) $NiO(OH) + H_2O + e^- = Ni(OH)_2 + OH^-$	0.49
(h) $2H_2O + 2e^- = H_2 + 2OH^-$	0.81
(i) $O_2 + 4H^+ + 4e^- = 2H_2O$	1.23
(j) $PbO_2 + SO_4^= + 4H^+ + 2e^- = PbSO_4 + 2H_2O$	1.69

2

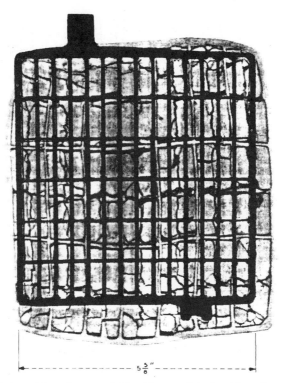

Fig. 2   Corroded positive plate overlaid with an original grid to demonstrate positive plate growth (after Refs. 5 and 6, reprinted by special permission. Copyright © 1970 AT&T).

where the positive (cathode) and negative (anode) electrodes correspond to the half cell reactions having the more positive and less positive reduction potentials, respectively. For a lead-acid battery the discharge reactions are

$$PbO_2 + 4H^+ + SO_4^= + 2e^- = PbSO_4 + 2H_2O \qquad \text{(positive)} \qquad (2)$$

$$Pb + SO_4^= = PbSO_4 + 2e^- \qquad \text{(negative)} \qquad (3)$$

From Table 1 and eqn. (1), it can be seen that the OCV is slightly over 2 volts. Table 1 also reveals what should be a fatal problem for the 2 volt lead-acid system, namely the decomposition of water into hydrogen and oxygen at only 1.23 V (from half cells (e) and (i) in Table 1). Fortuitously, high overvoltages for hydrogen and oxygen evolution slow the water decomposition kinetics sufficiently to allow the lead-acid battery to function quite well indeed.

As a standby power source in a telephone central office, the lead-acid battery has to be kept fully charged at all times. In this "float" mode, there is a continuous trickle charging current and, in addition to the reverse (charging) reactions of equations (2) and (3), the following reactions must be considered.

$$Pb + 2H_2O = PbO_2 + 4H^+ + 4e^- \qquad \text{(positive)} \qquad (4)$$

$$2H_2O = O_2 + 4H^+ + 4e^- \qquad \text{(positive)} \qquad (5)$$

$$2H^+ + 2e^- = H_2 \qquad \text{(negative)} \qquad (6)$$

Reactions (5) and (6) correspond to a loss of water if the oxygen and hydrogen do not recombine. However, corrosion of the positive grid via reaction (4) is the villain in our story.

3

Prior to 1950, the batteries in the old Bell Telephone System contained grids of lead alloyed with 3 - 12% antimony, which increases the tensile strength of lead and prevents sagging of the grids. However, antimony leaches out of the positive plate, travels through the sulfuric acid and redeposits on the negative plate [2, 3]. This results in depolarization sites for hydrogen evolution and for local self-discharge leading to irreversible sulfation and failure of the negative electrode. In the late 1940s, it was found [4] that 0.05-0.10% calcium hardens lead and is nondepolarizing. The reduced hydrogen evolution and water loss with Pb-Ca grids resulted in today's maintenance-free car batteries. Batteries with Pb-Ca grids were introduced into the Bell System beginning in 1950. Concurrently, the fragile glass or ceramic jar and cover materials were being replaced by injection moldable polystyrene. With lower float voltages, less frequent water additions and an expected 25 year life, over a million Pb-Ca cells were in service by 1970.

However, trouble with the Pb-Ca batteries emerged in the early 1960s. Serious fires in telephone battery plants were traced to positive grid growth (Fig. 2) via reaction (4), a reaction accommpainied by a 40% increase in volume. The grid growth caused stresses on the battery jars and on the jar-cover and post-cover seals, resulting in sulfuric acid leakage to grounded metal battery racks. Electrical arcing caused the explosion of the H_2-O_2 mixture in the battery and/or direct ignition of the flammable polystyrene jar and cover materials. Positive grid growth also led to a loss of effective contact between grid and active material with degradation of performance. Numbers of batteries were fading to 50% capacity in only 7 years, not the expected 25 years to 90% capacity.

In 1964, a group was formed at Bell Labs to design a new, reliable battery with a life of more than 30 years. In 1970, an issue of the Bell System Technical Journal was devoted to the result, a new "round cell" (Fig. 3). (This cell, originally marketed as the Bellcell, is now the AT&T Lineage 2000™ Round Cell.) The titles of the papers [5-15] indicate the scope of this effort. Special features of the round cell design included (a) pure lead grids, (b) a conical grid design, (c) a self-supporting cylindrical structure, (d) a tetrabasic lead sulfate positive paste, (e) flexible post seals and (f) a heat bonded polyvinyl chloride (PVC) jar-cover seal.

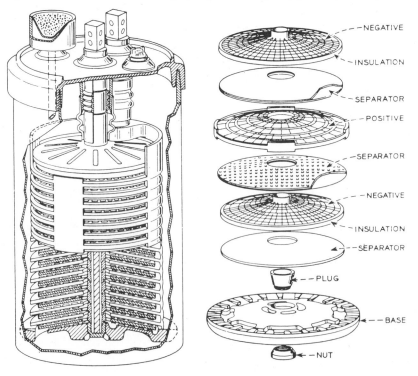

NEGATIVE
INSULATION
SEPARATOR
POSITIVE
SEPARATOR
NEGATIVE
INSULATION
SEPARATOR
PLUG
BASE
NUT

Fig. 3 Diagram of the lead-acid round cell (after Refs. 5 and 6, reprinted by special permission. Copyright © 1970 AT&T).

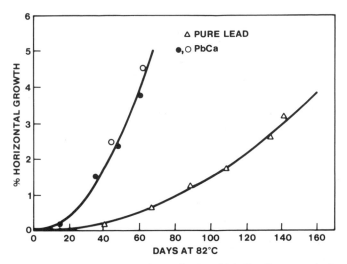

Fig. 4   Plot of the horizontal growth of pure lead and Pb-Ca alloy pasted plates as a function of time at 82°C (after Ref. 6, reprinted by special permission. Copyright © 1970 AT&T).

Why pure lead in place of alloy grids? The answer is the much lower corrosion rate of pure lead versus that of Pb-Ca (Fig. 4). How then to utilize soft pure lead in a grid? Taking a cue from Fig. 2, showing a rounding of the corroding rectangular grid, a cylindrical cell was proposed and, after the testing of a number of designs, conical grids stacked in a self-supporting structure were employed. The corrosion-induced grid growth as a function of time is given by

$$\text{growth} = k_1 k_2 t^2 \tag{7}$$

where k_1 is the rate constant for growth and k_2 a geometric factor equal to the ratio of the surface area to the cross sectional area of a grid member. By varying the grid member size with distance from the center to maintain a constant value of k_2, a "balanced" grid (Fig. 5a) was achieved such that the corrosion (growth) occurs in a cooperative, phased manner (Fig. 5b). This ensures

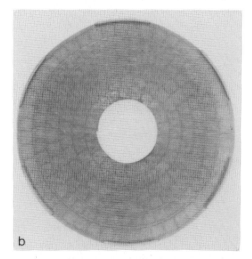

Fig. 5   (a) A positive grid for the present lead-acid round cell. (b) A positive plate after 114 days at 93°C (courtesy of A. G. Cannone).

5

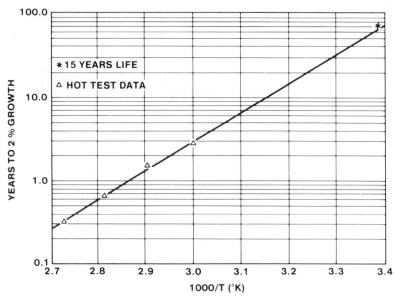

Fig. 6   Arrhenius plot of the temperature dependence of the time for 2% grid growth, taken as the point of failure of the battery. The 15-year point is a linear extrapolation of measured positive plate growth for round cells in the field for 15 years (courtesy of A. G. Cannone, R. V. Biagetti and R. E. Landwehrle).

effective grid-paste contact for the life of the battery. Indeed, the round cell capacity increases on aging due to the extra PbO_2 formed by grid corrosion. Data on grid growth at different temperatures were used to construct an Arrhenius plot (Fig. 6) which predicts a life of about 60 years at 25°C. An extrapolation of actual plate growth in round cells in the field for 15 years agrees very well with this prediction.

Another innovation was the use of tetrabasic lead sulfate ($4PbO \cdot PbSO_4$) in the positive paste. This compound, with a rodlike morphology, promotes an interlocking PbO_2 structure which minimizes any shedding or loss of grid-paste contact caused by the 92% volume change in the discharge reaction of equation (2).

Other new features included a high speed lead joining interconnection technique providing flaw-free positive plate multiple bonds and in situ casting of a lead-antimony connection rod providing mechanical and electrical interconnection of the negative plates. The flammable polystyrene jar material was replaced by a flame retardant PVC which provided improved impact and craze resistance and low cost fabrication. An infrared heat sealing technique (Fig. 7a) utilizing infrared radiation for localized melting and bonding resulted in a jar-cover seal (Fig. 7b) capable of supporting over 100 times the weight of the battery.

A new post-cover seal (Fig. 8a) included a rigid epoxy sheath to prevent leakage around the lead post, which in turn was flexibly coupled to the cover to allow stress-free movement of the cell element within the jar on handling or in use. The seal length was several times longer than in earlier seals to further discourage leakage. Another culprit in the battery plant fires, the metal battery rack, was replaced with a new plastic model.

With an expected life exceeding 35 years, the round cell was introduced into the Bell System. But corrosion was to rear its ugly head once more! In 1978, there were two telephone central office outages traced to electrical open circuits caused by severe positive post, not grid corrosion in round cells. Improper formulation of the epoxy in the post-cover seals was correlated with the corrosion and remedial actions were taken. However, even with proper formulation and further optimization of the epoxy mix, the post seals were still susceptible to corrosion, found to be related to the immersion of the expoxy in sulfuric acid. To remove the epoxy from the acid, a shorter post seal was designed (Fig. 8b) and the acid level in the round cell was lowered, at the

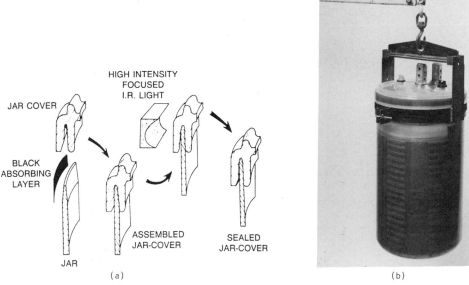

(a)　　　　　　　　　　　(b)

Fig. 7　　(a) Infrared heat sealing technique for the round cell jar-cover seal (after Ref. 12, reprinted by special permission. Copyright © 1970 AT&T). (b) Routine lifting of a round cell weighing over 150 kg by clamping under the cover, showing the strength of the jar-cover seal (courtesy of R. V. Biagetti).

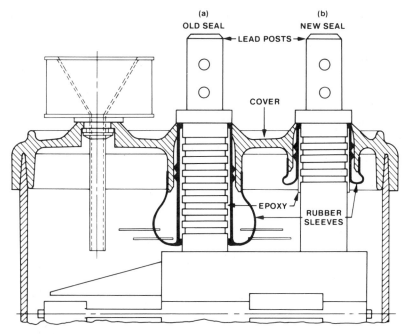

Fig. 8　　(a) Original round cell post-cover seal and (b) modified post-cover seal. The length is shortened and the rubber insert opened up to allow visual inspection (courtesy of R. V. Biagetti).

Fig. 9   (a) Oxide layer formed on a Pb grid taken from a cell which was stored without being floated. (b) Crystals of $PbSO_4$ formed on self-discharge from a similar cell.

cost of a slightly lower capacity (e.g., 1680 to 1600 Ah). Now, any contact of acid with epoxy is via creep of only a thin film of acid, greatly reducing the corrosion current. The life estimate for the new post seal design is 80 years and round cells with this design were first shipped in 1980.

A less traumatic problem involved the failure of some customers to float the round cells after delivery and storing them for prolonged periods on open circuit in warehouses in Arizona, where temperatures may be as high as 50°C! The result was again grid corrosion, flaking of active material and self-discharge. Examples of the lead oxide corrosion layer and of $PbSO_4$ crystals formed on self-discharge are shown in Fig. 9. The remedy here was simple, float as specified.

There was, however, an electrochemical problem even when the round cell was floated. In deciding on the float voltage to be specified for the battery plant, it was known that some 70-90 mV overpotential on the positive plate was required to prevent self-discharge due to local cell reactions. These result from the presence of Pb, PbO_2 and sulfuric acid in/on/at the charged positive plate and the high electrical conductivity of PbO_2 acting as a load through which these local lead-acid cells can discharge. A smaller overpotential on the negative plate was required to suppress corrosion and self-discharge. Too high a float voltage is undesirable due to increased power consumption, increased gas evolution, etc. Accordingly, to allow for variations in the round cells and still keep the float voltage low, a value of 110 mV was chosen as the standard overpotential. However, cells in the field were not delivering the expected capacities and actually were discharging under the recommended float conditions! Investigation of the polarization behavior revealed the situation shown in Fig. 10, which shows most of the 110 mV polarization on the negative, not the positive electrode. Hence, the self-discharge on float. Adding a very small amount of platinum, as chloroplatinic acid, shifts the polarization curve on the negative markedly to the left in Fig. 10, the bulk of the 110 mV now resides on the positive plate and the problem is solved. This is just one factor affecting float behavior [16].

Finally, keeping fingers crossed, a happy ending. Today, there are over 600,000 round cells in service, battery life greater than the design objective of 30 years is expected and not one jar-cover or post-cover seal has leaked! Today, some 40 years after the troubles with the Pb-Sb batteries, work still goes on to improve the lead-acid standby battery.

Nickel–cadmium and nickel–hydrogen batteries

The alkaline nickel-cadmium rechargeable battery is mentioned here not only in its own right but to show its influence on the course of development of other secondary batteries. The electrode reactions may be written as

$$2NiO(OH) + 2H_2O + 2e^- = 2Ni(OH)_2 + 2OH^- \qquad \text{(positive)} \qquad (8)$$

$$Cd + 2OH^- = Cd(OH)_2 + 2e^- \qquad \text{(negative)} \qquad (9)$$

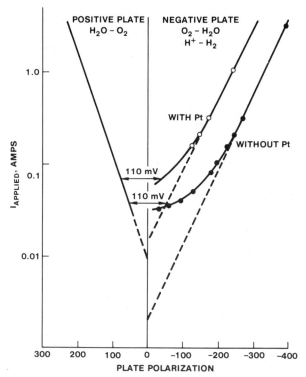

Fig. 10   Polarization behavior of floated round cell showing that, without addition of platinum, most of the 110 mV overvoltage was on the negative plate. Adding platinum resulted in the shift of most of the overvoltage to the positive plate (after Ref. 16, courtesy of M. E. Fiorino, F. J. Vaccaro and R. E. Landwehrle).

The 1.3 volt Ni-Cd battery is generally used in smaller sizes due to the higher cost of the active materials compared to lead. However, the weight advantage of Ni-Cd over lead-acid is significant for portable and space applications, e.g., in telecommunications satellites. Recently, Sanyo introduced a Ni-Cd AA cell with a capacity of 700 mAh, a significant advance. We shall see below that this improved Ni-Cd cell is a strong factor in the lithium rechargeable battery market.

The nickel-hydrogen battery, in which the life-limiting cadmium electrode of the Ni-Cd is replaced by a hydrogen electrode (half cell reaction (h) in Table 1), is extraordinary. Cells on test to date have yielded [17] over 6,000 and 30,000 cycles at 50% and 15% depths of discharge, respectively, making Ni–H$_2$ an obvious choice for long space missions and for satellites in low earth orbit. The challenge here is to bring the Ni–H$_2$ battery down to earth in an economical form by replacing the high pressure tank storage of the hydrogen with a material that will store hydrogen as a solid phase [18]. For example, LaNi$_5$ can store (and release) up to 6 hydrogens per molecule as the hydride LaNi$_5$H$_6$ at moderate pressures. However, the degradation of such materials on cycling has so far kept Ni–H$_2$ out in space.

Lithium primaries today – secondaries tomorrow?

With a reduction potential of -3 V (Table 1) and low equivalent weight, lithium is a natural choice as the anode in a high voltage, high energy density battery. Today, lithium primary batteries are everywhere, with solid electrolyte Li-iodine batteries in cardiac pacemakers, organic electrolyte Li–CF$_x$ and Li–MnO$_2$ batteries in cameras, watches, calculators, etc. and inorganic electrolyte lithium-sulfur dioxide or thionyl chloride batteries in military hardware. Lithium secondary batteries, however, have yet to make their mark despite two decades of intense effort [19]. Why the delay in commercializing rechargeable lithium? The answers include limited cycle life, cost, environmental concerns, limited rate capability, cathode limitations and safety. The

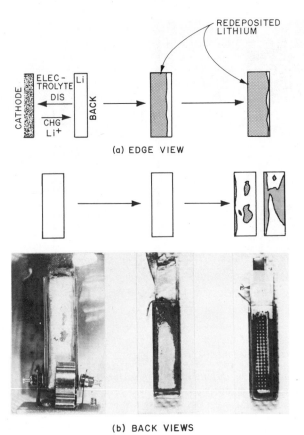

(a) EDGE VIEW

(b) BACK VIEWS

Fig. 11   Formation of particulate lithium as seen in a back view of a lithium anode cycling in a Li–NbSe$_3$ test cell. The lithium "mush" formation is observed through the polypropylene separator, which is transparent when wet by the electrolyte solution.

overriding concern is safety, which itself is tied into some of the other factors listed above, particularly lithium cyclability.

As with lead-acid, the existence of most lithium batteries depends on luck, this time as a passivating film (consisting, e.g., of oxides, carbonates, polymers and/or other organic or inorganic compounds, depending on the particular electrolyte solution). This battery-enabling film is also its curse, as demonstrated in Fig. 11. On discharge, lithium leaves the anode, typically leaving behind a pitted surface, and travels over to the cathode. The problem arises on charge, when lithium plates back onto the anode, where the passivating film prevents the epitaxial deposition of lithium on lithium to form a smooth coherent deposit. Instead, there is more or less random nucleation and growth resulting in a deposit of finely divided lithium which, on continued cycling, builds up into a "mush". Each particle has its own passivating film which interferes with the electrical contact to the main body of lithium and to the other particles. The film formation reactions also deplete the electrolyte and/or solvent. Open test cells of the type in Fig. 11 failed precipitously when the mushy lithium was seen to break the continuity of the bulk lithium anode.

The electrical isolation and film formation both lead to a low cycling efficiency for lithium. Although electrolyte solutions with cycling efficiencies approaching 100% have been found, there have been severe problems with each. The tendency of the lithium perchlorate-dioxolane solution to detonate is one example. Although propylene carbonate (PC) has been a favorite solvent for test purposes, cycling efficiencies are low. Yet, the Li–MoS$_2$ MOLICEL, the only rechargeable lithium cell now being marketed in significant quantities, apparently employs a PC-based electrolyte solution [20], as does the Li–NbSe$_3$ FARADAY cell [21]. It appears that in a sealed, tightly

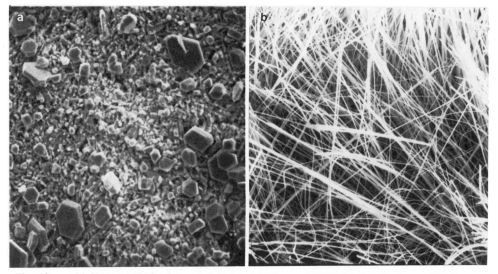

Fig. 12   (a) Hexagonal platelet morphology of $NbSe_2$. (b) Fibrous morphology of $NbSe_3$.

packed cell, the expansion of the lithium anode due to the mush formation (and possibly of the cathode) on cycling compresses the lithium particles together, maintaining electrical contact.

   . Initially, finding suitable rechargeable cathode materials was a problem but, in the early 1970s, various intercalation compounds were found to accommodate insertion and extraction of lithium. The cell reactions for one class of these compounds, the transistion metal chalcogenides MX_y , may be written

$$MX_y + ne^- + nLi^+ = Li_nMX_y \qquad \text{(positive)} \qquad (9)$$

$$nLi = nLi^+ + ne^- \qquad \text{(negative)} \qquad (10)$$

Two of these MX_y compounds, TiS_2 and MoS_2, have been used in commercial cells although only the MOLICEL™ , manufactured by the Moli Energy Corp., is now being marketed. The AA-size MOLICEL, with an amorphous MoS_2 cathode, is capable of a few hundred deep discharge cycles, has a mid-discharge voltage of 1.7 volts and a capacity of 700 mAh. However, the future of the

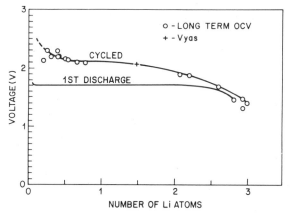

Fig. 13   Approximate open circuit equilibrium voltage behavior for cycled and uncycled Li–NbSe$_3$ cells (Ref. 22).

TABLE 2 – Characteristics of Selected Lithium Systems*

Cathode	Discharge reaction	Mid Voltage	Theoretical Wh/kg	Wh/L
CoO_2	$0.7Li + CoO_2Li_{0.3} = CoO_2Li$	4	766	2700
MnO_2	$0.5Li + MnO_2 = MnO_2Li_{0.5}$	2.8	415	1529
SO_2	$3Li + LiAlCl_4 + 3SO_2 + C = 3LiCl$ $+ LiAlCl_4 \cdot 3SO_2 \cdot C$ complex	3	650	983
V_6O_{13}	$3.6Li + V_6O_{13} = V_6O_{13}Li_{3.6}$	2.3	412	1243
TiS_2	$Li + TiS_2 = TiS_2Li$	2.1	473	1187
$NbSe_3$	$3Li + NbSe_3 = NbSe_3Li_3$	1.9	436	1600
MoS_3	$2Li + a-MoS_3Li = a-MoS_3Li_3$	1.9	494	1482
MoS_2	$0.8Li + \beta MoS_2Li_{0.2} = \beta MoS_2Li$	1.8	233	882

*From Gabano [23].

MOLICEL is clouded by the improved Ni-Cd battery discussed above. Although the Ni-Cd cell's voltage and energy density is lower, the capacity is the same and its lower cost and established position in the marketplace are formidable hurdles for a new lithium technology to surmount.

A more energetic lithium rechargeable system under study at AT&T Bell Labs for over 15 years is the Li–NbSe$_3$ system. The discovery of NbSe$_3$ as a cathode material was a serendipitous outgrowth of work on another Nb-Se compound, NbSe$_2$, with its hexagonal platelet morphology (Fig. 12a). NbSe$_3$ is a fibrous material (Fig. 12b) found as an intermediate phase in NbSe$_2$ film growth. To our surprise, NbSe$_3$ performed better than the NbSe$_2$ as a cathode material and studies on the latter were dropped.

For a Li–NbSe$_3$ cell, the equilibrium first discharge behavior is shown in Fig. 13, together with the (near) equilibrium discharge curve for cycled cells [22]. Initially, there is a voltage plateau out to a point corresponding to the insertion of 2 lithiums per NbSe$_3$ with a falloff to a lower voltage at 3 lithiums. The second and succeeding discharges, however, are at significantly higher voltages, resulting in a higher energy density cell. This fortuitous increase in voltage is associated with the NbSe$_3$ becoming amorphous to X-rays but the implied fundamental change in the nature of the NbSe$_3$ on lithiation (discharge) and delithiation (charge) has not been determined. The cell reactions may be written

$$3Li^+ + 3e^- + NbSe_3 = Li_3NbSe_3 \qquad \text{(positive)} \qquad (11)$$

$$3Li = 3Li^+ + 3e^- \qquad \text{(negative)} \qquad (12)$$

The ability to incorporate 3 lithiums gives a relatively high "theoretical" energy density of 1600 Wh/liter for the Li–NbSe$_3$ couple. Actually, this figure is higher, approximately 1800 Wh/liter, if one takes the open circuit voltage behavior for the cycled NbSe$_3$. The theoretical energy densities of a number of secondary lithium battery systems are summarized in Table 2. Table 2 contains a number of lithium systems with higher theoretical energy densities than Li–NbSe$_3$. However, the metallically conducting NbSe$_3$ does not need a conducting diluent such as carbon (graphite), generally used to increase the conductivity of and/or to "open up" the cathode structure. The fibrous nature of the NbSe$_3$ not only provides its own open structure but also has a "Velcro" morphology which eliminates the need for a binder, typically Teflon or some other polymer. Eliminating binders or diluents permits the attainment of a larger percentage of the theoretical energy density in a real cell. Furthermore, the fibrous NbSe$_3$ cathode supports high rates of discharge.

Coin cells (Fig. 14a) and AA-size cylindrical "FARADAY" cells (Fig. 14b) have been constructed using the Li–NbSe$_3$ couple. The coin cells [24] delivered up to 30 deep discharge-charge cycles with a typical precipitous decline in capacity at the end of life (Fig. 15). This

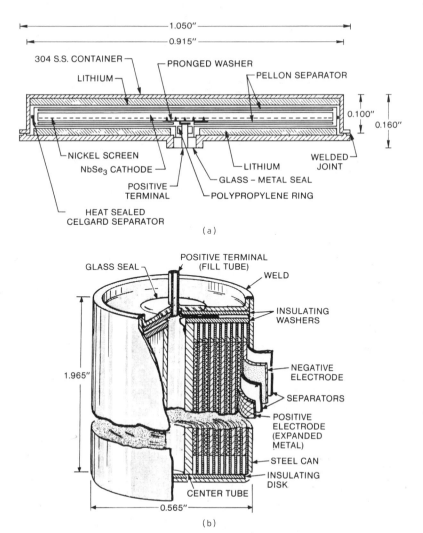

Fig. 14 (a) Diagram of a Li/NbSe$_3$ coin cell (courtesy of S. Basu). (b) Schematic diagram of the Li–NbSe$_3$ AA FARADAY cell.

modest cycle life and precipitous failure mode is similar to lithium failure in open cells of the type seen in Fig. 11. The tightly wound FARADAY "jellyroll" cells, on the other hand, have delivered [22, 25] in the range of 100-300 cycles depending on the voltage limits, discharge and charge currents, temperature, etc. Figure 16 demonstrates the cycle life and the rate capability of the Li–NbSe$_3$ AA cells. Note that the failure mode of these cells is not precipitous, indicating a gradual increase in the cell resistance, possibly related to buildup of reaction products from lithium reacting (to form films) with the electrolyte solution.

Lithium cells in general have a low rate of self-discharge due to the passivating film(s) on the lithium. The Li–NbSe$_3$ system is no exception, especially when compared with Ni-Cd (Fig. 17). An area where the Li–NbSe$_3$ system clearly excels is in the power the AA cells can deliver (Fig. 18).

Given the state of development of Li rechargeable battery technology, we return again to the question of commercialization. The introduction of any new battery technology is especially difficult in view of today's concerns with environmental, economic and safety issues. Indeed, it is doubtful whether the lead-acid or even Ni-Cd batteries could be introduced anew into the marketplace without years of delay. With Li, high rate capability is achieved at the cost of using

13

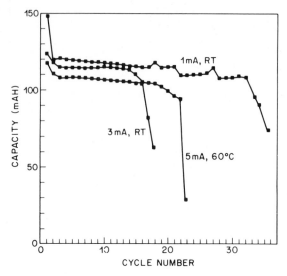

Fig. 15   Capacity as a function of cycle number for Li–NbSe$_3$ coin cells discharged at the
indicated currents. Li–NbSe$_3$ AA FARADAY cell.

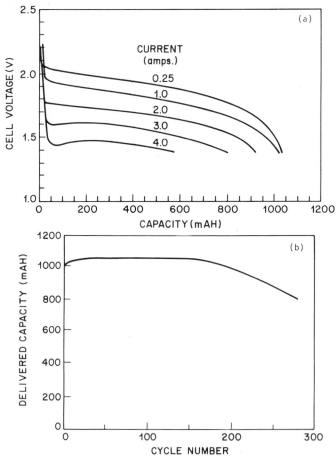

Fig. 16   (a) Discharge curves for a Li–NbSe$_3$ AA cell cycled 20 times and then successively
discharged at the currents indicated. (b) Capacity vs. cycle number for an especially
long lived Li–NbSe$_3$ cylindrical cell.

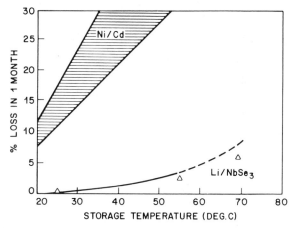

Fig. 17   A plot of the loss in capacity on open circuit stand typical of Ni-Cd AA cells and a similar plot for the Li–NbSe$_3$ cylindrical AA cells.

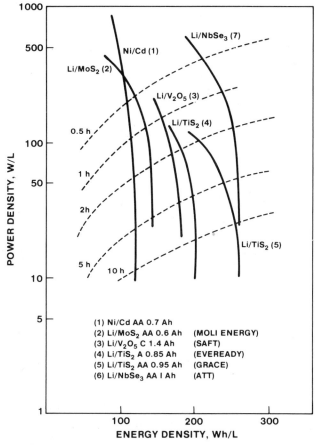

Fig. 18   Power density vs. energy density plots for a number of battery systems (courtesy of J. P. Gabano).

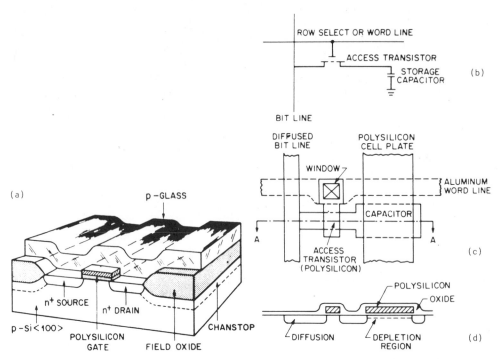

Fig. 19   (a) An n-channel MOSFET. By adjusting the drain and gate voltages, electrons are drawn across the channel from source to drain (after Ref. 28b, © 1985 John Wiley & Sons). (b,c) A simple NMOS DRAM circuit diagram and circuit layout incorporating an n-channel MOSFET and a storage capacitor. (d) Diagram of a silicon chip element corresponding to the DRAM NMOS circuit in (a). (after Ref. 28c, © 1981 John Wiley & Sons).

thin separators which in turn are more prone to internal short circuiting. A long cycle life is accompanied by an accumulation of finely divided lithium, which can melt under "hard" short circuits and lead to venting and flaming of the cell. As with the MOLICEL, the FARADAY cell has a potential competitor in the form of a Li–MnO_2 cell which, though lower in energy density (Table 2), has a voltage making it a direct one for two cell replacement of the Ni-Cd battery. MnO_2 is also inexpensive and a well known entity in the battery world. For high rate cells, however, it seems that all will have the same problems with safety and internal shorts, a problem that, hopefully, may be solved by stringent quality control or some breakthrough in technology, perhaps a high rate Li alloy anode. To answer the question posed in the heading for this section, there will be lithium secondary batteries tomorrow but whether "tomorrow" is next year or next decade is the question.

SEMICONDUCTORS

VLSI – the making of a silicon chip

Memory devices. In this book, Akridge [26] discusses small solid state batteries backing up semiconductor memories for decades. How is this possible? Typically, these memories are DRAMs (dynamic random-access memories) incorporating metal-oxide-semiconductor field effect transistors (MOSFETs). Figure 19 illustrates (a) an n-channel MOSFET, (b) its incorporation into a simple NMOS DRAM circuit element and (c) a silicon embodiment of this circuit. If both row and column lines of the DRAM element are brought to high values, e.g., +5 V, the MOSFET is on and a relatively large or small amount of charge will flow into the capacitor, depending on whether it was initially uncharged or charged, respectively. The state is read as a logic one or zero by the sensing circuit, depending on the amount of charge flow. The capacitor is then "refreshed" by

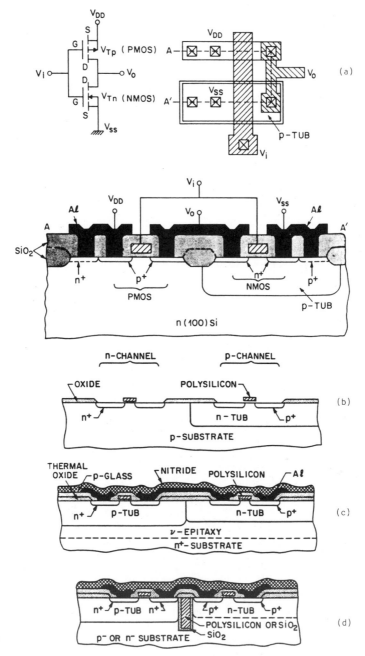

Fig. 20   (a) A CMOS DRAM circuit diagram, layout and device cross section indicating both
n- and p-channel MOSFETs (reprinted with permission from Ref. 28h, copyright 1985
Pergamon Press PLC). (b,c,d) Various CMOS structures. From top to bottom, n-tub,
twin tub and trench isolation structures. (after Refs. 28a and 28e, © 1982 IEEE). The
trench structure (d) is used to prevent "latchup", a potentially fatal condition arising
from interactions involving parasitic bipolar transistors in the other structures (e.g., the
n^+ drain/p-tub/n-tub combination in (c)).

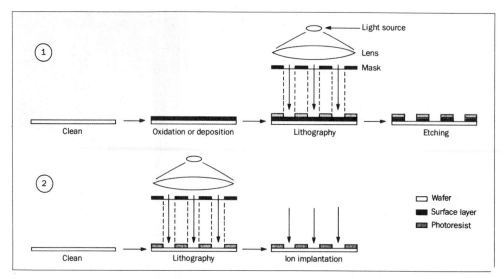

Fig. 21   Schematic diagram of two basic sequences used in VLSI technology (after Ref. 28f, reprinted by special permission. Copyright © 1987 AT&T).

charging it or discharging depending on its initial state. The leakage current of the capacitor mandates this refreshing every few milliseconds.

Today, the CMOS (complementary MOS) DRAM (Fig. 20) is the most common DRAM element, combining both n- and p-channel MOS devices. Why the preference for the CMOS memory, with its obviously more complex structure requiring additional processing steps to manufacture? One answer pertinent to microbatteries lies in the fact that CMOS memories consume significant current only during switching from one logic state to another. When switched, the leakage currents of the MOSFETs are only nanoamperes per device, the reason a small lithium primary battery can back up CMOS memories for many years.

The fabrication of VLSI structures such as the DRAM involves two basic process sequences (Fig. 21). The first is a photolithographic sequence in which oxidation or deposition is followed by patterning for further treatment, such as etching, metallization, etc. The second involves the introduction of impurities (ion implantation is shown in Fig. 21) in patterned areas to give the desired electrical properties. A silicon wafer may pass through some 100-200 steps to create hundreds of replicas of a device or "chip", each chip consisting of a network of thousands or hundreds of thousands of active and passive circuit elements. To accomplish this, there are many combinations of modern techniques such as vapor phase and molecular beam epitaxy, thermal oxidation, oxide or nitride deposition methods, optical, electron, X-ray or ion lithography, laser etching or annealing, plasma etching and plasma assisted deposition, diffusion, various metallization techniques, etc.

Space permits discussion of only a few of the various VLSI processes. Accordingly, we consider first the basic entity without which there would be no VLSI, the silicon wafer itself. The next topic is oxidation and again, without the protective oxide on silicon, VLSI would not exist as it is today. The deposition of dielectric and polysilicon films and metallization processes will be considered more briefly. Epitaxial growth will be considered in the concluding section on future and emerging developments, where molecular beam epitaxy and metal-organic chemical vapor deposition play such an important role. For detailed treatments of solid state devices and VLSI technology, see Refs. 27 and 28a, respectively.

Crystal growth [29a]. VLSI wafers are cut from silicon single crystals grown by pulling crystals from molten silicon (melting point of 1412°C) using the Czochralski technique. For pulling GaAs and GaP crystals, a high pressure puller (Fig. 22a) employing a layer of molten boron oxide over the melt is used. In this liquid encapsulation (LEC) method, the boron oxide

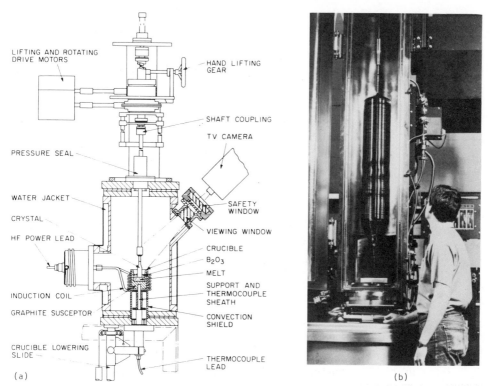

Fig. 22   (a) Schematic of a high pressure Czochralski crystal puller employing LEC (after Ref. 29b). (b) Photograph illustrating the size of today's silicon crystals (courtesy of the Kayex Corp., Rochester, New York).

prevents the escape of phosphorus or arsenic vapor as long as the pressure of an inert gas is higher than the partial pressure of the volatile component. With silicon, no high pressure or boron oxide is needed, one factor contributing to the lower cost of silicon vs. GaAs. In the early days of the transistor, crystals were grown by controlling the temperature manually and directly observing

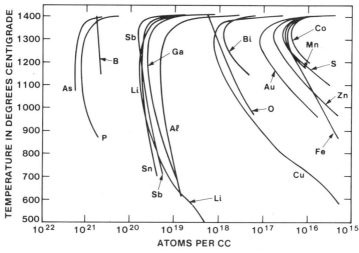

Fig. 23   Curves showing the solid solubilities of various impurity elements in silicon as a function of temperature (after Ref. 29c).

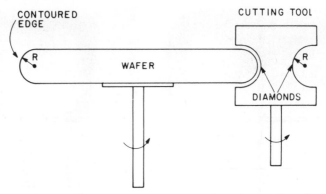

CONTOURED EDGE

CUTTING TOOL

R

WAFER

R

DIAMONDS

Fig. 24   Schematic diagram of edge contouring of a silicon wafer (after Ref. 29a).

the growth of the crystal, which was typically 2-3 cm in diameter and perhaps 5-10 cm long. Today, pulling is typically computer controlled and crystals 20 cm in diameter and 86 cm long weighing 75 kg may be grown in commercial pullers (Fig. 22b).

But size alone is not enough. The success of the VLSI process depends on the absence, control and/or removal of defects, on the crystal orientation, on the taming of unwanted impurities, control of doping level and resistivity both radially and along the length of the crystal, etc. The quality of the silicon crystal depends on factors such as the pull and rotation rates, precision of temperature control, purity of materials, thermal environment on cooling and removal from the hot zone, presence of a magnetic field, etc. Although the Czochralski technique has been used for pulling silicon for four decades, the challenge of growing the ever larger crystals of high perfection used in VLSI processing today is formidable indeed. This challenge is evident from the fact that the rejected material from a given crystal may be on the order of 50%, before any wafer fabrication begins.

Some defects, e.g., twins or grain boundaries, result in immediate discard of that part of the crystal. Dislocations are undesirable, acting as sinks for metallic impurities and altering diffusion profiles. Precipitates of dopants or other impurities can arise from supersaturation at lower temperatures due to the retrograde solubility behavior of most elements in silicon (Fig. 23). Oxygen, introduced from the silica crucible reacting with silicon, is always present in pulled crystals, typically at levels of $5 \times 10^{17} - 1 \times 10^{18}$ atoms/cc. Initially, about 95% of the oxygen is on interstitial sites, where it has the beneficial effect of increasing the yield strength (through solution hardening) by 25% over that of oxygen-free silicon. This beneficial effect can be offset by the formation at 400-500°C of oxygen complexes, which act as donors, and by the formation of SiO_2 precipitates, which generate defects such as stacking faults. These defects attract fast diffusing metallic impurities, and can cause large junction leakage currents in our MOS devices. On the other hand, we will see later that attraction of impurities to such defects can be put to good use to "getter" harmful impurities. Carbon is another ubiquitous impurity which may have detrimental effects if not handled properly.

Wafer preparation [29a]. After cutting away the unusable portions of the silicon crystal, the remaining ingot is ground down to a uniform diameter and flats used for identification and registry of the wafers in the VLSI processes are ground along the length of the ingot. The ingot is then cut into wafers, generally using an inner diameter slicing approach which results in 0.5-1 mm thick wafers, the thickness depending on the diameter, now typically 150-200 mm. After slicing and a two-sided lapping operation under pressure, which produces a wafer flat to within 2 μm, an edge contouring operation (Fig. 24) is carried out. Edge-rounded wafers are less susceptible to chipping, which can lead to fracture, dislocations and crevices for photoresist buildup. At this point the damaged and contaminated regions of the wafer are on the order of 10 μm deep and the wafer is then etched on both sides to remove some 20 μm per side. Etchants in the silicon field have never been very benign, consisting in the past of mixtures of HF, Br_2, H_2O_2, HNO_3, acetic acid, etc. More recently, alkaline etchants of sodium or potassium hydroxide have been introduced. The next step is polishing using a colloidal suspension of 100 Å SiO_2 particles in an aqueous solution of

(a)   (b)

Fig. 25   (a) L. Derick and the late C. J. Frosch receiving the 1969 New Jersey Council for
              Research and Development Outstanding Patent Award for their oxide-masking
              invention. (b) Original masks and transistor structures utilizing oxide-masking. "The
              End" was formed by oxidation, masking and etching and reoxidizing a 0.6 mm square
              Si chip and was used by Frosch in his first talk revealing the new invention. (Courtesy
              of L. Derick.)

sodium hydroxide. The NaOH oxidizes the silicon, while the SiO_2 particles abrade away the oxide.

After a cleaning to remove organic films and other remaining contaminants, one might think we're finally ready to make our CMOS DRAM structure. However, the low power consumption of the CMOS device depends on achieving low junction leakage currents. Metallic impurities, such as transition metals located at interstitial or substitutional sites, may act as centers for the generation and/or recombination of carriers and their silicide precipitates are usually electrically conducting. All contribute to high junction leakage currents. Therefore, we prepare our wafer to remove or getter these fast diffusing impurities, which are introduced in VLSI processing. This "pregettering" treatment can be achieved by deliberately introducing damage on the back surface of the wafer, away from the side on which the VLSI structure will be grown. Although mechanical means such as sandblasting have been used, a more sophisticated approach is to raster a laser beam across the back surface, creating an array of microdamaged areas. During processing at elevated temperatures, dislocations will radiate from the damaged areas and will be confined to the vicinity of the back surface where they serve as sinks for fast-diffusing impurities, isolated from the device action on the front surface.

Another method is "intrinsic" gettering involving the oxygen in the wafer. A high temperature (>1050°C) heat treatment serves to evaporate off oxygen near the surface, creating a "denuded" zone in which the oxygen will not supersaturate and precipitate at the lower temperatures. The high temperature heat treatment is followed by lower temperature heat treatment which leads to precipitation of oxygen in the interior of the wafer. These precipitates and the associated defects then act as getters for the impurities introduced in further processing, while the device action is confined to denuded zone of the wafer. Again, the inevitable trade-offs must be considered. During VLSI processing, the perfection of the wafer must be maintained in spite of the thermal stresses placed upon it on removal from hot furnaces. During removal the wafer edges cool rapidly by radiation but the center stays relatively hot. The yield strength of the wafer must be higher than the thermal stresses in order to prevent formation of dislocations. However, the yield strength (critical shear stress) of silicon can be reduced substantially (up to 5 times) by the oxygen precipitates. Thus the temperature of removal and oxygen precipitate formation must be carefully balanced, especially for today's large 200 mm wafers.

21

TABLE 3 – Diffusion Constants in SiO$_2$ [30]

Impurity	Diffusion constants at 1100°C (cm^2/sec)
B	3.4×10^{-17} to 2.0×10^{-14}
Ga	5.3×10^{-11}
P	2.9×10^{-16} to 2.0×10^{-15}
As	1.2×10^{-16} to 3.5×10^{-15}
Sb	9.9×10^{-17}

Yet another pregettering technique involves using the grain boundaries of a polysilicon (polycrystalline silicon) layer deposited on the back side of the wafer as the gettering entities. These gettering processes not only reduce junction leakage currents, but also contribute to improved quality of VLSI oxide growth. Metallic precipitates on the surface could interfere with the oxide growth and act as points of localized breakdown.

Oxidation [30] and oxide masking. In 1954, a faulty hydrogen regulator and two alert workers combined in the serendipitous invention of oxide masking, a key step toward the integrated circuit (IC) and VLSI. Frosch and Derick (Fig. 25a) were diffusing impurities into silicon at high temperatures but were stymied by pitting of the silicon. To avoid this pitting, they tried various highly purified, dry ambient gases such as nitrogen, carbon monoxide and hydrogen, without any success. One day, a faulty regulator allowed the hydrogen pressure to drop to zero. Instead of the expected disaster, there were beautifully colored, smooth wafers. They concluded that the colors were the result of interference in thin oxide layers grown by reaction with water vapor formed when the hydrogen burned back in the tube. The next day, they got the same result by adding water vapor to hydrogen. They soon demonstrated [31] masking of diffusing impurities by the thin oxide films and the foundation for the later IC and VLSI industry was laid. Memorabilia from these experiments are shown in Fig. 25b. The basis of oxide masking is seen in Table 3, with diffusion coefficients of a number of dopants in SiO$_2$. The oxide is seen to mask well against all the impurities in Table 3 but the more rapidly diffusing Ga.

Today, SiO$_2$ films serve as masks against implant or diffusion of dopants, provide surface passivation, isolate one device from another, are an integral component of MOS devices and provide isolation of multilevel metallization structures. The thermal oxidation of silicon proceeds by the diffusion of the oxygen through the SiO$_2$ layer to the Si–SiO$_2$ interface, where the oxidation occurs. For a given thickness of oxide d, there is a loss of $0.44d$ thickness of silicon which must be taken into account to get any desired registry with the surface. Attention must be paid to the stresses set up by the different thermal expansion coefficients of SiO$_2$ and silicon and by the volume increase in confined regions of VLSI structures. Fortunately, the viscous flow of SiO$_2$ at elevated temperatures can be utilized to attain stress-free growth at temperatures above 960°C. With increasing miniaturization of VLSI devices, the growth of oxides in the 50-100 Å range is demanded. Oxidation can also generate stacking faults, believed due to the presence of excess silicon interstitials or decreased vacancy concentration. These stacking faults can lead to increased leakage currents or decreased storage times in MOS structures due to decoration of the stacking faults by impurities which make the faults electrically conducting. Again, temperature plays a role. The number of faults can be reduced or even completely suppressed by, e.g., the use of high pressure oxidation below 950°C. In the DRAM structure, a dielectric layer consisting of a thin thermal oxide or a composite of a thermal oxide (<100 Å) and silicon nitride can serve as an active component of the storage capacitor, with the charge storage capacity depending on its thickness.

Oxide, nitride and polysilicon film deposition [32]. We have noted the use of SiO$_2$, polysilicon and silicon nitride in our CMOS DRAM. Silicon nitride Si$_3$N$_4$ is also used for both passivating silicon devices by acting as a barrier to the diffusion of water and sodium and masking to allow selective oxidation of silicon. Three common deposition methods used for these materials are atmospheric-pressure, low-pressure and plasma-enchanced chemical vapor deposition (APCVD, LPCVD and PECVD). Typical reactants are listed in Table 4, which also shows the wide range of temperatures available if low temperatures are needed for compatibility with earlier VLSI steps.

TABLE 4 – Selected Deposition Reactants [32]

Product	Reactants	Deposition temp. (˚C)
Silicon dioxide	$SiH_4 + CO_2 + H_2$	850-950
	$SiCl_2H_2 + N_2O$	850-900
	$SiH_4 + N_2O$	750-850
	$SiH_4 + NO$	650-750
	$Si(OC_2H_5)_4$	650-750
	$SiH_4 + O_2$	400-450
Silicon nitride	$SiH_4 + NH_3$	700-900
	$SiCl_2H_2 + NH_3$	650-750
Plasma silicon nitride	$SiH_4 + NH_3$	200-350
	$SiH_4 + N_2$	200-350
Plasma silicon dioxide	$SiH_4 + N_2O$	200-350
Polysilicon	SiH_4	575-650

Many of the gases used in film deposition are quite hazardous. Indeed, a fatal explosion in New Jersey was attributed in press reports to the presence of nitrous oxide in a tank of silane.

CVD SiO_2 films are used as insulating layers between multilevel metallizations, as ion implantation or diffusion masks, as capping layers to prevent outdiffusion and to increase the thickness of thermally grown oxide layers. Phosphorus-doped SiO_2 is used in the "P-glass flow" process at 950-1100˚C forming the insulator between the polysilicon gate and the top metallization, while at deposition temperatures under 400˚C it can form passivating layers. Doped SiO_2 is also used as a diffusion source.

An interesting problem occurs when trying to line a step or "trench" with oxide (Fig. 26). (Trench isolation and trench capacitors are standard VLSI features.) Three cases are shown: (a) rapid surface migration of the reactants or reactive intermediates (uniform coverage), (b) long mean free path in the gas phase and no surface migration and (c) short mean free path and no surface migration, both (b) and (c) giving nonuniform coverage. The uniform coverage (Fig. 26a) was obtained by decomposing tetraethoxysilane (TEOS), $Si(OC_2H_5)_4$, at 700˚C an a pressure of 30 Pa. The rapid surface diffusion of the TEOS results in the uniform coverage. Figure 28b shows a layer grown from silane and oxygen at reduced pressure with no surface migration. In both cases the mean free path was ~100 µm. Figure 28c shows growth from silane and oxygen at atmospheric pressure, where the mean free path was <0.1 umm. We see that choice of reactants and reaction parameters can be quite critical.

It is ironic that, after years spent trying to attain defect-free single crystals, two materials of interest these days are polysilicon and amorphous silicon. Amorphous silicon, actually silicon containing hydrogen, is a potential low cost solar cell material. Here, we are primarily concerned with the use of polysilicon as the gate electrode in MOS devices but it is also used for resistors, conductors, to ensure ohmic contact to shallow junctions and, with 25-40 atom % oxygen added, becomes a semi-insulating passivating coating for certain VLSI circuits. Polysilicon is formed by the pyrolysis of silane between 575 and 650˚C in a low pressure reactor. The structure depends on the temperature of formation and/or heat treatment. At the lower deposition temperatures (605˚C) amorphous silicon is formed, becoming columnar at a higher deposition temperature (630˚C) and on annealing at 700˚C both the amorphous and columnar phases become crystalline. The doping behavior is also related to the temperature and structure (Fig. 27). Amorphous silicon for solar cells is generally made by plasma-enhanced CVD. The deposition of semi-insulating polysilicon (SIPOS) is accomplished by the reaction of silane with nitrous oxide. By varying the N_2O/SiH_4 ratio, the resistivity of the deposit changes from 10^6 to almost 10^{11} ohm-cm, resulting in the passivating SIPOS, actually a multiphase mixture of crystalline and amorphous Si, SiO_2 and SiO.

This brief treatment of film deposition has shown a few of the possibilities for growth of different materials. Oxide deposition complements thermal oxidation in situations where the latter could, e.g., lead to defect generation, stresses, etc.

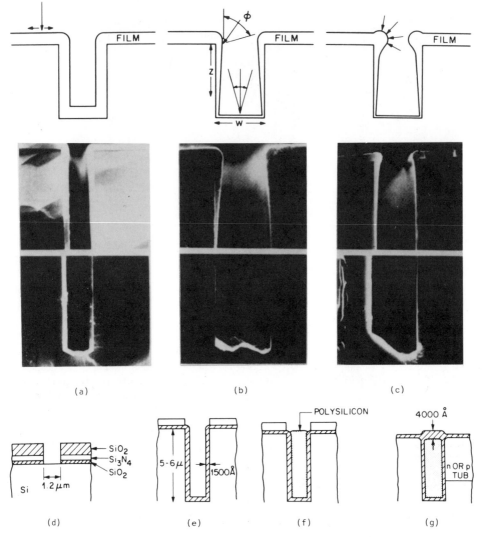

Fig. 26 Step coverage by CVD SiO_2 films. (a) Uniform coverage resulting from the decomposition of TEOS at 700°C and 30 Pa. (b) Nonuniform coverage by deposition of SiO_2 from silane and oxygen at a low pressure. (c) Nonuniform coverage from reaction of silane and oxygen at atmospheric pressure. (After Ref. 32.) (d-g) Utilization of uniform coverage in forming a trench isolation structure (reprinted with permission from Ref. 28g, copyright 1985 Pergamon Press PLC).

Metallization [33a]. In batteries, we saw that electrical contact with the outside world is crucial, witness the post corrosion in lead-acid batteries and the isolation of particulate lithium in the lithium batteries. In MOSFET technology, we are concerned with three groups of metallization: gate, contact and interconnection. We have seen that polysilicon is the gate metal in the central gate region of the MOSFET, separated from the silicon substrate by an oxide layer. Polysilicon serves also as the fine line metallization connecting the MOSFETs. Bilayers of refractory silicides (e.g., $MoSi_2$) on top of the polysilicon may be used to attain lower resistance metallizations. The gate and interconnection metallizations control the speed of the circuit via the resistance of the "runners", which determines the RC time constant. These metallizations also

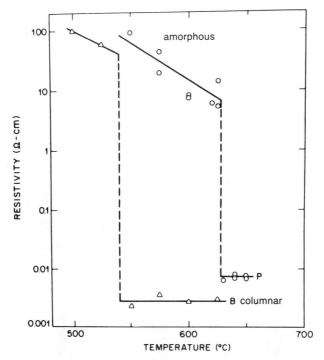

Fig. 27   Resistivity of polysilicon doped during deposition by adding diborane or phosphine. The sharp drop in resistivity in both the B- and P-doped polysilicon corresponds to the change from amorphous to columnar structures at the higher temperatures. (after Ref. 32).

contribute to the threshold voltage, which determines the gate-to-source voltage that will switch the MOSFET to the on condition.

The contact metallization is the metallization in direct contact with semiconductor. Aluminum is the choice here because of the ease of processing (deposition via evaporation), the ability to reduce the native oxide on silicon and its low resistivity. However, the low melting

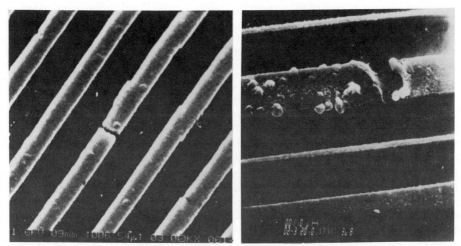

Fig. 28   Examples of the failure of metallization interconnections due to electromigration (after Ref. 33b, © 1980 IEEE).

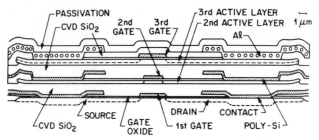

Fig. 29   Schematic showing three levels of stacking of active transistors (after Ref. 33c, © 1980 IEEE).

point of aluminum (660°C) and its lower eutectic temperature with silicon (577°C) limits the temperatures available for further processing. Furthermore, annealing of aluminum on silicon, e.g., at 450°C, results in pit formation due to the solid solubility of silicon in aluminum (1 atom%) and diffusion of aluminum into the silicon which can result in so-called spiking shorts. To solve this problem, silicon may be added to the aluminum or layers of silicides, TiC or TiN, e.g., may be deposited between the aluminum and the silicon to serve as diffusion barriers. Each of these approaches has its own advantages and disadvantages. The aluminum spiking problem is just one challenge in the metallization area and illustrates one of the simpler chemical incompatibilities which play a role in VLSI.

Aluminum also may serve in the second-level interconnections to the outside world. The main failure mode of aluminum lines is the breaking of continuity (Fig. 28) due to electromigration, the enchanced and directional movement of atoms caused by the electric field and by the collision of electrons with the atoms. This effect is strongly dependent on grain structure and size, possibly indicating the importance of grain boundary diffusion. The addition of copper to the aluminum or aluminum-silicon has been found beneficial in diminishing electromigration. Because of faulty metallizations, many large MOS memory chips are designed with extra metal rows and columns of bits which can be opened either electrically or with a laser and exchanged for any faulty links.

VLSI limitations. MOSFET devices get smaller and smaller as time goes by. However, fundamental limits are in sight. For example, gate oxide thicknesses less than 50 Å lead to tunneling. As the size (channel length) of the device is diminished, leakage currents become sufficiently high that the distinction between the on and off logic states becomes blurred. In addition, scattering of the electrons by impurities results in lowered conductivity and slower switching. At channel lengths of 0.1 μm quantum effects must be considered, as discussed later. Interconnections become a bigger problem. A 1 μm square wire carrying $10^5 A/cm^2$ is near the electromigration limit; the thinner the wire, the higher the resistance and RC time constant and the slower the device speed. Power considerations and heat dissipation requirements limit the number of gates on a chip, etc.

To continue packing more and more devices on the silicon chip, devices are already being stacked on top of each other. Figure 29 shows three levels of active transistors with two levels of deposited and recrystallized silicon. In such structures, care must be taken to maintain the integrity of the underlying devices during the additional deposition/annealing steps. Laser annealing avoids heating the whole chip to recrystallize the silicon.

Beyond these "conventional" VLSI structures lies the world of quantum devices with dimensions in the 0.01 μm range, of optoelectronic integrated circuits (OEICs) incorporating the light emitting and ultra-high speed capabilities of gallium arsenide and other compound semiconductors directly on the VLSI silicon chip, of neural networks whose ultimate form remains to be determined, of tandem solar cells, etc. All or most of these will rely to some extent on heteroepitaxy in that these structures will be composed of layers of different materials and composition. For example, the very concept of OEICs on silicon depends on heteroepitaxial growth of utmost sophistication to overcome the problems of gross defects and cracking of epilayers due to lattice mismatch (Fig. 30).

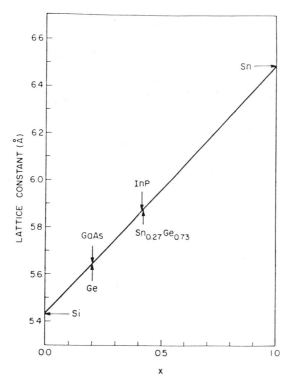

Fig. 30   Plot of the lattice constants of hypothetical Sn-Si solid solutions assuming Vegard's law. Also shown are the lattice constants of Ge, GaAs, InP and a hypothetical Sn-Ge alloy (Reprinted with permission from Ref. 47, copyright, CRC Press, Inc., Boca Raton, Florida).

Epitaxial growth techniques

LPE. To appreciate the sophistication of today's epitaxial techniques, a brief look at the use of liquid phase epitaxy (LPE) to grow light-emitting diodes (LEDs) and lasers is instructive [34]. LPE growth of GaP p-n junction LED material involved the elegantly simple "tipping" technique of Nelson [35] (Fig. 31a). A saturated solution of GaP in gallium is tipped onto a GaP substrate, the temperature is lowered and GaP precipitates out of solution as an epilayer, doped to form a light emitting p-n junction. The world's most efficient (1-2%) red-emitting LEDs at the time were grown in a small tipping furnace of the type shown in Fig. 31a.

A forward step in LPE was the use of a "slider" technique (Fig. 31b) to grow GaAs–$Al_xGa_{1-x}As$ LEDS and, later, laser hetrostructures [36a]. Figure 31c shows the original entry from M. Panish's notebook suggesting the use of the slider method to grow two epitaxial layers in a single run. The slider method later was modified to permit more layers to be grown by using multiple wells of liquid. The very close lattice match between GaAs and AlAs make this an ideal system for heteroepitaxy. The range of visible and infrared light emitting capabilities is illustrated in Fig. 32, which shows the range of bandgap energies available for selected ternary III-V alloys, some of possible interest for OEICs. Not shown in Fig. 32 are the quaternary $Ga_{1-x}In_xAs_{1-y}P_y$ alloys of practical importance in lightwave communications lasers, which have been fabricated using LPE.

MOCVD. Chemical vapor deposition (CVD) techniques are typically of a higher complexity than LPE. Figure 33 illustrates an early approach [37] to growing $Ga_{1-x}As_xP$ p–n junction material for LEDs. The process involves transporting Ga as volatile chloride(s) which react with As and P formed by the decomposition of AsH_3 and PH_3. The capability for control and grading of composition by changing the phospine and arsine pressures is apparent. However, the CVD process of greatest interest for heteroepitaxy is the metal-organic CVD (MOCVD) process, which

27

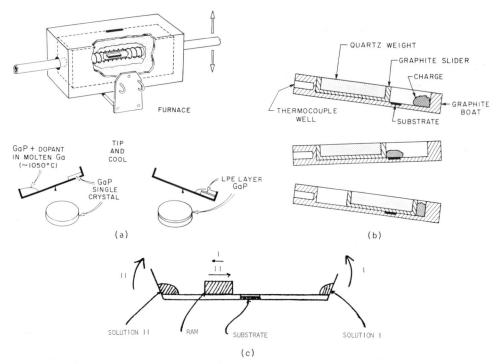

Fig. 31   (a) Schematic of the LPE tipping method applied to growth of GaP pn junction material for LEDs. (b) Schematic of LPE   slider'' (after Ref. 36b. This paper was originally presented at the 1968 Fall Meeting of The Electrochemical Society, Inc. held in Montreal, Canada) developed by Panish and Sumski (Ref. 36a). (c) Notebook entry describing the use of a slider for growing two epilayers in one run (courtesy of M. Panish).

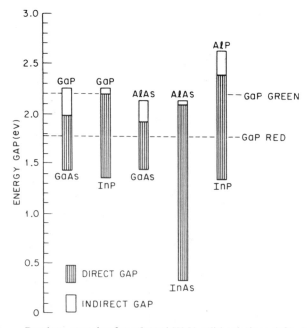

Fig. 32   Bandgap energies for selected III-V solid solutions (after Ref. 34).

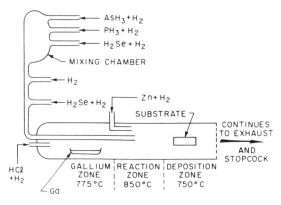

Fig. 33   An early system for the growth of GaAs$_{1-y}$P$_y$ epitaxial layers with Se and Zn doping capability (after Ref. 37, reprinted by permission of the publisher, The Electrochemical Society, Inc.).

involves transporting volatile organometallic compounds (e.g., trimethylgallium, triethylgallium, tetramethyltin, diethylzinc, etc.) over heated substrates where the vapors thermally decompose to form epitaxial layers of the III-V compounds/alloys (Fig. 34). Originally [38a], MOCVD was carried out at atmospheric pressure, but the development of low pressure MOCVD has catapulted it to becoming a most exciting epitaxial growth technique. The versatility of MOCVD is demonstrated by its capability for growing all of the common III-V compounds and alloys with doping control and, very importantly, to grow on dissimilar substrates as in the growth of GaAs on silicon. Selective growth on exposed substrate but not on SiO$_2$ masks has been demonstrated at low pressures. MOCVD is now used to grow abrupt structures, superlattice and other metastable structures at sufficiently low temperatures that atom migration is suppressed and at the highest purity levels.

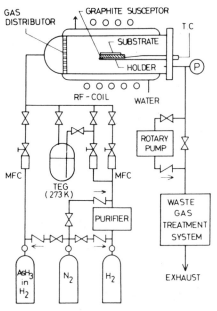

Fig. 34   Diagram of a MOCVD apparatus involving the growth of GaAs using arsine and triethylgallium. (after Ref. 38b, reprinted by permission of the publisher, The Electrochemical Society, Inc.).

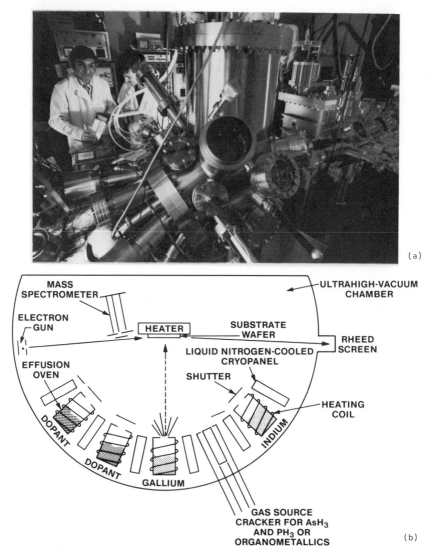

Fig. 35   (a) M. Panish (left) and R. A. Hamm shown with gas source MBE apparatus. (b) Diagram of a GSMBE apparatus incorporating gas source addition and RHEED capabilities. (Courtesy of M. B. Panish.)

MBE. The complexity of today's modern heteroepitaxial methods is sen by comparing Panish's notebook entry on LPE with his current activity (Fig. 35), gas source molecular beam epitaxy (GSMBE) [39]! Molecular beam epitaxy was initially an evaporative technique which depended on precise temperature control of ovens containing the source elements to control the pressures of these components, while shutters controlled the exposure of the substrate to the desired beams of vapor species emanating from the ovens [40]. This ultra-high vacuum (10^{-8}–10^{-11} Torr) technique permitted growth and control of doping and stoichiometry on the atomic level with dimensional control in the nanometer range. Although Si structures have been grown using MBE, these have been of a research or very specialized nature and MBE growth of Si has been of no practical significance in VLSI, primarily because of the expense and the low throughput capability of MBE machines. As with MOCVD, MBE has been most productive in the area of compound semiconductors, notably binary, ternary and quaternary allsoys of Ga, In and Al with As and P.

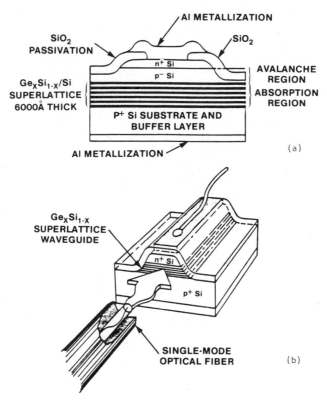

Fig. 36   Cross section and perspective view of a heterostructure lightguiding photodetector incorporating a $Ge_{1-x}Si_x$/Si superlattice (after Ref. 42a, © 1986 IEEE).

CVD, MOCVD and MBE have had their own special niches to fill in the epitaxial picture, each technique having its strengths and weaknesses. For example, the use of elemental phosphorus in MBE presents a problem due to the number of allotropic forms of P, the lack of precise temperature control capability at the low temperatures needed for the relatively volatile element, the low accommodation coefficient for P_4, etc. Such problems were addressed by introducing gaseous hydrides (e.g., PH_3, AsH_3) into the MBE apparatus, a marriage of MBE with CVD to give GSMBE [39]. The beam nature of the fluxes of entering reactants can be maintained, even at pressures as high as 10^{-3} Torr. Today, various combinations of MBE with CVD or MOCVD are used to perform heteroepitaxial feats of utmost complexity. One example is the use of GSMBE in growing lattice-matched $Ga_{1-x}In_xAs_{1-y}P_y$ alloys on InP demonstrating growth of quantum well structures with near monolayer abruptness, superlattice avalance photodiodes, heterostructure bipolar transistors with superior properties, etc. [39].

Heteroepitaxial GaAs and $Ge_{1-x}Si_x$ on Si [41].   The growth of GaAs on Si, with a 4% lattice mismatch, has long been one of the highest heteroepitaxy goals. Dramatic progress toward this goal has been made in recent years, with the help of the above epitaxial techniques and some key breakthroughs. Especially important has been the achievement of clean Si surfaces, relatively free of oxygen and carbon, with the aid of ultra-high vacuum techniques. Another breakthrough has involved tilting of the Si substrate off the standard VLSI <100> orientation to permit the growth of GaAs free of antiphase domains (APDs) and minimize "threading" dislocations penetrating the epitaxial material. The APDs arise from single atomic layer steps on the <100> orientation and the preference of As for Si, i.e., the growth of GaAs directly on Si starts (or is "seeded") with As. With single step domains on the surface there is not the appropriate registry of Ga and As atoms in adjoining domains; hence the APDs. By orienting off the <100>, double layer steps are introduced and the registry is correct. Another very important development has been the growth of intermediate strained layer superlattices, which tend to bend the threading dislocations. Annealing has also been found to reduce the number of dislocations.

An example of the use of strained layer epitaxy is in the light guiding photodetector shown in Fig. 36, which illustrates the growth of a metastable $Ge_{1-x}Si_x/Si$ superlattice combined with a silicon avalanche layer to amplify the signal [42]. The $Ge_{1-x}Si_x/Si$ heteroepitaxial structures can be grown without defects and without the autodoping problems encountered when, e.g., GaP is grown on Si where Ga and P and Si are donors an/or acceptors in Si and GaP [42b]. For a realistic critique of the prospects for a germanium silicide technology see Ref. 42c. One approach to growing GaAs on Si has been through the graded and strained layer epitaxial growth of Ge on Si, the Ge providing a lattice match to GaAs. However, recent efforts have concentrated on growing GaAs directly on Si without an intervening Ge layer. Extensive studies are in progress on the fabrication and study of artificially structured materials, e.g., alloys of Al-Ga-As consisting of alternate layers of AlAs and GaAs with the thicknesses of the layers adjusted to give the desired overall composition. In all these heteroepitaxial growths, the goal is that the active material be of the same quality as the homoepitaxial material.

Nothing is straightforward, however, especially when we consider the pros and cons of GaAs on Si. Indeed, why has GaAs been the "material of the future" for decades? The lack of a passivating oxide on GaAs has been a key fundamental obstacle to a "pure" GaAs VLSI technology. But practically, the answer lies in the persistent innovation in Si device technology and decrease in cost per bit on the Si chip. The cost of a GaAs wafer alone can be 25-30 times that of a Si wafer. It is also true that the GaAs epitaxial structures on Si may cost more than a GaAs wafer, depending on the particular devices and processes involved. A positive consideration is the fact that Si is 2.5 times less likely to fracture than GaAs, resulting in lower yield losses in handling for Si. The 8" diameter Si wafers, compared to the 4" GaAs wafers, also allow significant savings in manufacturing costs due to economies of scale. The low thermal conductivity of GaAs makes pulling of larger diameter GaAs crystals possible only with a significant sacrifice in quality [43]. The four times higher thermal conductivity of Si vs. GaAs also means better heat dissipation, especially desirable for power FETs or lasers, and allows a higher density of devices on a chip where heat generation is a problem. For space applications, there is a trade-off of silicon's lower density and the lower efficiency of solar cells of GaAs on Si vs. on GaAs. Although thin layer superlattice structures of the type in Fig. 36 may be grown defect-free, heteroepitaxial interfaces and the nearby material are generally associated with high dislocation densities, with relatively good quality material above that. This may be acceptable for FETs, where current paths are located near the surface, but not for the minority carrier devices where the current goes through the interface region. Another factor is the thermal expansion mismatch between GaAs and Si which can lead to wafer bowing with a concave GaAs surface and high tensile stress within the GaAs layer. This can be minimized by low temperature growth, by selective deposition of the GaAs or by etching into a pattern of islands of GaAs on the silicon chip.

Although GaAs-on-Si structures have been constructed with device characteristics similar to those in homoepitaxial GaAs devices, there remain the problems of reliability and life. Structures with good device characteristics may last only hours. Will unavoidable defects at the interface propagate dislocations or otherwise affect device performance and cause an early demise of even FETs? Obviously, many challenges remain before GaAs, on or off Si, becomes the material of the present.

Metastable materials. In Fig. 30, we saw the lattice constant of hypothetical Si-Sn alloys as a function of composition over the range from pure Si to pure gray tin. Actually, the solid solubility of Sn in silicon is only about 0.1 atom percent (Fig. 23). The hypothetical Sn-Ge alloy composition shown matching the lattice constant of InP would be a laser material if it could be prepared. The growth of such metastable alloys by MBE/MOCVD or other techniques now seems within the realm of possibility.

A spectacular example of the growth of metastable phases is the growth of diamond [44] and diamond like hydrocarbons (a-C:H) [45], on a variety of substrates including Si, by low pressure CVD and even at room temperature by ion beam deposition. Apparently, the success of these

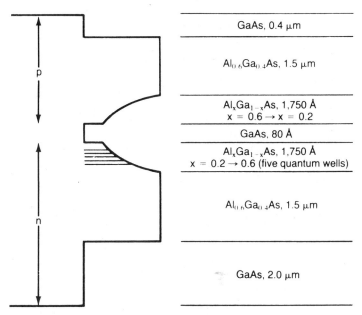

GaAs, 0.4 μm

Al$_{0.6}$Ga$_{0.4}$As, 1.5 μm

Al$_x$Ga$_{1-x}$As, 1,750 Å
x = 0.6 → x = 0.2

GaAs, 80 Å

Al$_x$Ga$_{1-x}$As, 1,750 Å
x = 0.2 → 0.6 (five quantum wells)

Al$_{0.6}$Ga$_{0.4}$As, 1.5 μm

GaAs, 2.0 μm

Fig. 37   Schematic view of a heteroepitaxial quantum well laser structure grown on Si by MBE
(after Ref. 48).

experiments is related to the preferential chemical erosion of graphite, presumed to codeposit with
the diamond, by hydrogen [46]. The challenge here is to grow diamond epilayers for possible
semiconductor devices, for heat sinking (diamond has the highest thermal conductivity), for optical
coatings, etc. The jewelry market is not yet threatened by the current micron size diamonds!
Metastable (or stable) new materials emerging from modern growth techniques could well spawn
new devices and technologies.

Emerging and future semiconductor technologies [47]

OEICs. The III-V and other compound materials offer a number of features unattainable in
silicon such as their optoelectronic properties and possibility of bandgap engineering (Fig. 32),
higher mobilities with higher speed device capabilities, radiation hardness, etc. OEICs, with
islands of these compounds grown on the silicon wafer or chip in the form of lasers or LEDs
driven by the silicon circuitry and combined with photodetectors, offer the possibility of inter- or
intrachip communication with the speed of light. The possibility of using a constant light source in
conjunction with transmission modulators controlled by the silicon circuits has been suggested, the
light source being external to the chip and acting like a battery. An example of a possible optical
interchip communications device is the heteroepitaxial quantum well laser (Fig. 37), which has
been grown on silicon by MBE [48]. Again, the future of OEICs on Si depends on achieving
reliability and life in these heteroepitaxial structures. Indeed, the OEICs of the future may not
involve Si but rather the lattice-matched Ga$_{1-x}$In$_x$As$_{1-y}$P$_y$/InP combination mentioned above.

Quantum dots. When the dimensions of a device approach the wavelength of the electron
in the semiconductor material (~100-200 Å in GaAs at room temperature), quantum effects come
into play (Fig. 38a). We have seen (Fig. 37) that one-dimensional quantum wells are already being
utilized in lasers. A concerted effort is underway to achieve and exploit the ultimate, truly
quantized 3-dimensional quantum dot. A driving force for this effort on such minute devices is the
quest for a new semiconductor revolution in which the cost per function might be reduced by a
factor of 10^3–10^4 and allow the circuitry of a supercomputer on a single chip [49]. Texas
Instruments is one company pursuing this technology and workers there have fabricated [50] a
quantum dot diode (Fig. 38b), a 1000 Å diameter diode operating at 1°K. The problems associated
with such extreme miniaturization in the areas of interconnection and all other aspects of a
quantum technology are formidable. To surmount these problems, a totally new chip architecture

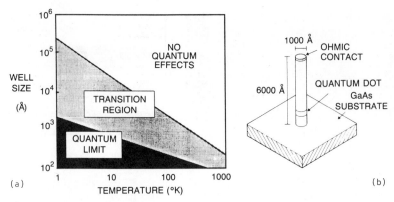

Fig. 38   (a) Temperature dependence of the size regime for quantum effects in GaAs/Al$_{1-x}$Ga$_x$As quantum-coupled structures. (b) A quantum dot diode (after Ref. 50).

is under consideration in which each device would communicate with only its neighboring devices and new rules would have to relate to assemblages of energy levels and tunneling between quantum dots. No field for the faint of heart!

Neural networks. Another area involving radically new approaches is that of neural networks, in which attempts are being made to mimic the operation of the brain. A simple illustration of the difference between the operating modes of the normal everyday computer and the brain was given by Hopfield [51],who used the example of a telephone book and the concept of local and nonlocal memory storage. If your name and number are listed on a page of the book and that page is torn out, you have become a nonperson. If that information has been written across the edges of the pages of the book, removal of one or even many pages of the book leaves the name and number still recognizable. This ability to remember and identify even a defective copy of an image is one feature distinguishing the brain and neural network approach from that of today's computer, which may be useless with one defective device.

There are many approaches to achieving a neural network, one of the problems being the huge number of connections for large networks. In this regard,the optically programmed network of Kornfeld, et al. [52] is of interest. As shown schematically in Fig. 39, their network consists of a photoconductive array of 120 x 120 photoconducting elements of amorphous Si (remember its solar cell potential). This approach requires only 240 electrical connections to access the 14,400 synapses, whose individual conductivities are determined by the amount of light falling on each

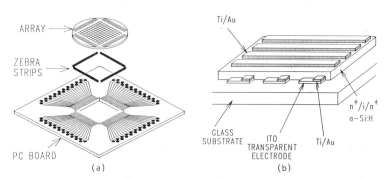

Fig. 39   Photoconductive neural network array. Light shining through the transparent glass substrate and indium-tin-oxide (ITO) electrodes strikes the (hydrogenated) amorphous Si photoconducting elements (Ref. 52, courtesy of R. C. Frye).

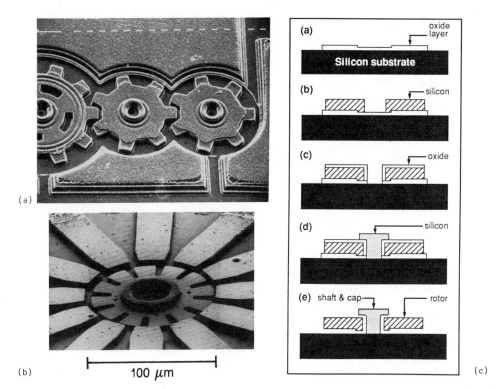

Fig. 40   (a) Miniature turbine in which the gears are turned by air entering through the channel at bottom (b) Motor run by static electricity. (c) A processing sequence to fabricate a miniature motor. (After Ref. 57, copyright 1988 by the AAAS; also courtesy of R. Muller and AT&T Bell Laboratories, M. Mehregany, K. Gabriel, W. S. N. Trimmer and J. Walker.)

element. The author has seen this array operate with rows of defective elements and, as with the telephone book example, a computer processed image is quite recognizable.

Superconducting devices/interconnections. Today, one cannot neglect superconductors or the improved performance when VLSI circuits are operated at low temperatures [53]. In a VLSI structure, the possibility of superconducting interconnects at operating temperature raises the possibility of higher speeds (lower RC time constants) and lower power requirements, of possible relevance to microbatteries. Other advantages of operating semiconductor devices at low temperatures include such things as lower noise, slower diffusion of impurities to degrade performance, lower leakage currents, slower electromigration to degrade interconnects, etc. Hybrid circuits involving Josephson junctions, with very high speed capabilities, and semiconductors are being investigated [54]. With the recent appearance of superconductors at temperatures above that of liquid nitrogen, the operating domains of superconductivity and semiconductor devices now overlap and at much less costly temperatures to maintain than those of liquid hydrogen or helium.

Solar cells. Solar energy has been a field of waxing and waning interest, depending on the price of oil at the time. However, steady progress toward lower cost, higher efficiency and more stable solar cells makes it more likely that solar energy will play an important role in the future. Of particular interest here are the "tandem" or composite solar cells. These cells contain layers of different materials such as amorphous silicon and/or other compounds such as III-Vs or certain chalcogenides of In and Cu, e.g., having different bandgaps. By covering more of the spectrum, the efficiency of conversion of solar energy to electricity is increased, compared to silicon or GaAs individually. These cells may not be composed of single crystal material, but may be formed using some of the deposition methods discussed above. The challenge is to achieve the conversion efficiencies (as high as 30%) which have been demonstrated in special structures [55], but in a less costly manner and in structures that do not decay significantly with extended use.

Micromachining. One final item for the future is an unusual offspring of silicon technology, micromachining and micromachines. An example is shown in Fig. 40a, which shows a miniature turbine with 3 gears micromachined from Si [56]. A motor running on static electricity (Fig. 40b) has been reported recently [57]. A processing sequence to build such a motor is illustrated in Fig. 40c. The oxide layers deposited along with silicon are etched away with acid, leaving the gear free to rotate around the hub attached to the Si substrate. Applications for these micromachines are postulated in the medical field ("smart pills", cutting away blockages in blood vessels, killing cancer cells, etc.?), in the handling of small items, fans to cool electronic circuits, etc. These are certainly enthusiastic predictions for this embryonic field but who knows, will microbatteries be needed to power these tiny machines?

CONCLUSION

This "overview" of batteries and semiconductors is perhaps more appropriately described as a "microview", considering the vast amount of work going on in both areas. Certainly, the excitement in the semiconductor arena is directed to smaller and smaller devices and growth on an atomic scale to allow hitherto incompatible materials to be merged. The power requirements of the new microdevices is certain to be followed with interest by the battery community, with its divergent needs for both smaller and larger batteries. The need for large batteries in electric vehicles, for power storage and for backup power in telecommunications is evident. Smaller, more energetic batteries are required to achieve the ultimate potential in portable applications, e.g., in portable cellular telephones. The need for microbatteries will be determined in large measure by future progress in some of the areas of semiconductor technology considered here.

ACKNOWLEDGMENTS

I am grateful to M. E. Fiorino, R. E. Landwehrle, V. A. Edwards, A. G. Cannone and especially to R. V. Biagetti for material and helpful discussions pertaining to the round cell and its recent history. The lithium battery material is drawn from the work of many past and present members of the "FARADAY Team" and earlier lithium efforts at AT&T Bell Labs. Unfortunately, proprietary considerations preclude naming of these individuals. I am indebted to Simon Sze for helpful material and discussions and for calling my attention to his 2nd edition of "VLSI Technology", which appeared just in time to provide a crash course in modern semiconductor technology after my 16 year absence from the field. L. Derick graciously supplied the memorabilia and history of the invention of oxide masking. M. B. Panish also supplied historical material and provided sobering comments on the manuscript and on the limitations of heteroepitaxy. I also thank S. Knight, R. C. Frye, W, S. Trimmer, J. C. Bean, B. A. ter Haar and L. W. ter Haar for their contributions and to H. J. Leamy and J. Broadhead for their support. This paper has also benefited from stimulating discussions with Max Schulz and other lecturers in the Silicon course at Erice.

REFERENCES

1. M. Faure, Faure Secondary Battery, Electrician 6:323 (May 7, 1881); J. S. Sellon, Improvements in Secondary Batteries or Magazines of Elctricity, British Patent 3987, September 15, 1881.
2. J. T. Crennell and A. G. Milligan, The Use of Antimonial Lead for Accumulator Grids: A Cause of Self-Discharge of the Negative Plates, Trans. Faraday Soc. 27:103 (1931).
3. H. E. Haring and U. B. Thomas, The Electrochemical Behavior of Lead, Lead-Antimony and Lead-Calcium Alloys in Storage Cells, Trans. Electrochem. Soc. 68:293 (1935).
4. U. B. Thomas and H. E. Haring, Corrosion and Growth of Lead-Calcium Alloy Storage Battery Grids as a Function of Calcium Content, ibid. 92:313 (1947).

5. D. E. Koontz, D. O. Feder, L. D. Babusci and H. J. Luer, Reserve Batteries for Bell System Use: Design of the New Cell, Bell System Tech. J. 49:1253 (1970).

6. A. G. Cannone, D. O. Feder and R. V. Biagetti, Positive Grid Design Principles, ibid. 49:1279 (1970).

7. R. V. Biagetti and M. C. Weeks, Tetrabasic Lead Sulfate as a Paste Material for Positive Plates, ibid. 49:1305 (1970).

8. P. C. Milner, Float Behavior of the Lead-Acid Battery System, ibid. 49:1321 (1970).

9. T. D. O'Sullivan, R. V. Biagetti and M. C. Weeks, Electrochemical Characterization of the Bell System Battery: Field Trials of the Battery, ibid. 49:1335 (1970).

10. T. W. Huseby, J. T. Ryan and P. Hubbauer, Polyvinyl Chloride Battery Jars and Covers, ibid. 49:1359 (1970).

11. A. D. Butherus, W. S. Lindenberger and F. J. Vaccaro, Electrochemical Compatibity of Plastics, ibid. 49:1377 (1970).

12. D. W. Dahringer and J. R. Schroff, Jar-Cover Seals, ibid. 49:1393 (1970).

13. L. H. Sharpe, J. R. Schroff and F. J. Vaccaro, Post Seals for the New Bell System Battery, ibid. 49:1405 (1970).

14. R. H. Cushman, Techniques for Bonding the Positive Plates, ibid. 49:1419 (1970).

15. H. J. Luer, Incorporating the New Battery into the Telephone Plant, ibid. 49:1447 (1970).

16. See, e.g., M. E. Fiorino, F. J. Vaccaro and R. E. Landwehrle, Factors Affecting the Polarization Curve of the Negative Plate of the Lead-Acid Battery and Their Impact on Float Performance, Proceedings of the Intelec Meeting, San Diego, Oct. 30 -Nov. 2, 1988.

17. See, e.g., D. K. Coates and R. M. Barnett, Nickel-Hydrogen Life Testing, Proc. 23rd Intersociety Energy Conversion Engineering Conference, 2:483 (1988).

18. See, e.g., S. Srinivasan and R. H. Wiswall, Hydrogen Production, Storage, and Conversion for Electric Utility and Transportation Applications, Proc. Symposium on Energy Storage, The Electrochemical Society, Inc., Pennington, New Jersey, 1976.

19. See, e.g. (a) J-P. Gabano, ed., "Lithium Batteries", Academic Press, London, 1983; (b) H. V. Venkatasetty, ed., "Lithium Battery Technology", John Wiley & Sons, New York, 1984.

20. K. M. Abraham, 4th International Meeting on Lithium Batteries (4IMLB), Vancouver, May 24-27, 1988.

21. L. E. Brand, I. Chi, S. M. Granstaff, Jr. and B. Vyas, U.S.Patent 4,753,859, June 28, 1988.

22. F. A. Trumbore, 4IMLB, Vancouver, May 24-27, 1988: J. Power Sources, to be published.

23. J. P. Gabano, Recent Work in Organic Electrolyte Secondary Lithium Batteries, in: "Primary and Secondary Ambient Temperature Lithium Batteries", J. P. Gabano and Z. Takehara, eds., Proc. Vol. 88-6, p. 311, The Electrochemical Society, Inc., Pennington, New Jersey, 1988; also revised version of this paper from J. P. Gabano.

24. J. Broadhead, F. A. Trumbore and S. Basu, J. Electroanal. Chem. 118:241 (1981).

25. J. Broadhead, Proceedings of the Battery Applications Conference, Long Beach, California, January 11-14, 1988: B. Vyas, Proc. Power Sources Symposium, Cherry Hill, New Jersey, June 13-16, 1988.

26. J. R. Akridge, this book.

27. M. Shoji, "CMOS Digital Circuit Technology", Prentice Hall, Englewood Cliffs, New Jersey, 1988; S. M. Sze, "Semiconductor Devices. Physics and Technology", John Wiley & Sons, New York, 1985.

28. (a) S. M. Sze, ed., "VLSI Technology", Second Edition, McGraw-Hill Book Company, New York, 1988; (b) W. E. Beadle, J. C. C. Tsai and R. D. Plummer, "Quick Reference Manual for Semiconductor Engineers", John Wiley & Sons, New York, 1985; (c) R. W. Hunt, Memory Design and Technology, in: "Large Scale Integration", M. J. Howes and D. V. Morgan, eds., Wiley, New York, 1981; (d) L. C. Parrillo, VLSI Process Integration, in: VLSI Technology, McGraw-Hill, New York, 1983; (e) R. D. Rung, H. Momose and Y. Nagakubo, Deep Trench Isolated CMOS Devices, in: Tech. Digest IEEE Int. Electronic Device Meeting, 1982, p. 237; (f) J. M. Neve, F .D. Ray and J. P. Sitarik, Improving the Performance of an Integrated Circuit Manufacturing Ling, AT&T Tech. Journal 66 (no. 5):39 (1987). (g) R. D. Rung, Trench Isolation Prospects for Application in CMOS VSLI, IEDM Tech. Dig. 574 (1984): (h) A. K. Sinha, S. M. Sze and R. S. Wagner, Silicon Devices of Integrated Circuit Processing, in "Encyclopedia of Materials Science and Engineering", M. Bever, ed., Pergamon, Oxford, 1985.

29. (a) See, e.g., C. W. Pearce, Crystal Growth and Wafer Preparation, in Ref. 28a; (b) S. J. Bass and P. E. Oliver, J. Crystal Growth 3:286 (1968); (c) F. A. Trumbore, Solid Solubilities of Impurities Elements in Germanium and Silicon, Bell System Tech. J. 39:205 (1960).

30. See, e.g., L. E. Katz, Oxidation, in Ref. 28a.

31. L. Derick and C. J. Frosch, U. S. Patent 2,802,760, August 13, 1957.

32. See, e.g., A. C. Adams, Dielectric and Polysilicon Film Deposition, in Ref. 28a.

33. (a) See, e.g., S. P. Murarka, Metallization, in Ref. 28a; (b) S. Vaidya, D. B. Fraser and A. K. Sinha, Electromigration Resistance of Fine Line Al, in: Proc. 18th Annual Reliability Physics Symposium, IEEE, New York, 1980, p. 165; (c) K. Sugahara, T. Nishimura, S. Kusunoki, Y. Akasaka and H. Nakata, SOI/SOI/Bulk-Si Triple-Level Structure for Three-Dimensional Devices, IEEE Electron Device Lett., EDL-7:193 (1986).

34. See, e.g., H. C. Casey and F. A. Trumbore, Single Crystal Electroluminescent Materials, Materials Science and Engineering 6:69 (1970).

35. H. Nelson, RCA Rev. 24:603 (1963).

36. (a) M. B. Panish and S. Sumski, J. Phys. Chem. Solids 30:129 (1969); (b) F. E. Rosztoczy, Electron. Div. Abstr. of The Electrochem. Soc. 17:516 (1968).

37. J. J. Tietjen and J. A. Amick, J. Electrochem. Soc. 113:724 (1966).

38. (a) H. M. Manasevit, Appl. Phys. Lett. 12:156 (1968); H. M. Manasevit and W. I. Simpson, J. Electrochem. Soc. 116:1968 (1969); (b) See, e.g., M. Sato and M. Suzuki, Growth Rate of GaAs Epitaxial Films Grown by MOCVD, J. Electrochem. Soc. 134:1540 (1987).

39. See, e.g., M. B. Panish, J. Crystal Growth 81:249 (1987).

40. A. Y. Cho and J. R. Arthur, Prog. Solid State Chem. 10:157 (1975).

41. J. S. Harris, Jr., S. M. Koch and S. J. Rosner, The nucleation and growth of GaAs on Si, in: "Heteroepitaxy on Silicon II", J. C. C. Fan, J. M. Phillips and B-Y. Tsaur, ed., Materials Res. Soc. Proceedings, 91:3 (1987); D. W. Shaw, Epitaxial GaAs on Si: Progress and potential applications, ibid., p.15.

42. (a) T. P. Pearsall, H. Temkin, J. C. Bean and S. Luryi, IEEE Elec., Dev. Lett EDL-7:104 (1986); (b) J. C. Bean, Silicon Based Semiconductor Heterostructures, in "Silicon-Molecular Beam Epitaxy", E. Kasper and J. C. Bean, eds., CRC Press, Boca Raton, Florida, 1988; (c) J. C. Bean, Technological Prospects for Germanium Silicide Epitaxy, Mat. Res. Soc. Symp. Proc. Vol. 126:111 (1988).

43. A. S. Jordan, A. R. von Neida and R. Caruso, J. Cryst. Growth 70, 555 (1984).

44. See, e.g., K. Kobashi, K. Nishimura, Y. Kawate and T. Horiuchi, J. Vac. Sci. Technol. A 6:1816 (1988); M. Kitabatake and K. Wasa, ibid. p.1793; B. Singh, Y. Arie, A. W. Levine and O. R. Mesker, Appl. Phys. Lett. 52:451 (1988); K. Kurihara, K. Sasaki, M. Kawarada and N. Koshino, ibid. p. 437.

45. See, e.g., J. C. Angus, J. Vac. Sci. Technol. A 6:1778 (1988); S. R. Kasi, H. Kang and J. W. Rabalais, ibid. p. 1788.

46. W. L. Hsu, J. Vac. Sci. Technol. A 6:1803 (1988).

47. S. Luryi and S. M. Sze, Possible Device Applications of Silicon Molecular Beam Epitaxy, in Ref. 42b.

48. H. Z. Chen, J. Palaski, A. Yariv and H. Morkoç, Research & Development, January 1988, p. 61. Copyright 1988, Research and Development.

49. R. T. Bate, The Quantum Effect Device: Tomorrow's Transistor?, Scientific American 258 (No.3): 96 (1988).

50. J. N. Randall, M. A. Reed, R. J. Matyi, T. M. Moore, R. J. Aggarwal and A .E. Wetsel, Nanofabrication of Quantum Coupled Devices, in: "Advanced Processing of Semiconductor Devices II", Harold G. Craighead, J. Narayan, eds., Proc. SPIE 945, 137-145 (1988).

51. J. Hopfield, Brian, Computer and Memory, Engineering and Science, California Institute of Technology, Vol. XLVI, No. 1, p. 2, (1982).

52. C. D. Kornfeld, R. C. Frye, C. C. Wong and E. A. Rietman, An Optically Programmed Neural Network, presented at the IEEE International Conference on Neural Networks, San Diego, California, July 24-27, 1988.

53. See, e.g., the proceedings volume "Low Temperature Electronics and High Temperature Superconductors", S. I. Raider, R. Kirschman, H. Hayakawa and H. Ohta, eds., The Electrochemical Society, Inc., Pennington, New Jersey, 1988.

54. See, e.g., B. A. Biegel, R. Singh and F. Radpour, Ultra high speed electronics based on

proposed high Tc superconductor switching devices, in: "High-Tc Superconductivity: Thin Films and Devices", R. Bruce van Dover, Cheng-Chung Chi, eds., Proc. SPIE 948:3 (1988).

55. T. Katsuyama, M. A. Tischler, D. Moore, N. Hamaguchi, N. A. Elmasry and S. M. Bedair, New Approaches for High Efficiency Cascade Solar Cells, Solar Cells 21:413 (1987): T.Baer, Volts from the Blue, High Technology, July, 1986, p. 26.

56. (a) M. Mehregany, K. J. Gabriel and W. S. N. Trimmer, Sensors and Actuators 12:341 (1987); (b) W.S.N. Trimmer and K. J. Gabriel, ibid. 11:189 (1987).

57. R. Muller, as reported by R. Pool in news note in Science, 242:379 (1988); see also R. S. Muller, Tech. Digest of the 7th Sensor Symposium, 1988, p. 7.

THIN FILM TECHNOLOGY AND CHARACTERIZATION:

THEIR USE IN MICROIONIC DEVICES

M. Ribes

Laboratoire de Physicochimie des Matériaux Solides U.A. 407
Université des Sciences et Techniques du Languedoc, Place E. Bataillon
34095 Montpellier Cedex 5, France

Most of the electrochemical generators manufactured today were discovered at the end of the 19th century. This is the case of the manganese dioxide cell (Leclanché, 1868), the lead accumulator (Planté, 1859) and the ferro-nickel accumulator (Edison, 1900). The characteristics of these batteries and accumulators were considered to be acceptable by users for a long time. For the past fifteen years or so, mass-production of industrial objects with their own sources of power (computers for the popular market, medicine, photography, watch-making, space, etc.) has led to a new growth in research on new essentially solid-state types of energy storage.

Solid-state systems have several advantages:
- simple design;
- natural seal;
- resistance to shocks and vibration;
- increased resistance to pressure and temperature variations;
- broad stability range of electrolyte, enabling the use of redox couples corresponding to large differences in voltage;
- very great selectivity of charge carriers resulting in the prevention of losses by self-discharge.

Finally, and most important of all, solid-state devices can be miniaturized.

These capabilities and an ever-increasing need to integrate electrical circuits using the miniaturization of components have led to the emergence of a branch of research called "microionics" by analogy with microelectronics.

The fabrication of ionic microcomponents requires mastery of thin layer coating of various constituents (electrolytes, electrodes, etc.). This field has profited from the considerable efforts made to reduce the size of integrated circuits. The VLSI generation of circuits is now coming on to the market. The spectacular progress achieved is the result of the improvement or devising of methods of obtaining thin films.

The purpose of this paper is a) to summarize briefly thin film technology using chemical methods, b) to recall the different techniques of characterization of thin films and c) to provide a number of examples of ionic microcomponents.

I THIN FILM TECHNOLOGY

Depending on the nature of the physical or physicochemical procedure used, deposition techniques for obtaining thin films can be placed in two main categories:
- physical deposition methods
- chemical deposition methods.

Physical deposition techniques are described by A. Levasseur (elsewhere in the present book). I shall describe the various chemical methods used to obtain thin films (thickness < 1 μm) or thick films (thickness > 1 μm).

A. CHEMICAL VAPOR DEPOSITION

This technique is used mainly in microelectronics and the examples given are drawn mainly from this field.

"Chemical vapor deposition" (CVD) covers a set of techniques which consist of triggering a chemical reaction between gases in a chamber, giving a solid product which is deposited on a substrate. The film is thus the product of a chemical reaction. The basic features are:

1. Reactants and by-products must be volatile and stable so that they can be transported to and from the deposition zone. The solid substance should have low vapor pressure in deposition condition.

2. The reaction can be stimulated by various energy sources:
   (i) thermal;
   (ii) glow discharge, plasma;
   (iii) electromagnetic radiation.

3. CVD reactions involve gas phase (homogeneous) reactions *and* surface (heterogeneous) reactions. The former can lead to gas phase nucleation, which is a major problem in CVD processes.

4. The effect of process parameters (reactor geometry, flow, gas mixture, temperature and pressure) on deposition rate, film uniformity and material properties depends on a combination of fundamental aspects:
   (i) thermodynamics and kinetics;
   (ii) transport phenomena (flow, heat and mass transfer);
   (iii) film growth nucleation and growth phenomena.

The various chemical reactions involved in CVD can be classified as follows:

- Pyrolysis
  $$SiH_4(g) \rightarrow Si(s) + 2H_2(g)$$

- Reduction
  $$WF_6(g) + 3H_2(g) \rightarrow W(s) + 6HF$$

- Oxidation
  $$SiH_4(g) + 4N_2O(g) \rightarrow SiO_2(s) + 4N_2 + 2H_2O$$

- Hydrolysis
  $$Al_2Cl_6(g) + 3CO_2(g) + 3H_2(g) \rightarrow Al_2O_3(s) + 6HCl(g) + 3CO(g)$$

- Disproportionation
  $$2GeI_2(g) \rightarrow Ge(s) + GeI_4(g)$$

- Organometallic reactions
  $$(CH_3)_3Ga(g) + AsH_3(g) \rightarrow GaAs(s) + 3CH_4(g)$$

42

- Chemical transport reactions
$$6GaAs(s) + 6HCl(g) \rightarrow As_4(g) + As_2(g) + 6GaCl(g) + 3H_2(g)$$

All CVD deposition involves gas phase (homogeneous) reactions and surface (heterogeneous) reactions. The most classic example is that of silane pyrolysis. An overall reaction is $SiH_4 \rightarrow Si(s) + 2H_2$. A possible mechanism is:

Gas phase	Surface
$SiH_4 \Leftrightarrow SiH_2 + H_2$	$SiH_4 \rightarrow Si(s) + 2H_2$
$SiH_2 + SiH_4 \Leftrightarrow Si_2H_6$	$SiH_2 \rightarrow Si(s) + H_2$
$SiH_2 + Si_2H_6 \Leftrightarrow Si_3H_8$	$Si_2H_6 \rightarrow Si(s) + H_2 + SiH_4$

The various stages of the deposition process can be broken down in a similar way as follows:

1. Gas phase reactions.
2. Mass transfer of reactants from the main body of fluid to the deposition surface.
3. Adsorption of at least one reactant at the surface.
4. Diffusion and reactions at the surface (this may involve several steps and influence the growth mode).
5. Desorption of gaseous products of deposition reactions.
6. Mass transfer of products from the deposition surface to the main body of fluid.

If one step is significantly slower than the others it will be *rate determining*.

- Low pressure chemical vapor deposition (LPCVD)

Like any classic chemical reaction, the CVD process can be used at different pressures. In a great number of cases, and particularly when very pure materials are sought, operating at low pressure is advantageous. The efficiency of the operation is then considerably improved:
- less gas used;
- less heating power required (loss by conduction is limited);
- better uniformity and less defects in the layers.

Working pressures can vary from 10^{-1} to 20 torr. As an example, excellent quality thin films of PSG (phosphorus silica glasses) passivation glasses have been obtained with the gaseous mixture $Si(OC_2H_5)_4/N_2\ PO(OCH_3)_3/O_2$. Deposition pressure and temperature were 3 torr and 740°C respectively. Under these conditions, the deposition rate was 350 Å/min.

Plasma-enhanced (or assisted) chemical vapor deposition. PECVD (or PACVD)

Here, thermal chemical activation is partially replaced by plasma activation. RF electrical (glow) discharges are used. These are low pressure (0.05-5 torr) *non*-equilibrium plasmas with:

electron density	$\approx$	$10^9 - 10^{12}$ cm^{-3}
neutral density	$\approx$	10^{15} cm^{-3}
electron energy	$\approx$	1-10 eV
ion energies	$\approx$	0.04 eV

The large energy difference between electrons and ions (and neutrals) is the key to plasma CVD reactions. The electrons gain energy from the electrical field applied and then by collision transfer it to reactant molecules, producing free radicals, ions, atoms, excited species and photons. Plasma chemical interactions create films at lower gas temperatures than is possible in thermal CVD. The properties of such films are strongly dependent on plasma conditions.

PECVD modelling is not as developed as that of conventional CVD because of difficulties in characterizing the electrical discharge structure and reactions.

Photo CVD

Selective absorption of a gas is used to initiate the deposition process. A mercury vapor lamp (2537 Å) is generally used as an excitation source for the mercury vapor which acts as a photosensitizer. Energy is transferred from excited mercury atoms (Hg^*) to reactants by collision; this generates free radicals which interact. For example, the mechanism proposed for Si_3N_4 is as follows:

$$Hg_0 + h^\nu \rightarrow Hg^*$$
$$Hg^* + SiH_4 \rightarrow \cdot SiH_3 + \cdot H + Hg_0$$
$$Hg^* + NH_3 \rightarrow \cdot NH_2 + \cdot H + Hg_0$$
$$\text{Free radical reaction} \rightarrow Si_3N_4$$

Laser CVD

Lasers can drive CVD by two different mechanisms:

(i) Thermal process in which the gas phase or the substrate (more commonly) is heated;
(ii) Photochemical processes in the gas phase in the adsorbed layer or on the surface of the substrate.

Apart from the advantages of spatial resolution and rapid energy transfer in the laser beam, the thermal process is not different from conventional CVD.

The features of physical deposition techniques (RF sputtering, vacuum evaporation) can be combined with those of various CVD processes to increase considerably the methods of obtaining thin films. Table 1 gives an idea of the possibilities.

Table 1. Combination of various methods of preparation of thin layers

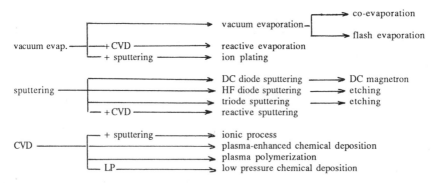

As the advantages of microionics have only emerged very recently, there are still very few examples of the utilization of CVD for the fabrication of electrolyte films and electrode materials. Several can nevertheless be mentioned:

TiS_2 films (1)

Titanium disulfide films (25 μm thick) were fabricated using PECVD with a substrate temperature of 450°C. RF power density was approximately $1W/cm^{-3}$ for 7.5 torr total gas pressure. The source gas was a mixture of 0.9% TiCl$_4$ and 7% H$_2$S diluted with Ar. The film microstructure consisted of small, narrow plate-like crystals with the crystallographic c-axis parallel to the substrate plane. The chemical diffusion coefficient of lithium ranged from 10^{-11} to 10^{-9} cm^2 s^{-1} depending on the Li concentration.

WO$_3$ films (2)

These films have interesting electrochromic properties. Polycrystalline tungsten oxide thin films were prepared using the following experimental procedure. Black tungsten (BW) films were deposited by pyrolysis at 400°C of W(CO)$_6$ in the presence of oxygen. Reflective tungsten (RW) thin films were obtained using the same CVD method but in the absence of oxygen. WO$_3$ thin films were then prepared by oxidation at 600°C in air for BW films and at 500°C in oxygen-enriched air for RW films.

GeSey vitreous films (3)

Good knowledge of GeSey glasses, which are also studied in our laboratory as inorganic resist, enabled us to make thin films with large areas ($\approx$ 80 cm^2) of ionic conductive Ag-Ge-Se glasses using a somewhat original method. Once sensitized by the silver and subjected to radiation (UV, X-rays, e$^-$ beam), the electrical properties of GeSey glasses become modified. They change from a very resistive semiconductor ($\sigma \approx 10^{-13}$ Ω^{-1} cm^{-1}, gap $\approx$ 2.2 eV) to pure ionic conductors ($\sigma \approx 10^{-5}$ Ω^{-1} cm^{-1} Ea $\approx$ 0.35 eV).

The procedure used to obtain the films is as follows:

- Low temperature ($\approx$ 100°C) deposition of GeSey glass by PECVD from germane (GeH$_4$) and hydrogen selenide (H$_2$Se) diluted in N$_2$. A schematic view of the reactor is shown in Figure 1.

- Sensitization of the film obtained by a solution of Ag$^+$ ions or by deposition of silver metal.

- Exposure to UV light or thermal treatment.

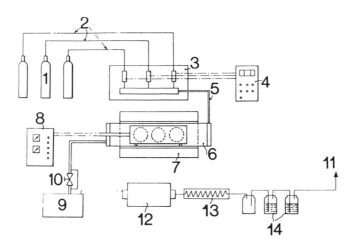

SYNOPTIC SCHEME

Figure 1. Diagram of a PECVD reactor

B. SOL-GEL PROCESS

This synthetic process, which can be used to obtain either ultra-fine powders which can be used for silk screen printing applications or monolithic materials (in bulk or film form), has only gained significant importance over the past fifteen years, and this has been mainly in the field of inorganic oxides. "Sol" and "gel" states must be defined briefly before describing the process and illustrating it with two examples in solid state ionics.

- **Sols**: these are stable colloidal dispersions in a liquid. Particle size ranges from 10 to 1000 Å and they contain 10^3 to 10^9 atoms. Distinction can be made between three main types:

- macromolecular sols;
- micellar sols or aggregation colloids which consist of amphiphile molecules;
- inorganic sols.

- **Gel**: this is a biphase medium formed by a porous solid whose pores are filled with liquid. The solid phase is formed on more or less reticulated polymers. Porosity is open.

Inorganic oxide gels

Two methods can be used to obtain inorganic oxide gels:

a) A colloidal solution can be destabilized by adding a solution of electrolyte which reduces the charge of the colloidal particles by raising the temperature or by varying the pH. A classic example is that of silica gel. The process can be summarized as follows:

b) Hydrolysis and polycondensation of organometallic compounds

After mixing suitable proportions of an organometallic compound M(OR)n, water and a solvent, the true single-phase system evolves towards a bi-phase system of metal oxide $MO_{n/2}$ with pores filled with solvent. The following simplified reaction mechanism has been proposed (4):

$$M(OR)n + nH_2O \rightarrow M(OH)n + n\ ROH$$
$$M(OH)n \rightarrow MO_{n/2} + n/2\ H_2O$$

Hydrolysis is not as easy when the length of the alkyl R chain increases. In addition, this reaction mechanism does not take into account the complexity of the process which leads to the intermediate compound $MO_{n/2 - x - y}(OH)_x(OR)_y$.

There are two ways of preparing multicomponent oxides:

- all the cationic constituents are added in organometallic form. The rates of hydrolysis of the organometallic substances are frequently different, and homogeneity can be maintained by:

46

- prehydrolysis of the slowest organometallic substance before mixing;
- introduction of some constituents in soluble salt form; this is generally used for nitrates or acetates because of their solubility and their low decomposition temperature.

Organic - inorganic polymers

Although the sol-gel process has been used for a long time for the synthesis of silicons by hydrolysis and polycondensation of organochlorosilane, it was only recently that Schmidt (5) developed synthesis of organosilicon compounds using the actual sol-gel process. The substances obtained are commonly known as ORMOSILS (ORganic MOdified SILicates). One of the advantages of the sol-gel method is the possibility of obtaining gels directly in various forms:
- fine, homogeneous powders which can be used for the fabrication of ceramics or applied in layers by screen printing;
- layers from a few μm to a few tens of μm thick obtained by dip-coating. A substrate is dipped in an alcoholate solution and withdrawn at a controlled, constant rate. This can be repeated several times. Polymerization takes place spontaneously in contact with atmospheric water vapor.
- monolithic gels obtained by slow hydrolysis. The solvent is generally removed under hypercritical conditions to prevent cracking during drying.

A general diagram of a sol-gel method is given in Figure 2.

From gel to material

Gels can be used as they are in certain cases, but it is usually necessary to apply thermal treatment to obtain material in the form desired. Two examples in the field of solid state ionics illustrate these two possibilities.

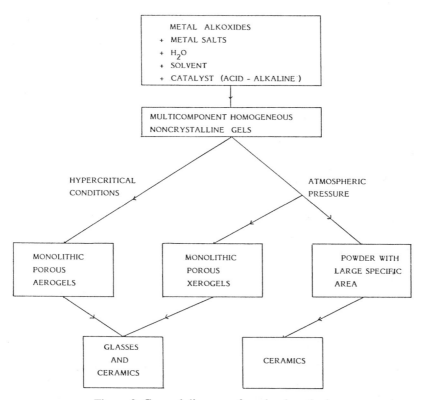

Figure 2. General diagram of a sol-gel method

a) Aminosils

A team in Grenoble, France (6), recently developed new solid electrolytes from ormosils by dissolving an ionic compound in the organosilicon matrix. The silica backbone provides mechanical strength and an amorphous state. The organic groups offer solvating properties with respect to the guest ionic compounds.

For example, hydrolysis and polycondensation of trimethoxiaminopropyl silane in the presence of a strong acid led to obtaining hard, transparent, non-porous films which were stable in ambient atmospheric conditions. The reaction diagram is as follows:

$$(CH_3O)_3 - Si - (CH_2)_3 - NH_2 \quad \xrightarrow[\text{hydrolysis}]{H_2O} \quad HO - \underset{\underset{OH}{|}}{\overset{\overset{OH}{|}}{Si}} - (CH_2)n - NH_2$$

$$\xrightarrow[\text{+ HX}]{\text{polymerization}} \quad \begin{array}{c} | \\ -Si - (CH_2)_3 - NH_2 \\ | \\ O \qquad\qquad (HX) \\ | \\ -Si - (CH_2)_3 - NH_2 \\ | \end{array}$$

Figure 3 shows the variations of conductivity Arrhenius diagram for an aminosil containing perchloric acid. Behavior obeys the free volume model. Proton conductivities of around 10^{-5} $\Omega^{-1}cm^{-1}$ were obtained at room temperature.

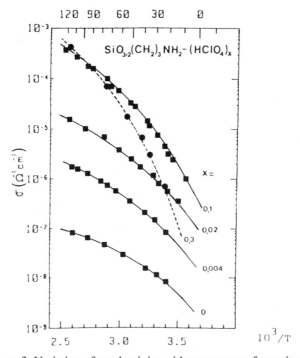

Figure 3. Variation of conductivity with temperature for aminosils

This new concept of solid electrolytes based on sol-gel process is opening up a very attractive field of investigation for the synthesis of new, high-performance materials in the field of solid state ionics.

b) Nasicons

Nasicons (an acronym for Na superionic conductors) were introduced as 3D solid electrolytes (7). Stoichiometry is $Na_3Zr_2Si_2PO_{12}$. Classic methods of synthesis of these materials leave traces of ZrO_2. These impurities collect at grain boundaries and affect the electrical properties. Sol-gel methods can be used to prepare perfectly pure, homogeneous fine Nasicon powders which will sinter at low temperature. There are several types of reaction mixture, as shown below.

zirconium
propoxide
$$\left\{\begin{array}{l} + \text{ sodium silicate } + NH_4H_2PO_4 \\ \\ + \text{ tetraethoxisilane } + NH_4H_2PO_4 \qquad + \qquad \text{Na butylate} \\ \quad \text{(TEOS)} \\ \\ + \qquad \text{``} \qquad\qquad + \text{ tributylphosphate } + \qquad \text{``} \end{array}\right.$$

These two examples show the potential of this "soft chemistry" technique. However, although it has given rise to industrial applications in certain fields such as vitreous materials, it is still at research laboratory stage in solid state ionics.

C. SILK SCREEN PRINTING

This technique is well-known in microelectronics since it is the basis of hybrid technology (screen-printing of calibrated junction resistances between different components). It is used to deposit layers which are generally several microns thick.

The material to be deposited is used in the form of a fine powder in a frequently complex mixture consisting of:
- an organic phase (solvent to facilitate screen printing);
- an organic binding phase (cellulose);
- a vitrifying glass phase (glass PbO, SiO_2, B_2O_3).

This mixture is frequently referred to as ink and is in the form of a thixotropic paste (viscosity lowered under the effect of pressure). The paste is applied with a squeegee through textile printed with perfectly defined patterns. The annealing phase is next, with three temperature phases:
- removal of solvent,
- removal of binder,
- sintering,
leading to obtaining a thick layer >10μm.

II THIN FILM CHARACTERIZATION

Here too we profit from technical facilities developed essentially for the characterization of semi-conductors. These techniques use the response produced by the interaction between a beam of particles or photons and the sample.

• **Scanning electron microscopy (SEM)**
SEM provides detailed information on the topography of a surface with magnification of up to x200 000.

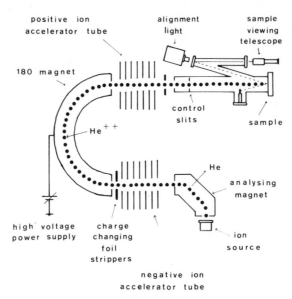

positive ion
accelerator tube

alignment
light

sample
viewing
telescope

180 magnet

control
slits

sample

He^{++}

He

analysing
magnet

high voltage
power supply

charge
changing
foil
strippers

ion
source

negative ion
accelerator tube

Figure 4. RBS configuration

• **Energy dispersive X-rays (EDX) and wavelength dispersive X-rays (WDX)**
Coupled with SEM, these systems analyse the X spectra emitted by interaction between the SEM electron beam and the sample. Depth analysed is in the order of 1 to 3 μm.

• **Auger electron spectroscopy (AES)**
AES can be used to detect all elements except hydrogen and helium by analysis of the energy of electrons ejected after excitation by an electron beam. Information is thus obtained on the first atomic layers.

• **Electron spectroscopy for chemical analysis (ESCA) (often called XPS for "X-ray photoelectron spectroscopy")**
X-rays cause photoemission of electrons whose binding energy is characteristic of their parent atom. Data is thus obtained on the first 10 atomic layers.

• **X-ray fluorescence spectroscopy (XRF)**
This is a simple technique which is rapid to use; it provides information on Z>10 elements.

• **Rutherford backscattering spectroscopy (RBS)**
A schematic diagram of an RBS spectrometer is shown in Figure 4. The probe particle is typically ^{4}He$^+$ with an energy of about 1 to 3 MeV. The helium beam strikes the sample, a small fraction collides with sample atoms and is backscattered. The energy lost depends on the mass of the "target" atoms and the amount of energy lost by the beam as it travels in the sample. Sample depth is usually between 0.5 and 0.3 μm and depends on the concentration and atomic masses of the elements in the sample.

• **Secondary ion mass spectroscopy (SIMS)**
The surface of the sample to be analysed is bombarded with an ion beam (1-15 keV) which sputters it. A small fraction of the atoms take the form of ions which are analysed with a mass spectrometer.

Table 2. Typical parameters for the materials analysis techniques

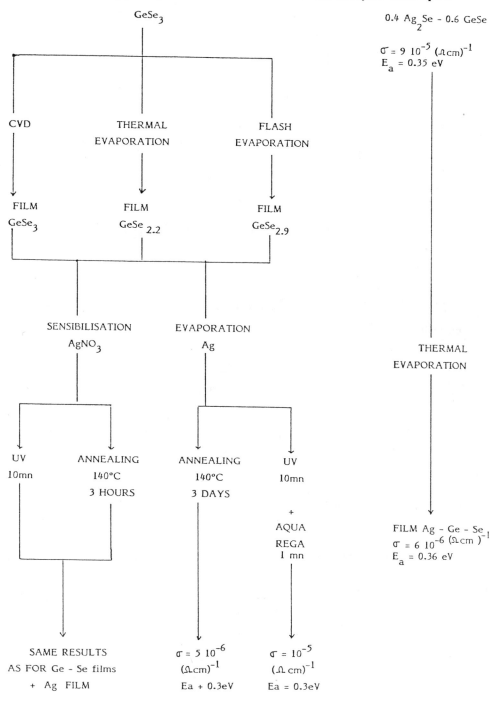

$$\text{GeSe}_3$$

$$0.4 \text{ Ag}_2\text{Se} - 0.6 \text{ GeSe}$$

$$\sigma = 9 \ 10^{-5} \ (\Omega\text{cm})^{-1}$$
$$E_a = 0.35 \text{ eV}$$

CVD   THERMAL EVAPORATION   FLASH EVAPORATION

FILM GeSe_3 FILM $\text{GeSe}_{2.2}$ FILM $\text{GeSe}_{2.9}$

SENSIBILISATION AgNO_3 EVAPORATION Ag

THERMAL EVAPORATION

UV 10mn   ANNEALING 140°C 3 HOURS   ANNEALING 140°C 3 DAYS   UV 10mn

+

AQUA REGA 1 mn

FILM Ag - Ge - Se
$$\sigma = 6 \ 10^{-6} \ (\Omega\text{cm})^{-1}$$
$$E_a = 0.36 \text{ eV}$$

SAME RESULTS AS FOR Ge - Se films + Ag FILM

$$\sigma = 5 \ 10^{-6} \ (\Omega\text{cm})^{-1}$$
$$Ea + 0.3eV$$

$$\sigma = 10^{-5} \ (\Omega \text{ cm})^{-1}$$
$$Ea = 0.3eV$$

- **Laser ionisation mass spectrometry (LIMS), also called laser microprobe mass analysis (LMMA)**

The principle of analysis is identical to that of SIMS. In this case, a high energy (10^7 to 10^{12} W cm^{-2}) finely focused (beam diameter $\approx$ 1 µm) laser pulse is used to eject ionized atoms or molecular species.

Table 2 gives a qualitative overview and comparison of several thin film analysis methods. It is important to note that in many cases complete characterization can only be achieved by using a combination of several techniques. In addition, electrical characterization, such as impedance measurement, is especially useful for solid electrolyte thin films.

AN EXAMPLE: THIN FILMS OF SOLID VITREOUS ELECTROLYTE CONDUCTING BY Ag$^+$ IONS. GLASS: Ag-Ge-X(X = S, Se)

Fast ionic conductors (FIC) by Ag$^+$ ions, and amongst them glasses, were among the first solid electrolytes to be used in all solid state electrochemical devices. Their advantages subsequently decreased and they have been replaced by lithium conductors. However, today the miniaturization of such devices combined with the possibility of constructing interdigital systems are opening up new perspectives for these glasses. The purpose of the example which follows is to illustrate the different possible ways of making thin films of Ag - Ge - X (X = S, Se) and above all to show the importance of the various analyses and physico-chemical treatments necessary to obtain homogeneous films.

Ag-Ge-S glasses

Thin films of these glasses have been made by vacuum evaporation, flash evaporation and by sputtering.

- **Vacuum evaporation**: as is shown by RBS results and electrical measurements (Figures 5 and 6), the film is not homogeneous. It consists of two layers: a lower layer of GeSx (1<x<2) and a surface layer rich in silver (composition close to Ag$_2$S). This non-homogeneity of composition is clearly shown by complex impedance: there are two arcs. Variations in conductivity in the Arrhenius equation confirms that the most conductive layer consists essentially of Ag$_2$S. Indeed, allotropic transformation is clearly visible at 170°C. It it thus necessary to apply homogenization treatment before using these films. As can be seen in Figure 7, annealing at 200°C for several hours enables the silver to diffuse to the substrate; it can also be seen that germanium is present as far as the surface of the film. However, there is still a small Ag gradient.

- **Flash evaporation**: RBS analysis shows that in this case Ag is present throughout the thickness of the layer (Figure 8). However, a high concentration of this element still remains near the surface of the film requiring thermal homogenization treatment here as well.

- **Sputtering**: this technique can be used to obtain homogeneous films which do not require any special treatment. However, deposition rate is very low ($\approx$ 800 Å/h) and the setting up of targets is very delicate.

Ag-Ge-Se glass

The PECVD procedure previously described gives perfectly homogeneous films ($\sigma25°C \approx$ 5 10^{-6} Ω^{-1} cm^{-1}, Ea $\approx$ 0.38 eV). The different possible ways of preparing Ag-Ge-Se glass thin films are shown in Figure 9.

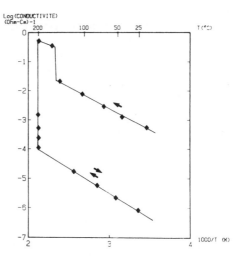

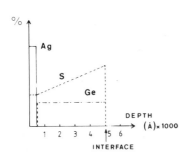

Figure 5. RBS analysis of a thin film of Ag-Ge-S glass obtained by thermal evaporation.

Figure 6. Electrical behavior of the film before annealing.

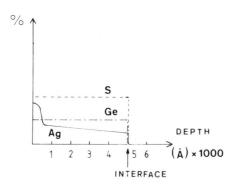

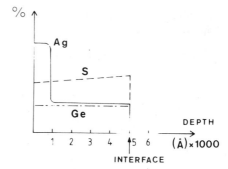

Figure 7. RBS analysis of a thin film of Ag-Ge-S glass obtained by thermal evaporation and annealed at 200°C for one week.

Figure 8. RBS analysis of a thin film of Ag-Ge-S glass obtained by flash evaporation.

III SOLID STATE MICROBATTERIES

One of the main advantages of solid state devices is the possibility of miniaturization. Although use of solid-state cells makes it possible to overcome a number of inherent disadvantages of liquid electrolytes, the "low" conductivity of solid electrolytes permits currents of only a few tens to several hundred microamperes.

Miniaturization thus has a two-fold advantage: reduction of the internal resistance of cells and above all enabling them to be integrated next to the circuits in electronic boards to

Technique name	SEM	EDX/WDX	AES	ESCA/XPS	XRF	RBS	SIMS	LIMS
Probe particle	Electrons Beam	Electrons Beam	Electrons Beam	X-Rays	X-Rays	Ions Beam	Ions Beam	Laser Beam
Detected entities	Electrons emitted or scattered	X-Rays emitted	Electrons emitted	Electrons emitted	X-Rays emitted	Energy of back scattered ions	Ions emitted	Ions emitted
Detectable elements	Surface topography	All>Be	All>He	All but H	All>O	All>He	All	All
Detection limits (atom %)*		1(EDX) 10^{-4}(WDX)	0.3-3	0.1-5	$10^{-4}-10^{-1}$	$10^{-2}-5$	$10^{-7}-10^{-1}$	$10^{-4}-10^{-2}$
Routine small aera analysis (diameter)	0.1 µm	1-3 µm	<1 µm	0.2-0.5mm	Sub µm to several µm	~ 1mm	Sub µm to several µm	~1 µm
Surface sensitivity		>1 µm	~10-40Å	~ 40Å	~1 µm	Depth resolution 25-200Å	~9-10Å	
Bulk sensitivity (>10 µm)	No	No	No	No	Yes	No	No	No
Depth profiling **	No	No	Yes (S.N)	Yes (S.N)	No	Yes (No S.N)	Yes	Yes (rough)
Analysis time (Routine)		2mn (EDX) 30mn (WDX)	5 mn	5 mn	15 mn	15 mn	5 mn	<1

* Detection limits are expressed as a range because differents elements have differents sensitivities

** Sputtering needed

Figure 9. Different ways of preparing Ag - Ge - S glass thin films

Table 3. Solid state cells realized for thin (or thick) films

Electrolyte	Deposition Techniques	Device (t = thickness)	Ref.
$RbAg_4I_5$	Liquid phase epitaxy	$Ag/SCI/I_2\text{-}RbI_3(NbSe_2)$ $t \sim 0,5\ \mu m$	8
$RbAg_4I_5$	Flash evaporation	$Ag/SCI/Au(Te)$ $t \sim 1\text{-}4\ m$	9
$RbAg_4I_5$ $(Ag_7I_4PO_4,\ Ag_4{,}5HgI_4,\ Ag_3SI)$	Chemical deposition	$Ag/SCI/I_2,C$	10
$Ag_{19}I_{15}P_2O_7$	RF Sputtering Flash evaporation	$Ag/SCI/Te$ $t \sim 2,5\ \mu m$	11
AgI	Evaporation	$Ag/SCI/Pt$	12
$AgBr$	Evaporation	$Ag/SCI/Pt$ $t \sim 5\text{-}12\ \mu m$	13
LiI	Electrolytic deposition	$Li/SCI/I_2$ $t \sim 1\ \mu m$	14
LiI	Evaporation of organic solution	$Li/SCI/AgI\text{-}Ag$	15
LiI	Evaporation	$Li/SCI/AgI$ $t \sim 15\ \mu m$	16
$LiI - LiBr$		Batteries for pacemakers	17
$Li_{3,6}Si_{0,6}P_{0,4}O_4$	RF Sputtering	$Li/glass/TiS_2$ $t \sim 5\ \mu m$	18
$Li - P - O$	RF Sputtering	$Li/glass/V_2O_5\text{-}WO_3$	19
$Li_2O - B_2O_3 - SiO_2$	RF Sputtering	$Li/glass/TiS_2$ $t \sim 0,2\ mm$	20
$Li_2O - B_2O_3$	Evaporation	$Li/glass + Polymer /$ $V_2O_5\text{-}P_2O_5$ $t \sim 10\ \mu m$	21
$Li_2O - P_2O_5 - B_2O_3$	Evaporation	$Li/glass/V_2O_5\text{-}TeO_2$ $t \sim 5\ \mu m$	21

protect read-write memories for example, which require practically only the application of voltage in case of power failure.

Another great advantage of miniaturized systems is the possibility of developing interdigital devices: assembly of microcells in series to supply high voltages, assembly in parallel for strong currents. In this case, the range of redox couples available would no longer be reduced to the Li/Li^+ combination alone, but other systems could probably be used (Ag, Cu, Na). Table 3 shows the different cells designed in the form of thin (≈ 1 μm) films using silver or lithium conductors.

In most cases, the devices use solid electrolyte and crystallized positive material. The problems encountered in electrochemistry are essentially those of charge transfer at the interfaces. Very high transfer resistances (compared to estimations) or even dendritic growth have been observed in some of the devices mentioned in Table 3. We have used vitreous electrolyte/electrode systems to avoid these serious disadvantages for rechargable systems. The glasses chosen have the same network former as far as possible. It can thus be hoped that there will be a larger area of contact at the interface and quasi-delocalization of the interface. Table 4 shows the results obtained since 1980 on thin films of vitreous ionic conductor.

Trials were carried out on lithium cells by Miyauchi (18). The electrochemical chain was as follows: $Li/Li_{3.6} Si_{0.6} P_{0.4} O_4/TiS_2$. Open circuit voltage was 2.45V. Cycling of this cell was carried out with currents of between 6 and 16 μA/cm^2 for discharge of 20 to 75%; rechargability was good (over 2000 cycles with less than 20% loss). Recently, the same authors (19) designed an "all glass" cell with the following chain: $Li/Li.Si.P.O/V_2O_5$-WO_3. The cathode film was deposited by reactive sputtering.

Table 4. **Thin films of solid vitreous electrolyte with conduction by Li^+ and Ag^+ ions**

Electrolyte	Deposition Technique	Conductivity $(\Omega\ cm)^{-1}$	Activation energy eV	Ref.
$-Ag^+$				
$AgI - Ag_2MoO_4$	Flash evaporation	10^{-2}	0.26-0.3	22
$AgI - Ag_2O - B_2O_3$				
$Ag_2O - P_2O_3$	RF Sputtering	$10^{-7} - 10^{-8}$	0.5	23
$Ag_2O - B_2O_3-P_2O_5$				
$Ag_2S - GeS_2$	Flash Evaporation	10^{-3}	0.3	23
$-Li^+$				
$Li_2O - B_2O_3$	Evaporation	$10^{-8} - 10^{-10}$	0.6-0.9	(24-25-20)
$Li_2O -B_2O_3 - X$ $(X=SiO_2, Li_2SO_4, LiI)$	RF Sputtering	10^{-6}	0.6	26
$Li_2O - SiO_2 - ZrO_2$	Magnetron sputtering	10^{-6}	0.57	27
$Li_{3,6}Si_{0,6}P_{0,4}O_4$	RF Sputtering	5.10^{-6}		18
$Li_2S - B_2S_3$	Evaporation	10^{-3}	0.4	28
$Li_2S - SiS_2$	Evaporation	10^{-5}	0.4	29

Table 5

Composition	:	Electronic conductivity Ω^{-1} cm^{-1} at 25°C
$0.6\ V_2O_5$ — $0.4P_2O_5$	:	$2\ 10^{-6}$
$0.6\ MoO_3$ — $0.4P_2O_5$	:	10^{-9}
$0.6\ WO_3$ — $0.4P_2O_5$	:	10^{-6}
$0.4\ Fe_2O_3$ — $0.6P_2O_5$	:	10^{-7}
$0.8\ V_2O_5$ — $0.2TeO_2$	:	$2\ 10^{-5}$

Levasseur et al. (20) designed a Li/Li_2O-B_2O_3-SiO_2/TiS_2 cell. In this case, thin film technology was only applied to the electrolyte. The active mass of the positive was formed by pressing.

As has already been mentioned, we used vitreous material for the cathode. The electronic conductivity in oxide glasses results from the presence of a transition element in two states of valence. Table 5 shows the composition and electronic conductivity of some glasses.

In bulk glasses it is possible to intercalate chemically or electrochemically up to 1 lithium per vanadium for example, with less than 4% change in volume (30). Like electrolyte glasses, these mixed ionic/electronic conductive glasses can be obtained in thin films. We thus made microcells (Figure 10) by successive deposition operations with the following electrochemical chain:

Li/Li^+ glass/V_2O_5 - TeO_2 glass; O.C.V. 3.2V.

The electrolyte glasses are oxide glasses of the Li_2O - B_2O_3 and Li_2O_3 and Li_2O - B_2O_3 - P_2O_5 systems. The lithium was protected by a film of hexamethyldisiloxane (1 μm thick) applied by plasma deposition. Typical discharge curves are shown in Figure 11. The internal resistance of the cells was greater than expected (loss of lithium doubtlessly occurred during evaporation of the electrolyte glass) and caused considerable ohmic drop.

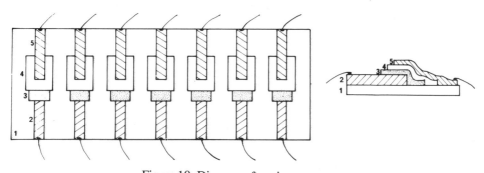

Figure 10. Diagram of a microgenerator.
1, insulating substrate; 2, branch connection; 3, cathode; 4, electrolyte; 5, anode.

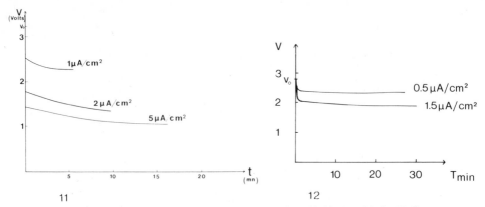

Figs. 11, 12. Intensiostatic discharge curve of a Li/Li+ glass/V_2O_5-TeO_2
Figure 11. Glass = Li_2O-B_2O_3-P_2O_5
Figure 12. Glass = Li_2S-SiS_2-P_2S_5

We have very recently started to make the same type of cell using sulfide glasses as electrolyte. The first intensiostatic discharge curves obtained are shown in Figure 12.

Although these studies are not very extensive, they have opened the way to the fabrication of microcells using ordinary integrated circuit construction techniques. All the work shows that it is difficult to achieve accurate control of film composition and texture; this accounts for the sometimes considerable deviation in electrical properties in comparison with the corresponding bulk materials.

THE PHYSICAL FORMATION PROCESSES OF THIN FILMS, THEIR CHARACTERIZATION
BY XPS, AES AND SIMS AND THEIR APPLICATIONS IN MICROBATTERIES

Alain Levasseur, G. Meunier, R. Dormoy and M. Menetrier

Laboratoire de Chimie du Solide du CNRS
Ecole Nationale Superieure de Chimie et Physique de Bordeaux
Universite de Bordeaux I, 33405 TALENCE CEDEX (France)

Deposition of thin films with controlled properties requires an operating environment that interferes as little as possible with the process of film formation. To minimize the interaction between residual gases and the surface of growing films one has to operate under vacuum.

I - VACUUM TECHNOLOGY

The term vacuum theoretically means a space without any residual matter. Usually it applies to an environment with a pressure lower than atmospheric. There are various degrees of vacuum (pressure in mbar) :

- low vacuum for $1 <$ Pressure < 1013 mbar

- medium vacuum for $10^{-3} <$ Pressure < 1 mbar

- high vacuum for $10^{-7} <$ Pressure $< 10^{-3}$ mbar

- ultrahigh vacuum for Pressure $< 10^{-7}$ mbar

The units that express the pressure remaining in a "vacuum" system are based on the force exercised by atmospheric air under standard conditions. This force is equal to 1.03 Kg cm^{-2}.

Solid State Microbatteries
Edited by J. R. Akridge and M. Balkanski
Plenum Press, New York, 1990

Gas pressure is expressed in terms of the height of a column of mercury supported by such a pressure in a barometer. Atmospheric pressure of 1.03 Kg.cm^{-2} will support a 760 mm high column of mercury.

1 torr = 1/760 of atmospheric pressure, i.e. 1 mbar = 0.75 torr = 10^2 Pascal The lowest pressures attainable reach about 10^{-13} mbar. The residual gas in a clean ultrahigh vacuum system consists mostly of hydrogen. It is the most universal contaminant, permeating nearly all materials. So far there is no known way to condense or trap it completely.

Two different methods may be employed to reduce the pressure in a vacuum enclosure. The first involves physical removal of gases from the vessel and exhausting of the gas load to the outside. The other evacuation method relies on condensation or trapping of gas molecules on some part of the inner surface of the vessel without gas exhausting.

I-A - VACUUM PUMPS

Five different types of pumps are mainly used to evacuate a vacuum system for the deposition of thin films:
- displacement pump (i.e., a mechanical pump)
- vapor stream pump (i.e., a diffusion pump)
- turbo-molecular pump
- cryogenic pump
- chemical getter pump

A-1) Displacement pump

Mechanical pumps move gases by cyclic motion of a system of mechanical parts. There are two main systems represented respectively in Figures 1 and 2, a rotary vane pump and a root pump. A definite volume of gas from the container can be exhausted, compressed and released toward the atmosphere. The lowest pressure obtained by a single stage rotary vane pump is about 10^{-2} mbar. A twin stage pump set up in series may reach about 2 x 10^{-3} mbar. The main advantage of the root pump is their ability to handle a large gas delivery. They offer economical advantages in industrial applications requiring 10 to 10^{-2} mbar.

A-2) Diffusion pump

When the pressure in the vacuum chamber decreased to values where the mean free path exceeds the dimension of this volume, the residual gas molecules collide more frequently with the walls than with each other. Conventional pumping process then became inefficient. It becomes necessary to wait until the gas molecules travel into the inlet opening of the pump and then give them a preferred motion direction by momemtum transfer. This principle has been described in 1913 by Gaede.

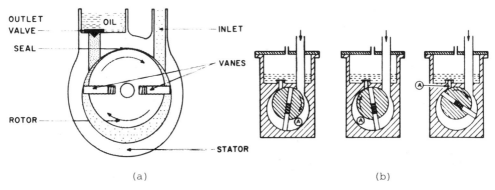

(a)   (b)

Figure 1. **Scheme of a rotary vane pump (a) and operating principle
(b) (1).**

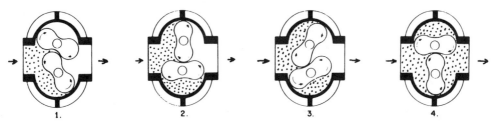

Figure 2. **Operating principle of a root pump (1).**

A work fluid such as a stable mineral oil is heated by a boiler, the hot vapor rising in a chimney. The direction of the flow is reversed at the jet cap, so that the vapor crosses a nozzle pointing away from the high vacuum side. During the exhaust from the nozzle the vapor expands. This expansion changes the normal molecular velocity distribution by creating a component in the expansion direction which is larger than the velocities associated with static gas at thermal equilibrium. The vapor jet moves with a velocity which is supersonic with respect to its temperature towards the water-cooled walls of the pump housing. Residual gas molecules from the high vacuum side diffusing to the jet are pushed downwards into the pump and compressed into the exit side. Thus a zone of reduced gas pressure is generated in the vicinity of the jet and more gas molecules from the high vacuum side diffuse towards this region. To enhance the pumping action, most diffusion pumps typically use three jets working in series. After the last stacking process the accumulated gas load must be removed by a backing pump (e.g., an oil sealed rotary pump).

The vacuum generated by a diffusion pump is not completely free of pump fluid molecules. It is the backstreaming effect which is impossible to eliminate completely despite several sophisticated additional systems.

The pressure obtained with such a system reaches about 10^{-8} mbar.

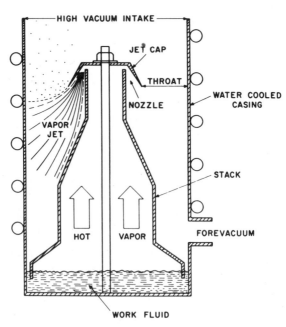

Figure 3. Diffusion pump (1).

A-3) Turbo-molecular pump

The turbo-molecular pump works like a diffusion pump. The principle of the molecular drag pump is based on the directional velocity imposed on gas molecules which strike a fast moving surface. If there is a second surface opposite to the first one this process is repeated a second time. A modern turbo-molecular pump contains alternate axial stages of rotating and stationary disks and plates with inclined coil channels. The preferential direction is given to the gas molecules by a fast moving turbine. With this system the pressure can reach 10^{-9} mbar without oil contamination, which is the main advantage of the turbo-molecular pump.

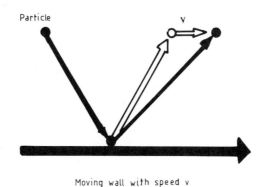

Figure 4. **Principle of the molecular pump** (2).

A-4) Cryogenic pumps

The action of cryopumps is based on the fact that the gas and the vapor molecules present in a vacuum chamber condense on a deep-cooled surface. The pump works at low pressure. It can be cooled by liquid nitrogen, hydrogen or helium. The working range of a cryopump is between 10^{-3} and 10^{-10} mbar. Among all high vacuum pumps, cryopumps have the highest pumping speed and also give rise to a very clean vacuum. Cryopumps are versatile: they can be used both as main pump and as an additional pump in conventional vacuum systems.

A-5) Getter pumps

Getter pumps are trapping devices which use reactive materials to fix residual gases on the inner surface. The trapping agents are freshly prepared metal films which catch gases using different mechanisms. This kind of pump does not utilize working fluids, they are free of organic contaminants and thus have an advantage over diffusion pumps. The most effective getter metal is tantalum, but for practical reasons molybdenum and titanium are the most used. As continued adsorption leads to saturation of the metal surface the trapping phenomenon stops at a certain moment. If the metal results from a sublimation of heated wires for example, it adsorbs all common gases except the rare gases and the adsorption products should be stable. In this case too, molybdenum and titanium are the most commonly used metals.

It is interesting to notice that the pumping systems equipped with diffusion pumps are the cheapest but that the cryopumping systems are only slightly more expensive. Turbo-molecular pumping systems, although cheap for small devices, become expensive for larger plants. Other more or less sophisticated pumping systems exist like evaporation pumps or ion sputtering pumps. They are used in ultrahigh vacuum technologies, but are generally not required in the thin film deposit processes.

II - FILM DEPOSITION PROCESS BY EVAPORATION AND CONDENSATION IN HIGH VACUUM

Thin films of materials with such different properties as metals, halides, oxides and sulfides can be obtained in crystalline or amorphous state by condensation of the vapor. The mechanism involved in film formation may be purely physical, a simple condensation, or may involve chemical reactions.

Evaporated films have been probably first carried out by Faraday in 1857 when he evaporated metal wires in an inert atmosphere. The deposition of metallic films in vacuum by resistance heating of platinum wires has been performed by Nahrwold in 1887 and only a year later this technique was used by Kundt to produce films for measuring the refractive indices of metals. In the following period, evaporated thin films led only to academic interest, although the vacuum evaporation of metal films by Pohl and Pringsheim in 1912 was performed under better developed technological conditions.

The technique of thin film formation by condensation at very low pressures has been developed parallel to industrial realization of techniques of high vacuum production in large volumes, using clean very high speed pumps.

Films obtained by condensation alone are usually made at pressures between 10^{-6} and 10^{-8} mbar down to the lowest values that modern vacuum technology can easily attain. This technique is combined with fast deposition rates to produce deposits in which only few foreign gas atoms are incorporated.

Of great technical importance was the discovery and the development of the reactive evaporation process by Auwarter in 1952 and Brinsmaid in 1953. Such processes which involve a chemical reaction between evaporated starting materials and the gaseous atmosphere are carried out mainly in the 10^{-4} mbar region and imply a slow deposition rate. The reaction is often activated either thermally or by UV radiation or by ionic or electronic bombardment.

The industrial development of those evaporation and condensation techniques results largely from the relatively easy obtaining of pure metal films or rigorously stoichiometric deposits of uniform thickness. The laws which determine the phenomena at low pressure are better defined than in condition close to atmospheric pressure. At very low pressures the mean free path of the vapor atoms or molecules exceeds the usual distance between evaporation source and substrate. Thus the substrate receives a vapor flux which moves in straight lines with no or only very few gas/vapor collisions and the spatial distribution of the condensate follows purely geometrical laws.

However, if the residual pressure has a value such that the source to substrate distance corresponds to the mean free path of the vapor atoms, only 37% of the atoms move without collision with the residual gas molecules. A residual pressures above 10^{-3} mbar, the propagation of the evaporated atoms between source and substrate is exposed to an isotropic diffusion effect due to increasing frequency of intermolecular collisions.

The atmosphere pressure in which deposition is optimally carried out depends on the utilized technology (1, 2).

II-A) Evaporation

The number of atoms or molecules evaporated from a liquid or a solid surface depends strongly on the temperature. As well known the equilibrium vapor pressure would be obtained in a thermodynamically closed system. However, in practical evaporation cases, no equilibrium is achieved as the environment of the vapor source plays the role of a vapor sink. The evaporated atoms condense on all walls that are at a temperature lower than the vapor source.

Systematic investigations of evaporation rates in vacuum have been performed mainly by Hertz, Knudsen and Langmuir. In those experiments it was found that a liquid has a specific ability to evaporate and cannot exceed a certain maximum evaporation rate at given temperature even if the heat supply is unlimited. The theoretical maximum evaporation rate is obtained only if the number of vapor molecules leaving the surface corresponds to that required for achieving the equilibrium pressure P_e on the same surface without return attain.

From such experimental considerations the following equation for the molecular evaporation rate can be formulated:

$$dN/A \; dt = \alpha_e \; (p_e - p_h)/\sqrt{(2\pi \; mkT)} \; cm^{-2}s^{-1}$$

where :
dN = number of evaporating atoms
 A = surface area
 t = time
α_e = evaporation coefficient
p_h = hydrostatic pressure
 m = atomic mass
 k = Boltzmann constant
 T = temperature (K)

With α_e = 1 and p_h = 0, the maximum evaporation rate is :

$$dN/Adt = p_e/\sqrt{(2\pi mkT)} \, cm^{-2}s^{-1}$$

Langmuir showed that this equation is also correct for evaporation from free solid surfaces. Multiplying by the mass of an individual atom or molecule yields the mass evaporation rate per unit area.

$$\Gamma = mdN/Adt = (m/2 \pi kT)^{1/2} p_e$$
$$\text{i.e. } \Gamma = 5.84 \times 10^{-2} (M/T)^{1/2} p_e \text{ gcm}^{-2} \text{s}^{-1}$$

where M is the molar mass and p_e is given in torr.

For most elements the mass evaporation rate Γ is of the order of 10^{-4} gcm^{-2}s^{-1} at $p_e = 10^{-2}$ mbar (1, 2).

Graphs of equilibrium vapor pressure vs. temperature are given for some elements in Figure 5.

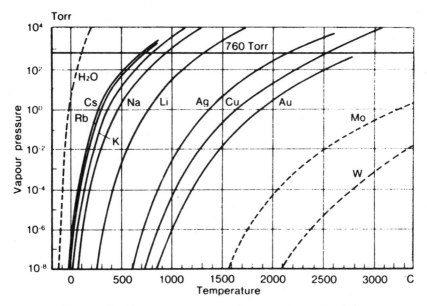

Figure 5. Vapor pressures of some elements (2).

II-B) Directionality of evaporating molecules. The cosine law of emission

An isothermal container with an infinitely small opening dAe related by vanishingly thin walls is shown on Figure 6. It is assumed to contain N molecules which have a Maxwellian speed distribution. In this case the total mass of evaporated material m_e is given by the expressions:

$$dm_e = m_e \cos \phi \, \frac{d\omega}{\pi}.$$

This is the cosine law of emission.

According to this law, emission of a material from a small evaporation surface does not occur uniformly in all directions but favors the directions approximately normal to the emitting surface where $\cos \phi$ has its maximum values (1, 2).

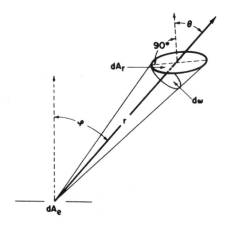

Figure 6. **Effusion from an isothermal container through a small area source and the surface element dAr receiving a deposit (1).**

The amount of material which condenses on an opposite surface depends also on the position of the receiving surface with regard to the emission source. The material contained in an evaporation beam of solid angle $d\omega$ covers an area which increases with the distance as well as with the incidence angle θ . The element of the acceptor surface corresponds to $dAr = r^2 d\omega / \cos \theta$.

The mass deposited per unit area is:

$$\frac{dm_r(\phi, \theta)}{dA_r} = \frac{m_e}{\pi r^2} \cos \phi \; \cos \theta$$

The extension of the cosine law to emission from liquid or solid surfaces is generally considered possible.

As it can be seen on Figure 7 the cross section in the symmetry axis of the spherical segment shows a sickle-shaped geometry of the deposited thin films. The mass M so deposited is:

$$M_s = \rho \, 2\pi \, X_s A_s, \; \text{where}$$

X_S is the distance between the mass center of the profile area A_s and the rotation axis (1, 2).

$m_s = \rho \cdot 2\pi \cdot x_s \cdot A_s$

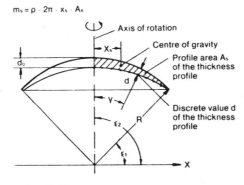

Figure 7. **Deposited film mass M_s on a spherical segment** (2).

$$X_s = 1/3A_s \left[R^3 (\sin \epsilon_2 - \sin \epsilon_1) - (\pi/180) \int_{\epsilon_1}^{\epsilon_2} (R-d)^3 \cos \epsilon \, d\epsilon \right]$$

$$A_s = 1/2 \, (\pi/180) \, R^2 (\epsilon_2 - \epsilon_1) - \int_{\epsilon_1}^{\epsilon_2} (R-d)^2 d\epsilon$$

II-C) Evaporation techniques

Various methods exist for performing an evaporation. However, for most materials there is only one optimum evaporation technique. It is related to a correct selection of the evaporation method, the evaporation source and the evaporation temperature. The technique to be applied depends primarily on the material used and the required film purity, but also on the plant and equipments used.

The most important technique is the indirect resistance heating. The evaporation material is placed in a container made of Mo, Ta, W or C which can have the shape of a boat crucible, a coil or a strip. Income cases, ceramic crucibles or inserts made of Al_2O_3, BeO or BN are also employed. The container is heated by current flow and the material is evaporated or sublimated from the various types of resistance-heated evaporation sources shown in Figure 8.

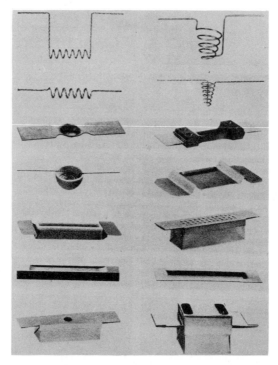

Figure 8. **Various types of resistance heated evaporation sources** (2).

Another convenient method is using an electron beam heating. The material is placed in a water-cooled crucible where it melts in its own environment and there is practically no chance of undesirable reaction with the crucible (Figure 9). A hot filament supplies current of the order of 1 ampere to the beam and the electrons are accelerated typically to a 10 kV

voltage and strike the surface to be evaporated. A magnetic field curves
the path of the electron beam scanned over the surface of the melt,
hindering the nonuniform deposition that would otherwise occur by formation
of a cavity in the molten source. Other methods may be employed such as
laser heating or high frequency heating.

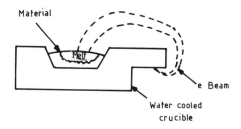

Figure 9. **Electron beam evaporation system.**

II-C-1) Flash techniques

Flash evaporation is another technique used for deposition of films
with constituents of different vapor pressures. In contrast to the two
source evaporation, it requires neither monitoring the vapor density nor
controlling the source temperature. Film composition checking is carried
out by full evaporation of small amounts of the constituents with the
required ratio. The temperature must be sufficiently high to allow
evaporating the less volatile material.

II-C-2) Reactive evaporation

For deposition of a stoichiometric oxide film by reactive evaporation,
a relatively high O_2 partial pressure and a slow metal-atom condensation
rate are required so that completely oxidized metal oxide films can be
formed. The partial pressure of the reactive gas component is usually a
few 10^{-4} mbar. The significant technology of reactive evaporation is
always applied when direct evaporation of a chemical compound is not
possible due to thermal dissociation or too low vapor pressure. In
practice oxide films are usually produced using suboxides or metallic

starting materials. However, basically it is also possible to obtain
sulfides, nitrides or other similar compounds by this technique.

II-D) Condensation and film formation

In a condensation process there is a relationship between a critical
deposition rate and the substrate temperature. An incident vapor atom has
a retention time on the surface which is proportional to reciprocal
substrate temperature and binding forces. Some atoms are reflected, which
means that they leave the surface by a process which directly results from
the collision against the surface. However, this effect will normally only
be observed at room temperatures when the binding forces between the
condensed atoms and the substrate are weak enough and the energy of the
arriving evaporated atoms is relatively low. An example of this condition
is Zn and Cd on glass surfaces. On the other hand, if the number of the
condensing Zn or Cd atoms per second is very high, a few large nuclei are
formed. The condensation can be improved by predeposition with small
amounts of other metals such as Ag or Cr.

Formation of a solid film by condensation is generally an irreversible
process. A certain relaxation time on the substrate and appropriate
surface diffusion processes of the atoms are responsible for the nuclei
formation. Heterogeneous nucleation is important as a first step in the
formation of a thin film by deposition from the vapor phase. Figure 10
gives a schematic representation of the various stages of metal film
deposition.

Most films deposited even at room temperature are in nonequilibrium
conditions and far from perfect. They contain vacancies, dislocations,
stacking faults and grain boundaries. A method for approaching equilibrium
is moving atoms in and on the surface layers. The most important parameter
controlling the mobility of atoms in a solid film is diffusion. If the
condensation process occurs close to the melting point of the film
material, a better ordered solid film is formed. This can be achieved for
instance by increasing the substrate temperature. In addition to
influencing surface mobility and ordering processes, the substrate
temperature will also affect the grain size. On the contrary, cooling of
the substrate can allow the obtaining of amorphous films.

Cathodic sputtering or more exactly sputtering of (usually solid) materials by bombardment with positive noble gas ions is the oldest vacuum process used for producing thin films. Sputtering, which resulted from the erosion of the cathode in glow discharge, often an undesirable effect, was discovered more than 130 years ago by Grove in 1852 in Britain and Plucker in Germany during gas discharge experiments.

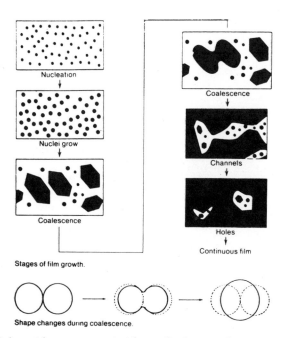

Figure 10. **Schematic representation of the various stages of metal film growth** (2).

In 1877 metal sputtering was applied to production of mirrors and later it was used for decorating various articles with noble metal films. Around 1930 it was applied to deposition of electrically conducting films of gold on the wax masters of Edison discs. Its importance decreased during following 30 years compared with the rapidly developing deposition of films by evaporation and condensation in high vacuum.

However, since approximately 1955 sputtering has undergone a renaissance. Intensive investigations of the phenomena occurring during the sputtering process and hence better control of the process together with the technical requirements for high quality films with very good adhesion and specific properties have certainly contributed largely to the development of sputtering (2).

III-A) General considerations

Without going into too many details, we will attempt here to discuss the parameters and system components of significance for cathodic sputtering and give a general survey of the fundamental correlations characterizing cathodic sputtering and the opportunities offered by this process.

Figure 11 represents a schematic drawing of a largely simplified device for cathodic sputtering. The process takes place in a vacuum chamber which has been evacuated as well as possible before the coating process. In order to prevent contamination of the films to be produced by incorporation of less defined residual gases, the starting pressure should be 10^{-6} mbar or lower. The working pressure is then achieved using the working gas. The sputtering process itself occurs in a gas discharge which is ignited at a pressure between 10^{-3} and 10^{-2} mbar, depending on the particular conditions required. Therefore, a final vacuum as high as possible and a high and constant pumping speed in the mbar range becomes a fundamental necessity. These requirements are met by diffusion pumps or even better by cryopumps or turbo-molecular pumps, which are actually often used in the practice. In order to maintain the gas discharge, a gas inlet, for instance in the form of a needle valve, has to be provided. The process itself is based on a flow-through principle. Noble gases, usually argon, are used as the working gas. However, for special applications, almost any other gas or gas mixtures can be envisaged (2).

Two electrodes are set up in the chamber. One of them, the so-called target, constitutes a material source for the films to be produced and is put at a negative potential. A substrate holder is situated just opposite to the target, which can be connected either to earth or to a floating potential. Furthermore, this holder can be heated or cooled. The positive ions provided by the gas discharge are then accelerated toward the negative target. Upon bombardment of the target, they cause ejection of mainly neutral particles by impulse transfer phenomena. The ejected particles move through the working gas and condense on the substrate. The energy range of the ions is usually between 10 and 5000 eV. A significant part of cathodic sputtering is bombardment of a solid surface with energetic particles with about 1 nm per 1 keV ion energy. The process of material erosion is definitely determined by momentum transfer to the impinging ions on the atoms of the upper layers of the lattice of the solid material, and shows an apparent resemblance to the behavior of billiard balls. This analogy was assumed long time ago, but could be confirmed only recently. The process is schematically shown in Figure 12 (2).

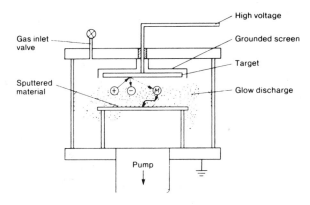

Figure 11. **Simple diode type sputtering system** (2).

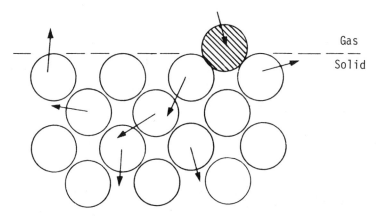

Figure 12. Collision cascade in a solid material during ion bombardment with ejection of two atoms (2).

75

In the areas near the surface of the solid target various complicated processes occur simultaneously. They are:

* knocking out of neutral atoms, compounds or fragmented species,
* secondary electron emission,
* ejection of positive and/or negative secondary ions,
* temperature increase,
* emission of radiation,
* chemical reactions and dissociation,
* implantation, solid state diffusion, crystallographic changes,
* reflection of incident and emitted particles.

All these processes may occur on both target and substrate, they determine the properties of the growing films. About 95% of the energy of the bombarding ions on the target are lost as emitted heat in the material and only 5% pass onto the secondary particles.

With this technique the deposition rate is generally low (20-100 Å/min) but the thin film fits perfectly with the substrate. It is, in fact, its main advantage.

III-B) Sputtering yield

The sputtering yield is measured by the number of emitted atoms per incident ion.

It is proportional to the effective section of showering of the particles and inversely proportional to the sublimation enthalpy of the solid.

The dependency of the sputtering yield on the atomic number is illustrated by Figure 13.

III-C) High-frequency discharges

Considerable changes arise in the discharge if high-frequency electric fields of more than 50 kHz are applied instead of dc. fields as just described. In practice, work is carried out in the frequency range from 5 to 30 MHz, usually with an industrial frequency of 13.5 MHz.

High-frequency cathodic sputtering differs mainly from direct voltage sputtering in two ways:

The electrons oscillate in the plasma and gain sufficient additional energy to produce more ionizing collisions. This results in the fact that the discharge is less dependent on the emission of secondary electrons (electrons emitted by the cathode) and that the "breakdown voltage" is reduced.

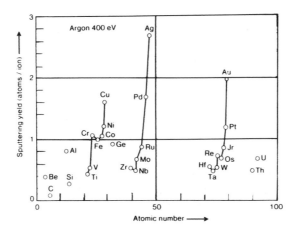

Figure 13. **Dependence of the sputtering yield on the atomic number of the target material** (2).

The target no longer needs to be electrically conducting as it must any way be coupled capacitively to the voltage supply. Due to the difference existing at those frequencies between the electron and ion mobilities and that the ions are relatively immobile, no ionic bombardment is expected to occur. This results in a high excess electron current on the target surface at beginning of the gas discharge. Because of the capacitive coupling of the target no voltage compensation can occur. It will lead to formation of a pulsing negative charge on the target surface, allowing only an efficient ion bombardment on this electrode. The method is mainly used for making thin film with insulating material such as glass for example (2).

III-D) **Discharge supported by magnetic field**

A normal gas discharge is a relatively inefficient source for ions, since only a few percent of the gas particles are ionized. However, a

77

higher portion of ionized particles is desirable for the sputtering process. To achieve this condition, first the required gas pressure can be reduced resulting in a simultaneous increase in erosion rate and growth rate. This can be achieved by applying magnetic field perpendicularly to the target surface. Due to these electron beam configurations, the electrons are submitted to spiral paths parallel to the target surface. It results in a considerable increase in ionization efficiency of the electrons.

The magnetron setup can be used for both dc and RF discharges. A possible geometry is the planar magnetron (Figure 14). This widely used technique generally gives a very high yield.

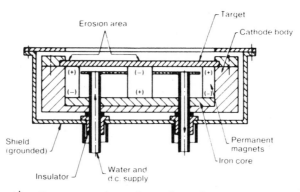

Figure 14. **Cross section view of a planar magnetron** (2).

IV - FILM DEPOSITION BY ION PLATING

Ion plating is a combination of the evaporation process and the sputtering technique. The apparatus is similar to that used in evaporation. The source to substrate distance is generally around 40 cm. The substrate holder, however, is electrically insulated and the substrates are biased negatively, so that an electric field exists between the source and the substrates. If the gas pressure is high enough and the voltage gradient appropriate a glow discharge is generated usually in an argon atmosphere. Evaporation is performed in the presence of a gas discharge and material ions are formed and accelerated in the electric field, so that condensation and film formation take place under the influence of the ion bombardment (Figure 15). This technique gives rise to high density films (often of an order close to that of the bulk material), a good adhesion of the film on the substrate and a high evaporation rate.

It should be mentioned here that one may substitute argon with a reactive gas (O_2, N_2,...) to perform "reactive sputtering" or "reactive ion plating" in a way similar to that used for the evaporation techniques.

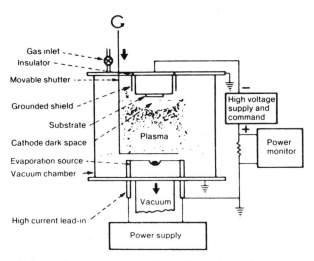

Figure 15. **Schematic representation of an ion plating system** (2).

IV-A) Film growth, grain size, crystallographic structure

The quality of thin films depends on many factors, such as substrate and film materials, substrate temperature, deposition speed, and sputtering pressure.

The nucleation density of sputtered films at the initial growth stage is larger than that of evaporated films and the grain size of sputtered films is usually smaller. Crystallites are often so small that one cannot obtain a clear pattern by X-ray diffraction and analysis.

Films of an unstable phase are occasionally obtained by sputtering. An interpretation of this phenomenon may be that the film formation process is a result of superquenching of a higher temperature phase. Furthermore, the impurity gas level affects the structure of the films.

IV-B) Microstructure

Usually vacuum-deposited films do not grow in a uniform manner as bulk material. Surface roughness of the substrate and geometrical shadowing will lead to preferential growth of the deposits, resulting in a columnar type structure of the films. This microstructure will be modified by substrate temperature, surface diffusion, impurity atom incorporation, incidence angle of the deposition adatom flux, deposition speed and sputtering pressure. Figure 16 shows the structural zone model for vacuum-deposited films. The model consists of three zones determined by the ratio of surface temperature to the melting point T_M of the deposited material:

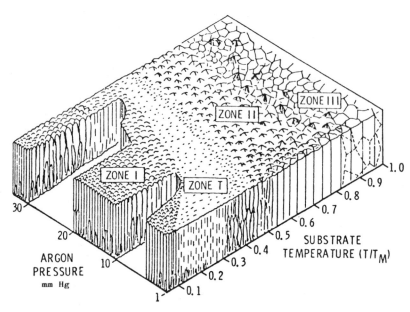

Figure 16. **Structural zone model for vacuum deposited film** (8).

- Zone 1 (T/Tm < 1/3) : forms when the adatom diffusion is insufficient to overcome shadowing effects and gives a columnar fine-grained microstructure with low density boundaries between columns. The individual columns are small size polycrystalline and are highly defective.

80

- Zone 2 (1/3 < T/Tm < 2/3) : the adatom surface diffusion prevails and the columnar structure consists of less defects and larger grains with high density boundaries between columns.

- Zone 3 (T/Tm > 2/3) : bulk diffusion and recrystallization dominates. The obtained material is similar to the bulk materials formed during the melting process.

- Zone T is a transitional zone proposed for sputtered films. It is constituted by fibrous dense columns.

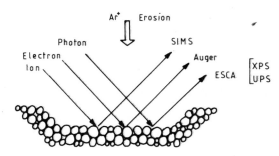

Figure 17. **Principle of the various techniques used for the characterization of thin films.**

V - CHARACTERIZATION OF THIN FILMS

Starting from a multicomponent material it is possible to obtain completely different materials by techniques developed in the previous items. The properties of thin films are frequently found to differ appreciably from those of the bulk material. It is absolutely necessary to characterize the thin films obtained. Two main difficulties remain : the very low thickness and the very small quantity of matter in the form of thin films. In this chapter only three techniques will be developed : Auger electron spectroscopy, SIMS and ESCA (XPS and UPS) (Figure 17).

V-A - AUGER ELECTRON SPECTROSCOPY (AES)

Auger Electron Spectroscopy (AES) analyzes a certain class of electrons, called Auger electrons which are produced when either an electron beam or a photon beam strikes the surface of the sample. Auger electrons are ejected from an atom as a consequence of core level ionization. As shown in Figure 18 an incident electron or photon of sufficient energy can expel an electron from the K shell of a target atom (silicon in Figure 18). An electron transition from L_1 to the empty position in the K level releases energy, which ejects an Auger electron from the $L_{2,3}$ level.

The incident electron beam usually falls between 2 and 10 keV and crosses only a small surface layer of the sample. Most Auger electron energies fall within 20 to 2000 eV, and appear in the spectral range between the low-energy (secondary electrons) and high-energy (primary backscattered electrons) peaks. Figure 19 shows for example an Auger spectrum of an oxidized TiS_2 sample. The ejection depth of Auger electrons is generally less than 50Å and even considerably less for the lower-energy transitions; thus, a chemical analysis of the surface regions can be realized from the AES data.

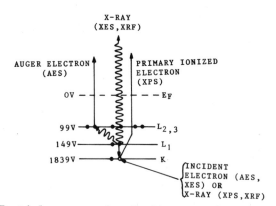

Figure 18. **Partial energy level diagram of the singly ionized silicon atom.**

Many diagnostic problems require depth analysis beyond the ejection depth of Auger electrons, and the sample must be ion milled in order to continuously give rise to a new surface which essentially moves through the sample during the AES analysis. Data are obtained either by interrupting the milling process at regular intervals or by continuously recording spectra during ion milling. The Auger peak heights can then be plotted as a function of milling time or milling depth, and a depth profile can be obtained.

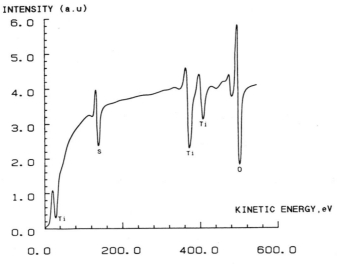

Figure 19. **Auger spectrum of an oxidized TiS$_2$ sample.**

For quantitative analysis, the concentration C_i of an element i in the host material (matrix) is calculated from the expression

$$C_i = \frac{\alpha_1 I_i}{\sum_j \alpha_j I_j}$$

where I_i is the intensity of the Auger peak of the element i and I_j the Auger peak intensity from the element in the matrix. The proportionality constants are most easily determined from well known standards (6).

In modern commercial instruments the resolution can exceed 0.1 µm.

V-B - SECONDARY ION MASS SPECTROSCOPY (SIMS)

In the SIMS method an ion beam sputters material from the surface of a sample, and the ionic component is mass analyzed and detected (Figure 20). Sputtered ions are extracted and the mass is determined with a magnetic prism or a quadrupole analyzer. In a system that uses a magnetic prism, a two-dimensional image of the distribution of an ionic species across the

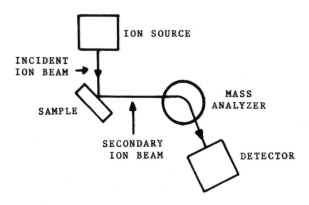

Figure 20. **Schematic diagram of a secondary ion mass spectrometer. An ion source creates a beam which strikes the surface of a sample and sputters material from the surface. The ion fraction of sputtered material is mass analyzed and displayed as a current intensity for a given mass or as a two-dimensional image of the distribution of that mass species.**

surface can be obtained by directing the secondary ion beam onto a channel plate. In quadrupole instruments the image is formed by recording the changing secondary ion-beam current as the primary beams scans the sample surface. The intensity of the detected signal is related to the mass concentration.

The detected signals can be displayed in a mass spectrum during ion milling, giving a depth profile of the chemical species.

Both positive and negative incident ions are used with a beam energy typically between 5 and 15 keV. Since only the ionic fraction of sputtered material produces a SIMS signal, ion beams are selected to produce the highest ion yields of the species investigated. Positive cesium ion beams are generally useful for producing high negative ion yields of electronegative species from a target, and O_2^+ ion beams are usually used for giving rise to high positive ion yields from electropositive species.

The incident beam scans a small area of the surface so as to create a crater with a nearly flat bottom. Mass analysis is performed on the ionic fraction of sputtered material only for a central portion of the crater. When very low primary ion currents are used, sputtering rates are lowered to the point where data can be collected from a few monolayers, and surface analysis can be achieved. Depth profiles are obtained by using higher primary ion currents. Vertical (depth) resolution is controlled by many factors, such as texture at the bottom of a crater, contribution of signals from the crater wall and impurity redistribution during ion milling.

Sensitivity limits are fixed by the factors described above and depend on mass interferences. The mass resolution of an experiment is defined by $M/\Delta M$, where ΔM is the minimum mass difference that can be detected at a mass level M. Typical values of $M/\Delta M$ are between 250 and 5000. Note that because of the complexity of the SIMS process, standards are required to apply it to a problem requiring quantitative analysis (6).

V-C - ELECTRON SPECTROSCOPY FOR CHEMICAL ANALYSIS (ESCA)
X-RAY PHOTOELECTRON SPECTROSCOPY (XPS)

X-ray bombardment of a sample can stimulate the emission of core-level electrons if the incident x-ray energy is high enough (Figure 18). If E_o is the energy of the incident X-radiation, E_b the binding energy of the emitted electron, $\Delta\phi$ the work function difference between the sample and spectrometer surface and E_{exp} the energy of the emitted electron (photoelectron kinetic energy):

$$E_{exp} = E_0 - E_b - \Delta\phi$$

With constant $\Delta\phi$ and E_o in a given experiment, electrons of different binding energies give rise to separate peaks in the photoelectron spectrum. The application of this phenomenon to chemical analysis of the bombarded surface is called X-ray photoelectron spectroscopy (XPS); this procedure is more recently also called electron spectroscopy for chemical analysis (ESCA).

The incident X-ray beam is usually generated by low-energy electron bombardment of an aluminum or magnesium anode. The K α radiation from magnesium has an energy of 1253.9 eV and a line width of 0.7 eV, and the Kα radiation from aluminum an energy of 1487.0 eV and a line width of 0.85 eV. Since $\Delta\phi$ is about 1 eV, photoelectron energies from aluminum or magnesium sources are sufficiently low, so that escape depths are less than 50Å. The two methods, XPS and AES, are therefore similar in providing chemical information from a region within a few monolayers of the surface.

The electron detection and analysis instrumentation used in XPS is similar and in some cases almost identical to that used for AES. As in AES, ion milling is used to obtain chemical depth profiles, and the depth resolution of XPS and AES is similar.

XPS is often used as a complement of AES because of following three advantages.

First radiation sensitive material can be nondestructively studied, because the scattering cross sections for X-ray induced desorption and dissociation are significantly lower than the corresponding cross sections for electron bombardment.

Second insulators can be studied with less surface charging, as a neutral incident beam is used.

Third information on chemical bonding can be obtained from the XPS data. The energy levels of core electrons are affected by the oxidation state and the nature of chemical bonding. The energy resolution of XPS peaks is typically around 0.5 eV, and since different chemical bonds often induce larger shifts in the binding energy, these shifts can be detected and the bond identified. As an example Figure 21 shows an XPS spectrum of an oxidized TiS_2 sample.

VI - THIN FILMS OF LITHIUM CONDUCTING GLASSES

A - BORON OXIDE BASED GLASSES

VI-A-1) Elaboration of the thin films

* Binary B_2O_3-Li_2O glasses

B_2O_3-Li_2O thin films have been prepared using a vacuum evaporation process (10^{-7} torr).

Glasses in the form of small lumps have been molted in a molybdenum boat at $600^{\circ}C$. Thin films (1 to 4μm thick) have been formed on a substrate heated at $250^{\circ}C$. The deposition rate, controlled by a piezoelectrical balance, was 2000 Å/min. After deposition, the substrate was progressively cooled with a $2^{\circ}C$/min. rate to avoid internal strains (3).

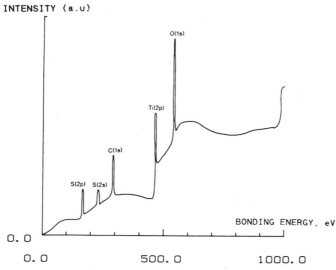

Figure 21. XPS spectrum of an oxidized TiS_2 sample.

* Ternary B_2O_3-SiO_2-Li_2O, B_2O_3-Li_2O-LiI and
B_2O_3-Li_2O-Li_2SO_4 glasses

It has been previously shown that on one hand the addition of SiO_2 to B_2O_3-Li_2O bulk glasses allows an increase in the Li_2O content and hence the conductivity. On the other hand a lithium salt (e.g., LiI or

Li_2SO_4) added to B_2O_3-Li_2O glasses has a similar influence on the Li^+ conductivity.

For those glasses a vacuum evaporation process was not appropriate. For borosilicate glasses the vapor pressure is actually very low at $1000^\circ C$. The halogenoborates show a large difference in the vapor pressures of LiI and of the other components: LiI evaporates at about $400^\circ C$. Li_2SO_4 in B_2O_3-Li_2O-Li_2SO_4 glasses decompose when they are heated in high vacuum. These unfavorable conditions do not allow the formation of homogeneous thin films. To overcome this difficulty RF sputtering has been used.

A bulky glass disc (of 50 mm diameter and 3 or 4 mm thickness) was used as a target. RF sputtering deposition was carried out under 10^{-2} torr argon pressure and with a power of 8 W/cm^2 on a substrate at room temperature. The distance between target and substrate was about 50 mm. Films of 1 or 2 μm thickness have been deposited thus with a rate of 40 Å/min.

VI-A-2) Thin film characterization

X-ray diffraction analysis has shown that the films obtained are amorphous, whatever the deposition method used.

Electrical measurements for a sandwich structure have been obtained by successive deposition of gold and glass (Figure 22). The geometrical arrangement was realized using several masks.

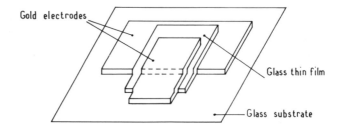

Figure 22. **Glass thin film sample between two gold electrodes.**

Ionic conductivity was measured by a classical complex impedance method using a Solartron 1170 Frequency Response Analyser in the frequency range of 0.1 mHz up to 10^4 Hz.

DC measurements were in good agreement with those obtained by ac and showed that the electronic conductivity is at least 10^5 times lower than ionic conductivity (3).

For all compositions considered plots of log σ against $1/T$ show a linear variation in the temperature range of the measurements $(25-300^{\circ}C)$.

In other words the conductivity data fit well an Arrhenius equation.

* $B_2O_3-Li_2O$ films

A sufficient amount of glass can be obtained by the evaporation process to allow a chemical analysis. As expected the lithium content in the amorphous films is higher than in the starting material (Figure 23).

Figure 24 shows the variation of log σ for $B_2O_3-Li_2O$ glasses with Li_2O content at $25^{\circ}C$. Conductivity values obtained with thin films fit well with those determined for bulk materials.

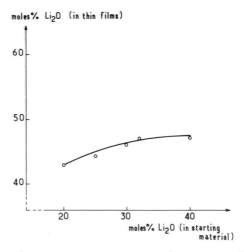

Figure 23. Variation of Li_2O content (in mole %) in the thin films with Li_2O content in the starting glasses.

89

A similar phenomenon is observed for the activation energies (Figure 25).

The ionic conductivity varies exponentially and the activation energy linearly with Li_2O content (Figure 24, 25). Similar behavior had already been observed for glasses belonging to the SiO_2-Na_2O and M_2O_3-Li_2O (M = Al, Ga, Bi) systems.

A scanning electron micrography of a thin film shows a surface which is quite uniform (Figure 26).

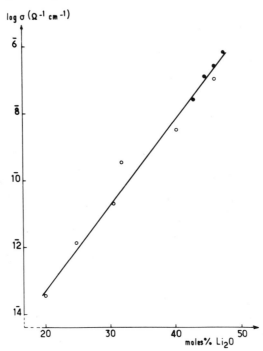

Figure 24. Variation of log σ of B_2O_3-Li_2O glasses (o : bulk glasses, ● : thin films) with Li_2O content at room temperature.

* B_2O_3-SiO_2-Li_2O, B_2O_3-Li_2O-LiI and B_2O_3-Li_2O-Li_2SO_4 glasses

It is difficult to obtain an amount of material large enough using RF sputtering for an efficient chemical analysis. However, qualitative energy dispersion analysis using a TRACORE NORTHEM TN 1705 apparatus coupled with S.E.M. showed the starting ternary glasses in the corresponding thin films.

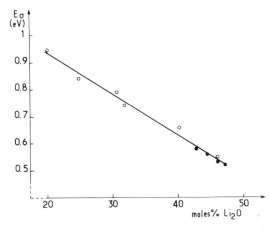

Figure 25. Variation of the activation energy of B_2O_3-Li_2O glasses (o : bulk glasses; ● : thin films) with Li_2O content.

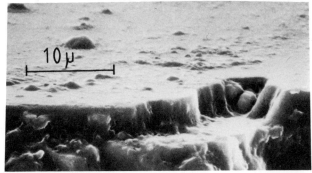

Figure 26. Scanning electron micrography of a B_2O_3-Li_2O film.

Electrical measurements seem to show that the thin film conductivity is similar to that observed in the starting bulk glasses as far as the composition of starting material and amorphous sputtering deposit does not change (3).

The best results have been obtained from borosilicate glasses. A 10^{-6} S/cm room temperature conductivity allows the attainment of a low electrical resistance (100 Ω cm for a 1 μm thickness and a 1 cm^2 at 25°C). This type of glass has been used for setting up microbatteries.

91

VI-B - BORON SULFIDE BASED GLASSES

VI-B-1)   Elaboration   of   binary B_2S_3-Li_2S and   ternary B_2S_3-Li_2S-LiI thin films

B_2S_3-Li_2S and B_2S_3-Li_2S-LiI thin films have been prepared using a high vacuum evaporation process (10^{-7} torr).

The starting materials and the obtained thin films are very hygroscopic. The deposition of such glasses requires a small evaporation apparatus, which has been designed and built up in order to be put easily into a glove box. This glove box is equipped with water vapor and oxygen purification devices.

The deposition rate controlled by a piezoelectrical balance is about 140 Å/sec, five times higher than that determined with oxide glasses. The obtained films are checked by SEM (4).

VI-B-2) Electrical measurements

The high ionic conductivity of the glasses will not allow use of a sandwich electrode structure for electrical measurements. The arrangement of the electrodes used for ionic and electronic conductivity measurements of B_2S_3-Li_2S and B_2S_3-Li_2S-LiI thin films is illustrated in Figure 27. The Au/thin film/Au cell is kept in a special airtight container during the electrical measurements.

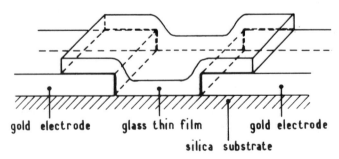

gold electrode   glass thin film   gold electrode

silica substrate

Figure 27. Arrangement of the electrodes for electrical measurements for sulfide glasses.

* The B_2S_3-Li_2S binary system

The conductivity measurements show that the thin film and the corresponding starting glass have similar conductivities (Figure 28). The ionic conductivity of this type of glasses varies exponentially with Li_2S content. Similar behavior had been observed for oxide glasses.

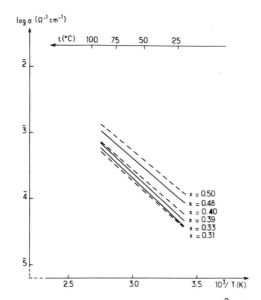

Figure 28. Variation of $\log \sigma$ vs. $10^3 \times T^{-1}$ for $(1-x)B_2S_3$-xLi_2S glasses (dotted lines : thin films, broken lines: bulk glasses).

* The B_2S_3-Li_2S-LiI ternary system

For B_2S_3-Li_2S-LiI films the conductivity measurements show two types of behavior according to the thermal treatment of the thin films.

During the first temperature increase from 25° to $90^\circ C$ the measured conductivity is similar to that of the starting glasses (Figure 29). This conductivity results essentially from the motion of the Li+ ions in the bulk of the thin film. Thus, despite the uncertainty of the exact iodine content, the films seem to have a composition and a behavior similar to these of the starting glasses.

When the films are heated for several hours at $90^\circ C$ the conductivity measures ten times higher and seems to be independent of the composition of the thin films (Figure 29). Such a type of annealing never produces any conductivity enhancement on similar composition bulk glasses.

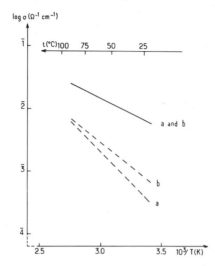

Figure 29. Variation of $\log \sigma$ vs. $10^3 \times T^{-1}$ for B_2S_3-Li_2S-LiI glasses (broken lines : thin films and starting materials; dotted lines: thin films after thermal treatment) a: $0.14B_2S_3$-$0.45Li_2S$-$0.41LiI$ b: $0.14B_2S_3$-$0.40Li_2S$-$0.46LiI$

This enhancement may be attributed likely to an interfacial conductivity due to rapid ion transport at the silica-film interface. The structural defects at the interface should be responsible for the low interfacial migration enthalpy as the thermal treatment increases the fraction of interfacial sites concerned.

VII - MICROBATTERIES

Miyauchi et al., prepared microbatteries according to the procedure (5):

(i) A titanium disulfide positive electrode film has been achieved on a silica glass substrate by low pressure CVD using $TiCl_4$ and H_2S as a source gas.

(ii) A solid electrolyte $Li_{3.6}Si_{0.6}P_{0.4}O_4$ amorphous film was rf-sputtered on the TiS_2 film at a pressure of 3 Pa, using a mixture of $Li_{3.6}Si_{0.6}P_{0.4}O_4$ and Li_2O powders as target pellets, and a gas mixture of $Ar/O_2 = 60/40$ as a sputtering gas. The substrate temperature was kept below 100°C to avoid pollution by oxygen of the TiS_2 film.

(iii)   Finally a Li negative electrode film was vacuum evaporated at a pressure of 10^{-4} Pa.

Discharge and charge curves for cell (A) and cell (B) at 3-16 $\mu A/cm^2$ are shown in Figure 30. The open circuit voltages for both cells were 2.5V. The quick voltage drop in the initial stage is due to ohmic polarization. A gradual decrease follows due to discharge. These discharge curves are similar to those found for a TiS_2 cathode in organic solvent electrolyte cells. The discharge capacities of cell (A) from 2.5-1.5 V at 3 $\mu A/cm^2$ and 6 $\mu A/cm^2$ were about 45 $\mu Ah/cm^2$. These capacities represented about 80% of the theoretical value for a TiS_2 single crystal. The discharge capacity of cell (B) at a 3 $\mu A/cm^2$ current density was 150 $\mu A/cm^2$, which was also about 80% of the theoretical value. The increase in discharge capacity was created by an increasing lithium diffusion coefficient and thickness. There was no significant difference between the discharge and charge curves except the potential drop due to internal resistance. This indicates sufficient reversibility for the investigated cells.

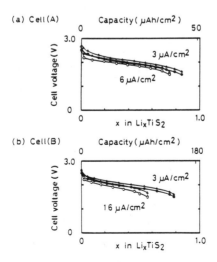

Figure 30. **Discharge and charge of thin films cells (5) (o, o: discharge; $\triangle$, $\triangle$: charge).**

Meunier et al., from a TiS_2 target, obtained a thin film of a positive electrode with a RF sputtering magnetron technique. A 1 μm thick film of an oxide glass $(B_2O_3-Li_2O-Li_2SO_4)$ was deposited over the previous one. Finally lithium is evaporated on the glass layer (7). These microbatteries can deliver up to 30 μ A/cm^2 and more than 100 cycles have been obtained without trouble (Figure 31).

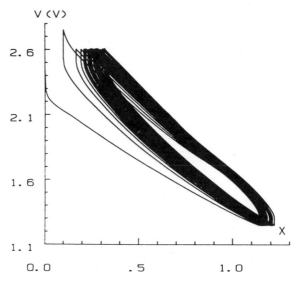

Figure 31. Cycling of a thin film microbattery between 1.25 and 2.5 V at 10 μ A/cm^2 (X is the lithium intercalation content in the positive electrode).

ACKNOWLEDGEMENTS

The authors wish to thank Prof. Paul Hagenmuller for fruitful participation in discussions and also Commission of European Communities for the financial support of the work on microbatteries.

Some figures and formulas are reproduced from reference 2 with the kind permission of H.P. PULKER and Elsevier Science Publishing Company.

REFERENCES

1. Handbook of thin film technology (edited by L.I. Maissel and Reinhard Glang) (McGraw - Hill Book Company).
2. "Coating on glass", H.P. Pulker, Elsevier, Amsterdam (1984).
3. A. Levasseur, M. Kbala, P. Hagenmuller, G. Couturier and Y. Danto, Solid State Ionics, 9&10, 1439 (1983).
4. M. Kbala, M. Makyta, A. Levasseur and P. Hagenmuller, Solid State Ionics, 15, 163 (1985).
5. K. Kanehori, K. Matsumoto, K. Miyauchi and T. Kudo, Solid State Ionics, 9&10, 1445 and 1469 (1983).
6. VLSI Technology (edited by S.M. Sze, McGraw - Hill Book Company).
7. G. Meunier, private communication. Patent no. 88-14435.
8. B.A. Movchan and A.V. Demchishin, Phys. Met. Metall. (USSR), 28, 83 (1969).

STUDY AND CHARACTERIZATION OF THIN FILMS OF CONDUCTIVE GLASSES
WITH A VIEW TO APPLICATION IN MICROIONICS

J. Sarradin, R. Creus, A. Pradel, R. Astier and M. Ribes

Laboratoire de Physicochimie des Materiaux Solides U.A. 407, Universite' des Sciences et Techniques du Languedoc, Place E. Bataillon, 34060 Montpellier Cedex, France.

INTRODUCTION

Application of ionic or mixed conductive glasses is without a doubt most promising in the field of microionics. The development of microsolid state energy storage devices and the design of ISFET type iono-sensitive sensors can be envisaged using thin films of glasses. Oxide and sulphide glasses used as electrolyte and cathode material, and an aluminosilicate glass which is sensitive to alkaline ions (Na^+) were characterized their electrical properties examined.

I - STUDY OF OXIDE AND SULPHIDE TYPE GLASSES FOR USE IN MICROBATTERIES

a) Glasses used as electrolyte

There are numerous Li^+ conductor glasses with interesting conductive characteristics. This is the case of $Li_2O-B_2O_3-P_2O_5$ glass, whose conduction properties were investigated both in bulk and thin film form [1]. Good quality films were obtained by flash evaporation in spite of considerable differences in the melting points of the different constituents. Figure 1 shows the variation in conductivity obtained in a thin film (~1 m) of $0.4Li_2O-0.2B_2O_3-0.4P_2O_5$ glass. The reason for the lower conductivity and higher activation energy than those observed in the initial bulk glass is doubtless a deficit of Li_2O, since its melting point is much higher than those of the other components.

Much work [2,3] has shown that substitution of oxygen with sulphur leads to a considerable increase in conductivity. Such glasses have already been used in bulk form in solid-state power storage devices (button cell type) [4]. Thin films have also been made [5]. Lithium conductive sulphide glasses of the Li_2S-SiS_2 system [6] were studied as well as various ternary and quaternary glasses, in particular $0.14SiS_2-0.09$ $P_2S_5-0.47Li_2S-0.30LiI$ whose stability with regard to lithium has been recently investigated [7]. These thin films were obtained by evaporation. Their high hygroscopic properties necessitated the construction of an evaporation unit (made in the laboratory) in a glove box. Their electrical properties were determined from complex impedance diagrams. Figure 2A shows the characteristics of a $0.66Li_2S-0.34SiS_2$ glass in bulk and thin film form.

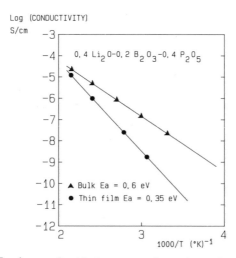

Figure 1. **Ionic conductivity as a function of temperature.**

Likewise, Figure 2B shows the electrical characteristics of a thin film of $0.14SiS_2-0.09P_2S_5-0.47Li_2S-0.30LiI$. As in most cases in which films are obtained, the material displayed lower conductivity than that of the same material in bulk.

b) **Glasses used as cathode material**

Use of glass as cathode material enables continuity at the electrolyte/cathode material interface and thus improves its quality. The glass chosen must also possess good electronic conductivity. $V_2O_5-TeO_2$ glasses were selected which are among the best electronic conductors known to date. The making of thin films of $(0.6V_2O_5-0.4TeO_2)$ obtained by evaporation often results in the formation of a phase characterized by lack of homogeneity of composition which can be observed in the complex impedance diagrams (2 more or less well defined arcs). Although this is not harmful for applications in microionics, it was preferable to obtain these layers by sputtering using a bulk target with the composition above. This method has the advantage of producing thin films that are more homogeneous, but deposition times are much longer.

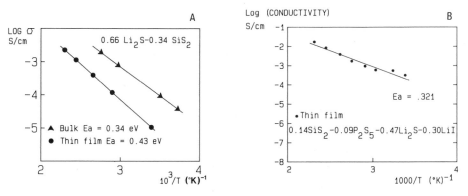

Figure 2 A and B. **Ionic conductivity as a function of temperature.**

II - STUDY OF AN ALUMINOSILICATE GLASS USED AS A SENSOR

ISFETs (Ion Selective Field Effect Transistors) are the sensors of the future in many domains; however, an important limitation is their lack of sensitivity. In order to improve this feature, use of a solid electrolyte specific to the species to be assayed could be used as a selective membrane. The $0.71SiO_2-0.18Al_2O_3-0.11Na_2O$ glass was studied which is generally used as a membrane in electrodes sensitive to Na^+ ions [8].

Given the high melting point of this material, sputtering had to be used to obtain thin films. Stabilization annealing is required before thin films are used. As can be seen in Figure 3, the electrical properties of the material became stable after annealing at $370^{\circ}C$ for 48 hours. Neither conductivity nor activation energy changed after annealing in this way. These characteristics were comparable to those of bulk glass.

CONCLUSION

Conductive glasses have large possibilities of application in microionics. This requires a good knowledge of the formation of thin films.

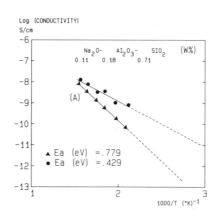

Figure 3. **Effect of annealing.**

REFERENCES

1. Delord, V., Dec. 1986, Thesis, Montpellier, France.
2. Souquet, J.L., Robinel, E., Barrau, B. and Ribes, M., 1981, Solid State Ionics 3/4: 317-321.
3. Malugani, J.P. and Robert, G., 1980, Solid State Ionics 1: 519.
4. Akridge, J.R. and Vourlis, H., 1986, Solid State Ionics 18/19: 1082-1087.
5. Kbala, M., Makyta, M., Levasseur, A. and Hagennmuller, P., 1985, Solid State Ionics 15: 163.
6. Pradel, A., June 1988, Thesis, Montpellier, France.
7. Kennedy, J.H. and Zhang, Z., in press, Solid State Ionics.
8. Koryta, K. and Stulik, K., 1983, "Ion Selective Electrodes" (2nd Edition) Cambridge University Press, Cambridge, UK.

CHEMISTRY, PHYSICS AND APPLICATIONS OF POLYMERIC SOLID
ELECTROLYTES TO MICROBATTERIES

Bruno Scrosati

Dipartimento di Chimica

Universita di Roma

"La Sapienza", Rome, Italy

INTRODUCTION

In recent years various types of new conductive polymers have been discovered and characterized. One category of this new class of conductors is characterized by an electrical transport mainly due to ion carriers, and it may be considered a group of 'polymer electrolytes' (1,2).

Other types are those polymers that acquire a high electronic conductivity after a reversible electrochemical doping. This group may be considered 'polymer electrodes' (3,4).

Consequently, these discoveries have now opened the possibility of realizing polymer-based electrochemical devices with unique characteristics of great importance in the presently developing technology.

In particular, and as it will be stressed later, these polymers may be easily and efficiently prepared in the form of thin films; therefore, conductive polymers are especially suitable for the development of new types of plasticlike thin-layer batteries.

This interesting aspect of advanced electrochemical science will be described in this paper. Before illustrating their applications in batteries, the properties and the characteristics of the polymer electrolyte and the polymer electrode materials will be briefly described.

Solid State Microbatteries
Edited by J. R. Akridge and M. Balkanski
Plenum Press, New York, 1990

For more detailed information the reader is referred to more extensive reviews presently available in the literature (1,3,5,6).

POLYMER ELECTROLYTES

Ion conducting polymers were first studied by Wright and his co-workers in the early seventies (7), but their potentialities as practical electrolyte materials in electrochemical devices were recognized by Armand and his co-workers (8) in 1978. These materials are complexes between metal salts and high molecular weight polymers containing solvating heteroatoms. The most common examples are complexes formed by the dissolution of alkali salts MX, in poly(ethylene oxide), PEO. The heteroatom, here oxygen, acts as donor for the cation M^+ while the anion X^-, generally of large dimension, stabilizes the PEO-MX complex. An oversimplified but commonly shown picture of this type of complex is illustrated in Figure 1.

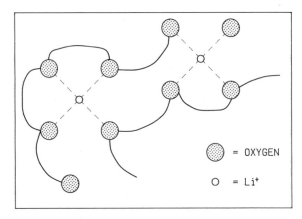

= OXYGEN

O = Li+

Figure 1. **Schematic illustration of the interactions between the PEO chains and lithium ions.**

Since the ionic mobility in the complexes may be interpreted on the basis of a hopping mechanism between coordinating sites (1,5), local structural relaxations and segmental motions of the polymer chains are essential to confer high conductivity in the electrolyte (Figure 2).

Generally, polymer electrolytes have a multiphase nature consisting of salt rich crystalline phases, pure polymer crystalline phases and amorphous phases with dissolved salt. This is shown by Figure 3 which illustrates the phase diagram of the PEO-LiClO$_4$ system (9). As is common in polymer electrolyte practice, the concentrations are reported in terms of number of heteroatoms (oxygens) per cation, here as O/Li. The diagram clearly reveals the presence of various phases, such as the (PEO)$_3$LiClO$_4$ and the (PEO)$_7$LiClO$_4$ crystalline complexes, the crystalline PEO polymer, and the liquid amorphous phase (L).

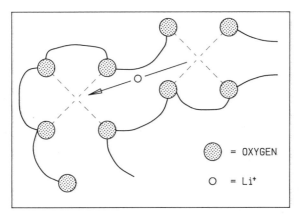

Figure 2. **Schematic illustration of the mechanism of the ionic transport in a polymer electrolyte.**

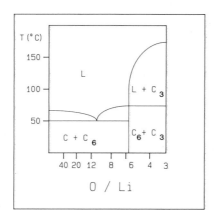

Figure 3. **Phase diagram of the PEO-LiClO$_4$ system.**

In the crystalline phases the polymer chains are rigid. The ion mobility is almost negligible and the overall conductivity is very low. Above the glass transition temperature, where the amorphous region progressively increases, the polymer chains acquire faster internal modes in which bond rotations produce segmental modes. This in turn favors the hopping interchain and intrachain ion movements and the conductivity of the polymer electrolytes becomes appreciable. The conductivity dramatically increases above the crystalline to amorphous transition temperature, as typically shown by Figure 4, which illustrates the conductivity of the $(PEO)_8LiClO_4$ electrolyte.

One of the most important characteristics of polymer electrolytes is that their conductivity is a property of the amorphous, elastomeric phase. Even above their glass transition temperatures, high molecular weight polymers may still exhibit mechanical properties which are similar to that of a true solid, and the result of chain entanglement and cross-linking of various types. At microscopic level, local relaxation processes take place which provide liquidlike degrees of freedom similar to those observed in ordinary molecular liquids (1).

In general, polymer complexes may be regarded as unusual forms of electrolytes with properties which lie between those of a solid and those of a highly viscous liquid. In the amorphous region, the polymer host (e.g., PEO) resembles a nonaqueous solvent capable in principle of dissolving a wide variety of salts; however, the complexes behave like a solid crystalline electrolyte, since not all the ionic species present are necessarily mobile.

A large number of electrolytes can be obtained. The only practical restriction is to select polymers with proper coordinating atoms and inorganic salts with large anions and delocalized charge in order to assure the stability of the entire complex.

Emphasis was initially placed on complexes between the readily available commercial PEO and lithium salts, such as $LiClO_4$, $LiCF_3SO_3$ and $LiBF_4$.

Figure 4 shows the temperature behavior of the conductivity of these complexes. Notice the jump in conductivity around 60^oC, which relates to the crystalline to amorphous phase transition. PEO-based polymer electrolytes with monovalent, divalent, and possibly multivalent ions can be obtained as well (10). As a typical example, Figure 5 illustrates the conductivity behavior of a family of polymer electrolytes formed by combining PEO with a divalent copper salt (i.e., $Cu(CF_3SO_3)_2$). Other

106

electrolytes may be obtained by replacing PEO with different polymer chains. Indeed, this is an important route since the proper choice of the polymer host may lead to the characterization of electrolytes capable of offering the required amorphous conductive condition at temperatures lower than $60^{\circ}C$ to $70^{\circ}C$.

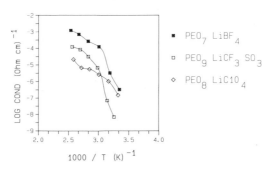

Figure 4. **Conductivity of polymer electrolytes based on the dissolution of $LiClO_4$, $LiCF_3SO_3$ and $LiBF_4$ in PEO.**

One possible strategy in this respect is to select polymers having a lower glass transition temperature (Tg) than PEO. An example is shown in Figure 6 which reports the conductivity of complexes between a lithium salt, $LiClO_4$, and an ethoxy-ethoxy-ethoxy-vinyl-ether (EEEVE) polymer chain (12). Indeed, the room temperature conductivity of this electrolyte is considerably higher than that of the more 'conventional' $(PEO)_8LiClO_4$ electrolyte, also reported in Figure 6 for comparison purposes.

Finally, another important characteristic of the polymer electrolytes, again related to their liquidlike property, is that both cations and anions can be mobile even if to a different extent. The solvation processes in these electrolytes are regulated by enthalpy changes. Salt dissolution may take place only if exothermic ion-polymer interactions compensate for lattice energy. Anion solvation is known to arise as a result of hydrogen-bonding while cation solvation is primarily a result of electrostatic interactions. Since the most common polymer hosts in polymer electrolytes have no hydrogen bonding capabilities, the enthalpy of solvation is mainly related to electrostatic interactions between the positive charge on the cation and the partial sharing of a lone pair of electrons on the coordinating atom in the polymer (e.g., oxygen in PEO). This leads to the formation of a coordinating bond (1).

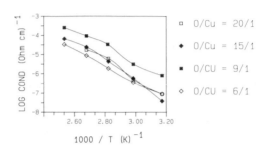

Figure 5. **Conductivity of polymer electrolytes based on the combination of PEO with the $Cu(CF_3SO_3)_2$ salt at various concentrations.**

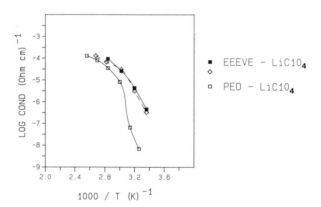

Figure 6. **Conductivity of EEEVE-LiClO_4 polymer complexes. Data of PEO-LiClO_4 are also reported for comparison purposes.**

The mobility of the anions is mainly influenced by steric effects, that of the cation is strongly related to the bonding character. In fact, the stronger the ability of the cation to form coordinating bonds with the polymer, the lower will be its mobility.

In this fashion, highly polarizable cations, such as Mg^{2+} or Ca^{2+}, are essentially immobile in PEO-based complexes. Cations such as Li^+ or Na^+, which react weakly with the hard ether oxygens, have a much higher mobility (10).

In conclusion, the main requirement for high mobility seems to be the facile dissolution of the cation by the coordination group of the polymer. In reality, the mechanism of ion mobility is more complicated since it certainly involves not only ion-polymer interactions but also ion-ion interactions. The detailed analysis of these effects, however, is beyond the scope of this paper. The purpose is mainly that of providing an overview of the properties of the polymer complexes in view of their application in thin-layer electrochemical devices.

PREPARATION AND ADVANTAGES OF POLYMER ELECTROLYTES

The preparation of polymer electrolytes involves simple casting procedures. The polymer host and the inorganic salt are dissolved in adequate reciprocal compositions in suitable solvents (e.g., acetonitrile). The two solutions are then mixed, and after stirring, the solvent is slowly evaporated to finally obtain a thin film of the desired polymeric complex (Figure 7). By proper control of the conditions of the synthesis, it is possible to obtain polymeric films having thicknesses between 25 and 100μm, still retaining good mechanical properties. This and other unique specific characteristics give polymer electrolytes significant advantages over conventional electrolytes in the area of high energy batteries. The thin film fabrication process allows the realization of polymer layers, which while acting as electrode separators, permit the fabrication of cells with variable geometry and high electrode/electrolyte interfacial area. Figure 8 pictorially illustrates some examples of advanced designs of multilayer cells, as originally proposed by Hooper and co-workers (13).

Finally, the viscoelasticity of polymer electrolytes (i.e., their ability to accommodate volume changes) minimizes the problems generally associated with all solid-state secondary batteries which arise from

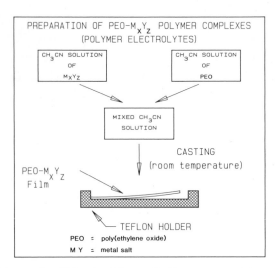

Figure 7. Schematic illustration of the preparation procedure of thin films of polymer electrolytes.

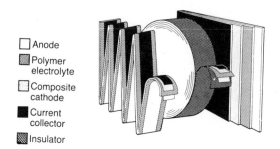

Figure 8. Thin film configurations of electrochemical devices based on polymer electrolytes. (Courtesy of Dr. A. Hooper, AERE Harwell, U.K.).

deformations and deteriorations of the interfaces during charge/discharge cycles.

ELECTROCHEMICAL CHARACTERIZATION OF POLYMER ELECTROLYTES

Three main parameters are of key importance in the electrochemical characterization of polymer electrolytes: the ionic conductivity, the ion transport number and the electrochemical stability.

Ionic conductivity

The most common and rapid method for determination of ionic conductivity is based on complex impedance analysis. The method consists of stimulating a cell formed by the electrolyte under test sandwiched between two 'blocking' electrodes (i.e., electrodes of such a nature as to avoid any charge transfer at the interfaces) and examining the response in a complex imaginary (-jZ") and real (Z') plane (14). Under these conditions, the cell can be described by the equivalent circuit (Figure 9) formed by a capacitance C_{dl} (double layer capacitance), and a parallel RC combination consisting of C_g (the electrolyte geometric capacitance) and R_e (the electrolyte bulk resistance).

Figure 9 also shows the theoretical response of this circuit in the -jZ" - Z' plane to an ac signal of variable frequency. It may be noted that the response basically assumes the form of a semicircle (related to the RC parallel combination) followed at low frequency by a line parallel to the complex axis (related to the blocking capacitance). From the intercept with the real axis, the resistance of the electrolyte and its conductivity may easily be determined.

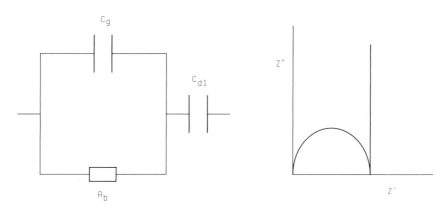

Figure 9. **Equivalent circuit of a cell with blocking electrode and its response in the complex impedance plane to an ac signal.**

A practical example is shown in Figure 10 which illustrates the impedance plot at $90^{\circ}C$ of the $(PEO)_9Cu(CF_3SO_3)_2$ complex in a cell using two stainless steel blocking electrodes. As previously indicated,

111

the intercept with the real axis easily gives the bulk resistance, R_e, of the $(PEO)_3Cu(CF_3SO_3)_2$ electrolyte. By repeating the measurement in a heating cycle, the temperature dependence of the conductivity, and Arrhenius plots similar to those reported in previous figures can be obtained. The ac impedance method is widely used since it readily provides reliable values of the ionic conductivity for the majority of the polymer electrolytes.

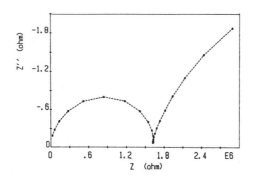

Figure 10. **Impedance plot at $90°C$ of a cell formed by the $(PEO)_9Cu(CF_3SO_3)_2$ electrolyte sandwiched between two stainless steel blocking electrodes.**

Ionic transport number

More complicated and less precise are the measurements of the fraction of charge transported by each type of ion in a polymer electrolyte. Due to the quasi-liquid nature of the polymer complexes, both the ionic species are mobile. For instance, in PEO electrolytes formed with lithium salts having monovalent anions, both anions and cations conduct. The transport characteristics of PEO solutions of divalent salts vary more widely. Those formed with halides of highly polarizing cations, such as Mg^{2+} or Ca^{2+}, are essentially pure anion conductors. Others formed with salts of more polarizable cations, such as Pb^{2+} and Cd^{2+}, conduct via both cations and anions (15). The determination of the ionic transport number is, in general, of great importance in the characterization of polymer electrolytes.

A variety of methods have been used for this purpose, including NMR (16), low-frequency complex impedance analysis (17,18), dc polarization (19) and the Tubandt method (20). These methods are not easily applicable and the results may be affected by various unpredictable side reactions and/or by practical difficulties. Indeed, different measurement techniques applied to the same polymer electrolyte frequently do not yield the same

TABLE 1

CATION TRANSPORT NUMBER OF THE $(PEO)_8 LiClO_4$ POLYMER ELECTROLYTE AS OBTAINED WITH DIFFERENT MEASUREMENT TECHNIQUES

METHOD	TEMPERATURE $^\circ C$	t_{Li+}	REFERENCE
Impedance analysis	20 - 170	0.54	(18)
" "	70 - 150	0.30	(21)
" "	> - 100	0.25	(22)
dc polarization	50	0.5	(19)
Pulse magnetic			
Field gradient	50 - 120	0.17 - 0.29	(23)
Tubandt	70 - 120	0.19 - 0.37	(20)

values for the ionic transport number. A typical example, concerning the $(PEO)_8-LiClO_4$ electrolyte, is reported in Table 1. These discrepancies are due to the fact that each of the listed methods suffer some limitations. The Tubandt method (the most popular technique used in crystalline solid electrolyte) cannot easily be used in the case of the polymer complexes when linear high molecular weight PEO electrolytes are considered, because of the difficulties in separating the various layers forming the measuring cell. The dc polarization methods are affected by the phase heterogeneity of the polymer complexes which introduces concentration gradients, and induce some uncertainties in using the classical diffusion equations (23).

The most common technique is again based on the perturbation by a small signal ac voltage, this time on a cell using nonblocking electrodes (i.e., electrodes of such a nature to allow charge transfer at the interface). The cell can be described by the equivalent circuit of Figure 11 consisting of a sequence of two parallel RC combinations. One is formed by

113

the electrolyte geometrical capacitance C_g and the electrolyte bulk resistance R_e, and the other formed by the double layer capacitance C_{dl} and the charge transfer resistance R_{ct}. Figure 12 shows the theoretical response of this circuit on the Z" - Z' complex impedance plane, which is basically formed by two semicircles related to the two RC combinations.

A practical example is shown in Figure 13 which illustrates the response of a cell formed by the $(PEO)_9Cu(CF_3SO_3)_2$ electrolyte sandwiched between two copper electrodes. The appearance of the two

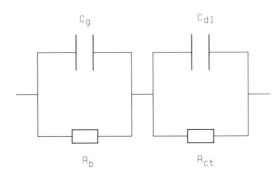

Figure 11. **Equivalent circuit of a cell with nonblocking electrodes.**

semicircles, in agreement with the theory, is already a preliminary indication that copper ions are mobile in the electrolyte; otherwise, the copper electrodes would have behaved as blocking electrodes with a response similar to that previously shown in Figure 10. The bulk resistance of the electrolyte, R_e, and the charge transfer resistance, R_{ct}, of the given electrode/electrolyte interface can be obtained from the intercepts of the real axis as illustrated in Figure 12.

The determination of the transport number value requires the extension of the ac complex analysis to very low frequencies. Under these conditions, the cell may be described by the equivalent circuit of Figure 14, which includes the two previously defined RC combinations followed by a so called diffusion impedance, Z_d. The theoretical response, also reported in Figure 14, shows at medium to high frequencies the two semicircles related to the two RC combinations already illustrated in

Figure 12. At very low frequencies, the amount of charge exchanged at the electrode interfaces during half cycles is sufficient to produce concentration gradients in the electrolyte and this is reflected in the impedance plot by a linear part of an arc having a slope of 1. At still lower frequencies, concentration waves and quasi-steady state profiles may develop, and if the electrolyte thickness is low enough to become comparable with the diffusion layer, the cell impedance curve closes again to the real axis (18).

Such an analysis, applied to a $Cu/(PEO)_9Cu(CF_3SO_3)_2/Cu$ cell having a thin-layer (~50μm) electrolyte, is shown in Figure 15. According to the above described method, originally proposed by Sorensen and Jacobsen (18), the transport number of the cation, here Cu^{2+}, may be obtained by the relation:

$$t^+ = \frac{R_e}{R_e + Z_d(0)} \qquad (1)$$

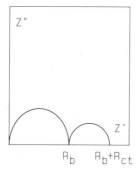

Figure 12. **Response in the complex impedance plane of the circuit of Figure 11 to an ac signal.**

R_e is the bulk resistance of the electrolyte and $Z_d(0)$ is the intercept at zero frequency (obtained by extrapolation of the experimental data).

The method, even if basically correct, may in practice be affected by some side effects related to the specific properties of the materials involved. For instance, salt concentration gradients may form at low

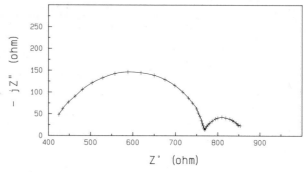

Figure 13. **Impedance plot of Cu/(PEO)$_9$Cu(CF$_3$SO$_3$)$_2$/Cu Cell at 85°C.**

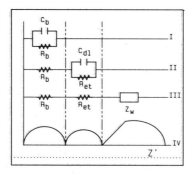

Figure 14. **Equivalent circuit of a cell with nonblocking electrodes and its response in the complex impedance plane extended to very low frequencies.**

frequency involving the nucleation of nonconducting crystalline clusters in the vicinity of the anode; furthermore, passivation reactions may also occur at the electrode/electrolyte interfaces (23). Both effects are not taken into account in the equivalent circuit model proposed by Sorensen and Jacobsen (18) and this may explain the discrepancies among the values obtained by different workers. This is clearly shown in Table 1. Three independent measurements, based on the same ac technique, have given results which differ up to 50% one from the other.

Possibly, these discrepancies result from a fact that generally is not considered; there is a basic difference between the underlined transport number t^+ and the underlined transference number T^+ (24).

For an electrolyte which dissociates in single cationic and anionic species, the electric transport number, say t^+, is defined by the net number of Faradays of charge carried by the cations across a reference plane (fixed with respect to the solvent) when a total of 1 Faraday of charge passes across that plane i.e.:

$$t^+ = \frac{i_+}{i_+ + i_-} = \frac{i_+}{i} = \frac{u_+}{u_+ + u_-} = \frac{D_+}{D_+ + D_-} \tag{2}$$

In this diagram i_+ and i_- are the currents carried by the cations and the anions, i is the total current, and u and D are the ion mobility and the ion diffusion coefficient, respectively.

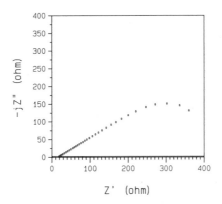

Figure 15. Impedance plot of a $Cu/(PEO)_9Cu(CF_3SO_3)_2/Cu$ cell at $80^\circ C$ extended to very low frequencies.

Consider now a salt M^+A^- dissolved in a polymer with low permittivity. Various equilibria are likely to exist in which associated species are formed:

$$M^+ + A^- \xrightarrow{\hspace{1cm}} MA \xleftarrow{\hspace{1cm}}$$

$$M^+ + MA \xrightarrow{\hspace{1cm}} M_2A^+ \xleftarrow{\hspace{1cm}}$$

$$A^- + MA \xrightarrow{\hspace{1cm}} MA_2^- \xleftarrow{\hspace{1cm}}$$

etc.

The cation constituent M, is carried towards the cathode not only by M^+ but also by M_2A^+, $M_3A_2^+$, etc. The anion constituent A is carried toward the anode by A^-, MA_2^-, etc. in such a way the only quantity that can be measured experimentally is the difference between the fluxes of M^+ containing species directed towards the cathode and the anode (i.e. the net number of moles of M being transferred in one direction).

Accordingly, we may now define the transference number T^+ as the net number of Faradays of charge transferred across the reference plane by the cation constituent in the direction of the cathode when one Faraday of charge passes across that plane.

In any transference number experiment (very often erroneously called 'transport number' measurement) the quantity being measured is the net charge of cation constituent or anion constituent. When ion association exists (as in most polymer electrolytes) it is usually impossible to divide the charge passed into the amount carried by each ionic species. Consequently, in most polymer electrolyte systems it is not the transport number that is measured but, rather, the transference number (24).

This may account for the large discrepancies over the values of transference numbers reported for the same system, since different preparation methods, differences in the thermal history and similar variations may greatly influence the association equilibria.

Vincent and co-workers (24,25) have proposed a method of measuring the transference number which, being based on a combination of ac and dc polarizations. It should make allowance for electrode effects and hence give more reproducible measurements of the transference number of polymer electrolytes.

The method is applied to a symmetrical cell using nonblocking electrodes. By ac impedance analysis, the initial resistance R_0 of the cell is determined by the intercept of the second semicircle with the real axis (see Figure 12). Then the cell is submitted to a small dc bias, and the initial current I_0 is measured and monitored with time until it reaches a steady value I_s. Under dc polarization a decline in current is expected in a symmetrical cell using a polymer electrolyte because of a combination of factors, which include growth at the electrodes of a passivation layer and the establishment of concentration gradients in the electrolyte. Once the steady state conditions have been reached, the resistance of the cell, R_s, is again determined by ac impedance. The transference number, say T^+, is then given by (24,25):

$$T^+ = \frac{I_s(DV - I_0 R_0)}{I_0(DV - I_s R_s)} \tag{3}$$

As a practical example, this analysis is applied to a Li/(PEO)$_7$LiBF$_4$/Li symmetrical cell. From the complex impedance response of the cell to an ac signal of 10 mV peak to peak, a value of R_0 = 5900 ohm is found by the abscissa intercept of the second high-frequency arc. A small dc polarization of 30 mV (DV) is then applied to the cell and the resulting current is monitored with time until stabilization (steady state) is reached. From the current time behavior the values of I_0 and I_s are determined to be 1.18 µA and 0.93 µA, respectively. Finally, the cell resistance is again determined by ac impedance. The high frequency arc has expanded, giving an R_s value of 8210 ohm. From the cited numerical values, the cation transference number, T^+, was determined by equation (3) to be 0.8 at 75°C. As already stressed, this indicates that in the (PEO)$_7$LiBF$_4$ electrolyte there is a preeminence of cation constituent transport but this does not necessarily mean that the current is fully transported by single Li$^+$ ions.

Electrochemical stability

The determination of the voltage range in which a given polymer electrolyte is electrochemically stable is an important parameter in view of its application in electrochemical devices.

Polymer electrolytes which are based on simple polyethers are likely to have a wide range of electrochemical stability. As an example, Figure 16 shows the sweep voltammetry of a $(PEO)_7LiBF_4$ electrolyte at $100^{\circ}C$, using a Pt working electrode and a Li reference electrode. The sweep defines the electrochemical domain (often also indicated as 'stability window'), which is limited at the cathodic side by lithium deposition (usually a reversible process) and at the anodic side by the oxidation of the anion (usually an irreversible process). The two related voltages define the stability window, which in the case of Figure 16, extends over 4 volts. This in turn indicates that the electrolyte may be used with electrodic couples having average open circuit voltages up to 4 volts. Similar stability windows have also been found for other common lithium polymer electrolytes, such as the $(PEO)_8LiClO_4$ and the $(PEO)_9LiCF_3SO_3$ complexes (26,27).

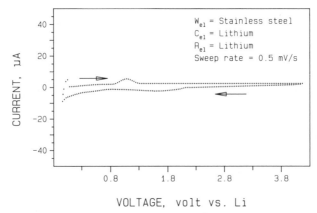

VOLTAGE, volt vs. Li

Figure 16. **Sweep voltammetry at $100^{\circ}C$ of the $(PEO)_7LiBF_4$.**

THIN-LAYER POLYMER/ELECTROLYTE RECHARGEABLE BATTERIES

As discussed in the previous sections, the polymer complexes may be considered plastic electrolytes, that is, a compromise between liquid and solid crystalline ionic conductors. The specific advantages of this new class of quasi-solid electrolytes include the viscoelasticity (which favors accommodations of volume changes at the interfaces during cell operation), the chemical and electrochemical stability (which allows the use of various high voltage electrodic couples) and the simplicity of preparation (which allows the formation of thin film batteries).

Common structures for this type of battery involve the sequence of a strip of lithium metal (anode), a thin film of the polymer electrolyte, and a thin layer of the active cathodic material (generally an intercalation electrode) mixed with graphite (to assure adequate electronic conductivity) and the polymer (to favor the electrode/electrolyte interfacial contact). Figure 8 gives typical examples of such layered lithium batteries.

Among the various possible candidates, the $(PEO)_x LiClO_4$ and the $(PEO)_x LiCF_3 SO_3$ complexes have so far been the most commonly used polymer electrolytes (26).

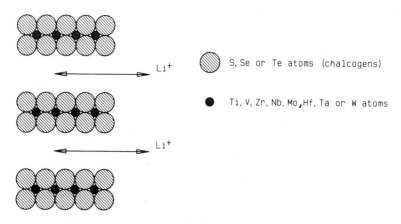

Figure 17. **Schematic illustration of the electrochemical insertion of Li ions in the chalcogenide sulphide host lattices.**

Properties of the intercalation electrode

The intercalation electrodes are based on materials that have an open structure of a layered or tunneled type, in which ions from the electrolyte can be easily inserted via an electrochemical process. An example is titanium sulphide. Figure 17 schematically illustrates the mechanism of the electrochemical insertion of Li^+ ions in the titanium sulphide host lattice.

The basic concept of the polymer electrolyte, lithium battery is the following: In discharge, lithium is dissolved at the anode as Li^+ ions are transported through the polymer electrolyte to the cathode where they are inserted into the intercalation electrode. The charging process is obviously the reverse. Accordingly, the electrochemical scheme is the following:

$$xLi + IE \underset{charge}{\overset{\overset{discharge}{\text{--------->}}}{\xleftarrow{\hspace{1cm}}}} Li_x IE \qquad (4)$$

Typical intercalation electrodes (IE) are TiS_2, V_6O_{13}, or LiV_3O_8.

Prototypes of these thin-layer, polymer electrode batteries are presently under test (28-31) and their performances are encouraging. It appears interesting to examine their behavior in detail by considering first the properties at the two electrodic interfaces and then the performance of the entire cell.

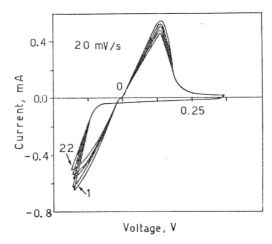

Figure 18. Cycling voltammetry of the lithium plating and stripping process on a nickel support from the $(PEO)_8LiClO_4$ polymer electrolyte at $85°C$. Li reference electrode.

The lithium/polymer electrolyte interface

The lithium electrode is basically reversible in polymer electrolytes and good lithium cyclability has been found in the most common PEO-based lithium complexes (27). The cyclability of lithium in the $(PEO)_8LiClO_4$ electrolyte is revealed by Figure 18 which illustrates the cyclic voltammetry of the lithium plating and stripping process on a nickel support.

On the initial cathodic scans, the voltage required to deposit lithium metal on the nickel surface is some 50 mV negative, probably because of electrocrystallization overvoltage (mismatch of lithium and nickel lattice spacings). Once a monolayer has been formed, further plating can occur at any potential below zero volts (vs. Li) and the stripping peak appears in the anodic scan. The cyclability is good and the recovery of lithium is high.

An as yet unsolved general problem is the passivation of the lithium electrode in PEO-based polymer electrolytes. There are various items of evidence (26,32) which indicate that lithium is effectively passivated in these electrolytes. One is the growth of a resistive layer at the interface similar to the phenomenon observed in liquid organic electrolytes (33).

As an example of this effect, Figure 19 shows the evolution of the ac complex impedance response of a $Li/(PEO)_7 LiBF_4/Li$ cell at progressively longer periods of storage at $80^{\circ}C$. It may clearly be seen that the low frequency semicircle increases with storage time. By circuit analysis (see Figure 12) this semicircle may be associated with the lithium/electrolyte interface. The result of Figure 19 reveals that the passivation of the lithium effectively takes place at the electrode interface with growth of a nonconductive layer, possibly as the result of a decomposition reaction of the electrolyte. The nature and the mechanism of this passivation process have not yet been clarified and they appear to depend in an unpredictable manner upon the type of polymer used and the temperature of operation.

Similarly, it is still not clear to what degree the passivation phenomenon can influence the long term rechargeability of practical, polymer electrolyte, lithium cells. Possibly, the passivation effects in these cells may be not as critical as those observed in liquid organic electrolyte batteries (33), since the nature of the polymer electrolyte allows thin film constructions which result in large extended surface areas (see Fig. 8). This may considerably reduce the value of the current density, and as a consequence, reduce the chances of building up points of nonuniformities upon cycling. Even under this circumstance, problems in cell cyclability cannot be excluded, as will be stressed later.

The intercalation electrode/electrolyte interface

Intercalation compounds are currently considered the most promising positive electrodes for lithium, organic electrolyte batteries (33,34). It is not surprising that also in lithium polymeric systems, the attention has been focused on these type of electrode material, and in particular on the most common of them, titanium sulphide and vanadium oxides.

Indeed, polymer electrolyte batteries based on the Li/TiS_2 couple are currently under study by the Hydroquebec Laboratories in Canada (29), those based on the Li/V_6O_{13} coupled by the Harwell Laboratories in UK (28), and at the University of Minnesota, USA (30), while the feasibility of the Li/LiV_3O_8 couple in polymer cells has been demonstrated by us (31) and other authors (35). The basic characteristics of the three cited electrodic couples are summarized in Table 2. The cyclability of the related intercalation processes is demonstrated by Figure 20 which shows their cyclic voltammetries in the $(PEO)_9LiCF_3SO_3$ polymer electrolyte. Quite good reproducibility and peak definition are obtained for the three intercalation compounds.

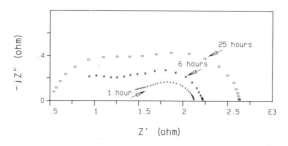

Z' (ohm)

Figure 19. Evolution of the ac complex impedance response of the $Li/(PEO)_7LiBF_4/Li$ cell at progressively longer periods of storage at $80^\circ C$.

Higher energy density (Table 2) and better structure retention upon cycling (36,37) are expected for the LiV_3O_8 vanadium bronze; therefore, this material appears to be a very promising positive electrode in lithium, polymer electrolyte rechargeable systems.

The basic structural elements of the LiV_3O_8 vanadium bronze are octahedra and trigonal bipyramids which are arranged to form puckered layers with Li^+ ions situated in between by occupying octahedral sites (Figure 21). The unit cells comprise 6 empty tetrahedral sites where

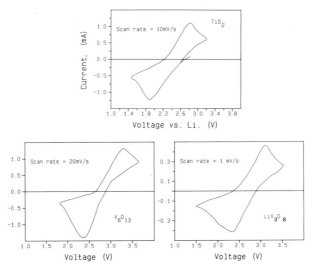

Figure 20. Cyclic voltammetry of the TiS_2, V_6O_{13} and LiV_3O_8 electrodes in the $(PEO)_9LiCF_3SO_3$ electrolyte at $100^{\circ}C$. Li reference and Li counter electrodes.

TABLE 2

THEORETICAL ENERGY OF RECHARGEABLE POLYMER ELECTROLYTE LITHIUM BATTERIES
(ELECTRODES ONLY).

CATHODIC MATERIAL	ELECTROCHEMICAL REACTION	SPECIFIC ENERGY Wh/Kg	VOLUME ENERGY DENSITY Wh/dm^3
V_6O_{13}	$8Li + V_6O_{13} = Li_8V_6O_{13}$	800	1900
TiS_2	$Li + TiS_2 = LiTiS_2$	480	1200
LiV_3O_8	$3Li + LiV_3O_8 = Li_4V_3O_8$	730	2500

excess of lithium may be accommodated with excellent structure retention (37). The maximum reversible lithium uptake corresponds to about 3 eq/mol, as shown in Figure 22 which illustrates the potential composition curve of

$Li_{1+x}V_3O_8$. This curve was obtained by intercalation and deintercalation galvanostatic steps in a lithium cell using the $(PEO)_9LiCF_3SO_3$ polymer electrolyte. A single phase behavior, typical of solid solution formation, is only visible up to an x value of about 1.5. At higher Li^+ content, a structural rearrangement of the vanadium bronze takes place. Indeed, repulsion among Li^+ in the unit cell causes ordering of Li^+ in specific sites so that a new phase appears, which is in equilibrium with that corresponding to the upper limit of the solid solution. The potential tends to remain constant, as is typical of a two phase region.

Detailed X-ray analyses (36,37) have shown that these changes in structure are highly reversible and that the vanadium bronze electrode can be repeatedly cycled over a composition range of 0<x<3. This is confirmed in Figure 22 by the matching between the potential values obtained in discharge and those found in recharge.

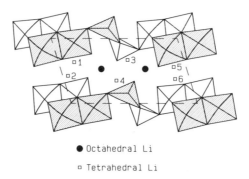

● Octahedral Li

□ Tetrahedral Li

Figure 21. **Structure of LiV_3O_8 projected** onto (010). **The unit cell is dashed.**

The data of Figure 22 also allows one to estimate the energy content of the system. An average voltage of 2.83V, and a maximum energy density of 730 Wh kg^{-1} can be derived for the $Li/Li_{1+x}V_3O_8$ couple. This value compares well with those related to other intercalation electrodes (Table 2).

The favorable structural characteristics illustrated in Figure 21 suggest a fast and highly reversible lithium intercalation process in the

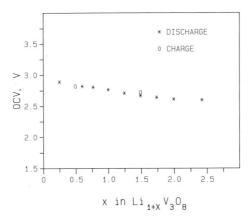

Figure 22. Potential composition curve at $100^{o}C$ for the $Li_{1+x}V_3O_8$ electrode in a $(PEO)_9LiCF_3SO_3$ cell. Li reference electrode.

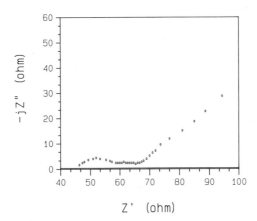

Figure 23. Alternating Current complex impedance diagram of the $Li/(PEO)_9LiCF_3SO_3/ LiV_3O_8$ cell at $100^{o}C$.

vanadium bronze. This is confirmed by experimental results obtained by ac impedance and cyclic voltammetry.

Figure 23 shows the impedance diagram of a $Li/(PEO)_9LiCF_3SO_3/ Li_{1+x}V_3O_8$ cell. The $Z' - Z''$ diagram reveals three relaxation effects. Proceedings from high to low frequency, a first semicircle is noticed which may be associated with the Li/electrolyte interface, followed by a second semicircle which may be related to the vanadium bronze/electrolyte charge transfer process, and finally a $45^{o}C$

Warburg diffusion line, the latter suggesting kinetics controlled by the diffusion of the intercalated lithium in LiV_3O_8.

A relatively fast diffusion of lithium is expected for this system, at least at the initial stage of the electrochemical process. In fact, large diffusion coefficients (i.e., of the order of 10^{-7} - 10^{-8} cm^2/s^{-1}) have been found (37) for the intercalated lithium in the $Li_{1+x}V_3O_8$ vanadium bronze.

Properties of the Li-LiV₃O₈ polymer/electrolyte, thin-layer battery

A thin-layer battery having the basic following sequence:

$$Li/polymer\ electrolyte/LiV_3O_8 \qquad\qquad\qquad (5)$$

was realized.

As lithium polymer electrolyte, the $(PEO)_8LiClO_4$, or the $(PEO)_9LiCF_3SO_3$ or the $(PEO)_7LiBF_4$ complexes, respectively, can be successfully used: A LiV_3O_8:C:PEO 40:20:40 weight percent ratio was used for the cathodic mixture. For optimized cell operations, the cathodic mixture can be cast (using a procedure basically similar to that illustrated in Figure 7) into a thin elastic, plasticlike layer. With this procedure, homogeneity in the cathodic mixture and good electrode/electrolyte interfacial contacts are realized. Figure 24 pictorially illustrates this thin-layer, polymer electrolyte battery.

Figure 25 shows the voltage trend of the first ten consecutive deep ($x\sim3$) discharge cycles of the battery. It may be clearly seen that the theoretical capacity 0.28 Ah/g, corresponding to an intercalation level $x\sim3$, is almost totally achieved in the first cycle, while a certain decline in capacity may take place upon prolonged cycling. It is now clear (38) that this effect which has also been observed by other authors in different polymer electrolyte batteries (35,30), is not due to irreversible degradation of the structure of the LiV_3O_8 vanadium bronze, but rather to the deformation and/or contact losses at the electrode/electrolyte interfaces. These interface nonuniformities may concern both the positive (isolation and contact losses of the active electrode material in the cathodic mixture) and the negative (passivation of the lithium electrode) sides of the cell.

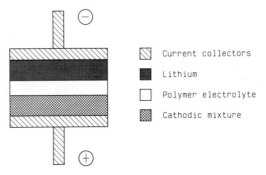

⬚ (hatched)	Current collectors
⬛	Lithium
☐	Polymer electrolyte
▨	Cathodic mixture

Figure 24. **Pictorial illustration of a Li/LiV₃O₈ thin-layer, polymer electrolyte battery.**

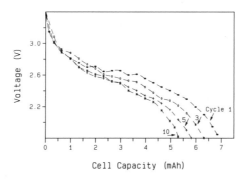

Figure 25. **Consecutive deep discharge cycles of the Li(PEO)$_7$LiClO$_4$/LiV$_3$O$_8$ cell at 100°C.**

Under these circumstances a fraction of the charging current may be lost. This may in turn produce an 'apparent' decline in discharge and capacity, especially under prolonged cycling conditions.

Various authors (29,30) including ourselves (38) have shown that these interfacial problems can be successfully overcome by using proper cell structures and geometry. These include optimized casting preparation procedures for the cathode mixture (to assure a high degree of homogeneity of the components) and extended electrode surface areas (to reduce values of current densities and thus reduce the chances of building up points of nonuniformities upon cycling). Therefore one may conclude that with appropriate cell geometry, high energy, long cyclable polymer electrolyte, thin-layer batteries can eventually be realized.

In the case of 'conventional' PEO-based electrolytes, an operation temperature of about $100^{\circ}C$ is generally required to assure low internal IR drops and fast kinetics. This may represent one of the drawbacks of the polymer electrolyte batteries if they are directed to the electronic consumer market.

Various types of new polymer electrolytes having improved low temperature performance are presently under development and are being characterized. The availability of these electrolytes will allow the realization of polymer-based batteries of more general interest.

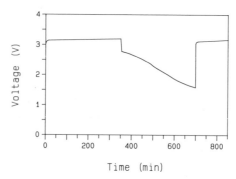

Time (min)

Figure 26. **Typical charge/discharge cycle at 0.01 mA/cm^2 of the Li/EEEVE-LiBF$_4$/LiV$_3$O$_8$, C cell at 45°C.**

An example of these improved electrolytes are the cited EEEVE-based complexes, whose electrical behavior is shown in Figure 6. Figure 26 illustrates a typical charge/discharge cycle of a Li/EEEVE-LiBF$_4$/LiV$_3$O$_8$, C cell at 45°C, run with a current density of $10\,\mu A/cm^2$. Under these conditions, the cell could be cycled several times without showing any appreciable degradation. Even after lowering the temperature of operation to 35°C, the cell could still be cycled at the same rate. Even more promising are the results obtained at Hydroquebec Laboratories in Canada, where a battery using an improved polymer electrolyte is currently under test at ambient and subambient temperature (39).

POLYMER ELECTRODES

The applications of ionically conducting polymers were described in the previous paragraphs. Another class of polymers having a predominant electronic conductivity has been characterized recently.

These are mainly unsaturated polymers with pi-electrons that can easily be removed or added to the polymeric chains. Typical examples are conjugated polymers, such as polyacetylene (PA), and heterocyclic polymers, such as polypyrrole (PPy), polythiophene (PT) and their derivatives.

By exposing these polymers which are intrinsically nonconductors to an oxidizing agent X or reducing agent M, one obtains a positively or a negatively charged, highly conductive, polymeric complex, and a counterion which is the reduced X^- or the oxidized M^+ form of the oxidant or the reductant.

These oxidation (or reduction) reactions, which induce high conductivity in the polymers, have been termed 'p-doping' (or n-doping) processes by a direct transfer of semiconductor terminology. It is clear that these reactions are quite different from the typical doping reactions in inorganic semiconductors, in that they closely resemble an intercalation process, and involve the 'insertion' of the counterion in the polymer matrix.

An interesting aspect is that the processes which modify the transport characteristics of the unsaturated polymers can be driven electrochemically by polarization in suitable cells. A cathodic polarization will induce n-doping processes accompanied by the insertion of cations from the electrolyte solution, according to the general scheme:

$$P + yM^+ + ye \xrightleftharpoons{\hspace{2cm}} P(M)y \qquad (6)$$

where P is the selected polymer and M^+ is the cation from the electrolytic solution.

131

Accordingly, an anodic polarization will induce p-doped processes accompanied by the insertion of anions from the electrolytic solution, according to the general scheme:

$$P + yX^- \xrightarrow{\hspace{1cm}} \xleftarrow{\hspace{1cm}} P(X)y + ye \qquad (7)$$

where X^- is the anion from the electrolytic solution.

The electrochemical 'doping' processes are reversible and this allows the use of the related polymers as a new class of electrode materials for the development of interesting, if not revolutionary, rechargeable thin-layer electrochemical devices.

To understand the properties of these polymers as electrodes, it seems necessary to briefly illustrate their electrical and optical properties. Here again, for detailed information, the reader is referred to more extensive reviews (3,6).

ELECTRICAL AND OPTICAL PROPERTIES OF DOPED POLYMERS

The electrochemically induced doping processes greatly modify the properties of the polymers. In describing the evolution of the electronic structure upon doping, the chemical approach elaborated by Bredas and Street (40) will mainly be followed here.

Let us consider as a typical example the oxidation (p-doping) process of polypyrrole, which in the undoped state is a poorly conductive material with an energy gap of 3.2 eV between the conduction and the valence bands. The removal of the first electron in the p-doping process leads to the formation of a charge localized in the polypyrrole chain, accompanied by a local distortion of the lattice. The localized charge is termed 'polaron' and in chemical terms it represents a radical cation associated with a relaxation of the structural geometry of the polymer towards a quinoid form which extends over four pyrrolic rings.

The presence of a localized electronic state in the gap is referred to as a polaron state situated at 1/2 spin polaron level at 0.5 eV from the band edges (Figure 27a).

When a second electron is removed from the polymer chain, a 'bipolaron' is formed, which is defined as a pair of like charges associated with a strong localized lattice distortion, which again extends over four pyrrolic rings. Obviously, the lattice relaxation is now stronger than that associated with a single polaron and thus the spinless bipolaron electronic levels are moved to within 0.75 eV from the band edges (Figure 27b).

At high doping levels (the case of polypyrrole may extend up to 33% of dopant (X) per polymer repeat unit, see equation (7)), the overlap between the bipolaronic states leads to the formation within the gap of two 'bipolaronic bands' approximately 0.4 eV wide (Figure 27c).

The described band evolution accounts for the unique transport properties of polypyrrole and of conducting polymers in general. Under an applied electric field, the spinless bipolarons become mobile, and especially at high doping levels when the coulombic attraction with the counterions is largely screened, they can easily transport the current.

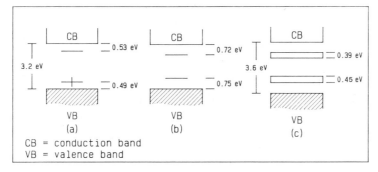

Figure 27. **Evolution of the electronic structure of polypyrrole following the p-doping (oxidation) process.**

It is worth noting that since transport takes place via mobile bipolarons, the conduction mechanism also requires the movements of the negative counterions which must migrate along the polymeric chain to compensate for the transport of the positive bipolaronic charges.

Since the electric transport implies both polaronic and ionic charge movements, the kinetics of the doping mechanism remain controlled by the diffusion of the slower carrier (i.e., the ionic specie) in and out of the polymer structure. This aspect is of crucial importance in the practical use of conducting polymers, since it may seriously affect their response when used in electrochemical devices.

Finally the presence of bipolaronic states in the polymeric chains allows optical absorptions prior to the gap transition. For example, referring to the band structure evolution of Figure 27, it is clear that the transitions may take place from the valence band to the lowest bipolaronic level as well as to the highest bipolaronic level. Furthermore, a third optical absorption corresponding to the transition between the two bipolaronic levels is possible.

These intergap transitions induce marked color changes on passing from the pristine to the doped state of the polymer, with resulting chromic effects of relevant technological interest.

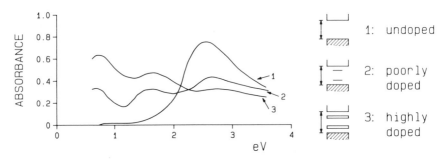

Figure 28. **Optical absorption spectra of** ClO_4^- **doped polythiophene as a function of dopant concentration.**

As a typical example, let us consider the case of polythiophene, a polymer with electronic properties similar to those of polypyrrole. Figure 28 shows the optical absorption spectra as a function of dopant concentration. In the polymer pristine state, only the gap absorption at about 2.5 eV is shown and the polymer has a red colour. In the doped state, two absorptions are present below the gap, at about 1.8 eV and 0.7 eV, respectively, ascribable to the bipolaron states, the polymer assumes a deep blue color.

134

PREPARATION AND ADVANTAGES OF POLYMER ELECTRODES

Most promising for electrochemical applications are the heterocyclic polymers, such as polypyrrole, polythiophene and their derivatives. These materials may be very conveniently polymerized and doped by a single electrochemical operation, carried out with a simple cell consisting of two plane electrodes (the substrate and the counter electrode) immersed in a generally nonaqueous solution (e.g., an acetonitrile solution) containing the monomer (e.g., pyrrole) and a supporting electrolyte (e.g., lithium perchlorate). Following an applied galvanostatic polarization, a voltage is reached at which the first polymerization takes place (usually via a-a' links) and then the polymer ionization (positive in the case of oxidation or p-doping and negative in the case of reduction or n-doping processes), followed by its deposition as a conductive film on the selected substrate.

The electrochemical preparation offers many specific advantages. First, because there is no need for catalyst, the electrodeposited polymers are essentially pure. Then, by controlling the time of electrodeposition, the thickness of the polymer film may be easily varied from a few Angstroms to many microns, and by changing the nature of the counterions M or X (see equations (6) and (7)) in solution, the electrical and physicochemical properties of the polymer may be greatly changed.

Furthermore, different substrates may be used for the electrodeposition, allowing different uses of the polymer films. Platinum or nickel substrates may be used for battery applications, while tin oxide coated glass may be used for optical tests.

Finally, the conditions of the electrochemical synthesis may greatly influence the physical, morphological and electrochemical properties of the final polymer film (41).

THE LITHIUM POLYMER/POLYMER THIN-LAYER BATTERY

Conductive polymers, such as polypyrrole, polythiophene and their derivatives, have been proposed and effectively tested as new electrode materials in lithium, liquid organic electrolyte, rechargeable batteries (42-45).

The results tend to indicate that polymer electrodes are of great potential interest in advanced battery technology. The rate capability and the coulombic efficiency of the electrodes are sensibly affected by the slow diffusion of the dopant ions in the bulk of the polymer (47). This appears to be an intrinsic limitation of polymer electrodes which may prevent their use in high rate, high power batteries.

Consequently, it seems at the moment more promising to direct the application of conducting polymers to the development of low rate, thin-layer electrochemical devices, possibly designed for the consumer market. In this respect, it would be desirable to replace the liquid electrolyte with a solid electrolyte. An entirely solid configuration would in fact assure higher reliability and versatility to the device.

PEO-based complexes, having a conductivity due to both anion and cation transport (1), appear to be appropriate solid electrolytes for lithium batteries using conductive polymers as positive electrodes. In fact, these are characterized by electrochemical doping processes which require anion (X) insertion in the polymer chains (see equation (7)).

This interesting concept may be exploited by studying the behavior of solid-state cells, using a lithium anode, a polymer electrolyte and a polymer electrode. A typical example may be that based on the following sequence:

$$Li/PEO-LiX/PPy \qquad (8)$$

where X can be ClO_4^-, BF_4^-, or $CF_3SO_3^-$ and PPy is an abbreviation for polypyrrole, $(C_4H_5N)_x$.

The electrochemical process of the cell involves the oxidation (p-doping) of PPy:

$$(C_4H_5N)_x + xyLiX \xrightarrow[\text{discharge}]{\text{charge}} (C_4H_5N(X)_y)_x + xyLi \qquad (9)$$

which, being reversible should assure a good cycle life to the cell. Furthermore, since both polymer electrolytes and electrodes may be easily prepared in the form of thin films, advanced thin-layer cell designs which may provide flexibility and plasticity to the entire structure can be envisioned. Figure 29 illustrates a possible example.

Two main factors may considerably influence the behavior of these solid-state, lithium polymer cells: the conductivity of the electrolyte and the nature of the dopant.

As already stressed, the conductivity of the most common PEO-based polymer electrolytes is quite low at room temperature, becoming appreciable only above $60^{\circ}C$ (i.e., above the transition into the amorphous phase). Cells based on these electrolytes may operate only at temperatures above

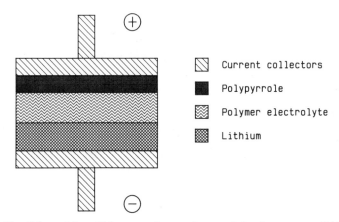

Current collectors

Polypyrrole

Polymer electrolyte

Lithium

Figure 29. Schematic illustration of a thin-layer, solid-state, lithium polymer battery.

ambient. This is clearly shown by Figure 30 which illustrates the cyclic voltammetry of a Li/PPy cell using a PEO-LiClO$_4$ complex as polymer electrolyte. No effect is revealed at $25^{\circ}C$, while at $60^{\circ}C$, the response is appreciable, and above the transition the response is evident with well-defined doping/undoping peaks.

The kinetics of the electrochemical process (9) are controlled by the diffusion of X^- in the bulk of the polymer (46). This is evidenced by Figure 31 which reports the ac impedance analysis of a Li/PPY solid-state cell using the (PEO)$_7$LiBF$_4$ complex as polymer electrolyte. Three relaxation phenomena can be identified in the frequency response. The first, revealed by a high-frequency arc, is related to the Li/polymer electrolyte interface. The second, at medium frequencies, may be

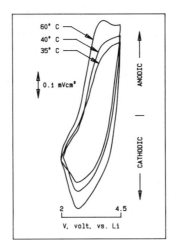

Figure 30. Cyclic voltammetry at various temperatures of a lithium/ polypyrrole cell using the $(PEO)_7LiBF_4$ polymer electrolyte. Scan rate 10 mV/s.

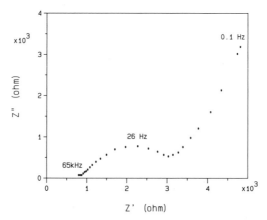

Figure 31. Complex impedance plot of a lithium/polypyrrole cell using the $(PEO)_7LiBF_4$ polymer electrolyte.

associated with the PPy/polymer electrolyte interface. The arc is here followed by a $45°$ Warburg line, which typically characterizes diffusion-controlled kinetics (14). Finally, at very low frequencies, the $45°$ line approaches a $90°$ line, indicating that in these conditions the polymer assumes a capacitive behavior (47,48).

Due to the diffusion effects, the nature of the dopant ion X^- becomes a critical factor. Detailed work should be devoted to the identification of the most suitable counterion-optimized battery applications. At the present preliminary stage of development, the majority of the studies have been limited to the $(PEO)_x LiClO_4$ electrolyte.

The thin-layer battery:

$$Li/(PEO)_x LiClO_4/PPy \qquad\qquad (10)$$

has an open circuit voltage (OCV) of 3.15V at $100°C$. The operating capabilities of the battery are demonstrated by Figure 32 which confirms the good cyclability, expressed in terms of delivered capacity upon prolonged cycling.

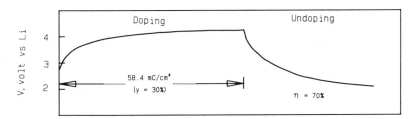

Figure 32. **Charge/discharge cycle of a $Li/(PEO)_8 LiCF_3 SO_3/PPy$ solid-state battery at $100°C$.**

CONCLUSIONS

Polymer-based, thin-layer lithium solid-state batteries are a feasible concept since preliminary, nonoptimized prototypes offer promising operating capabilities, such as high voltage and good rechargeability.

High·energy, solid--state, polymer electrolyte, rechargeable lithium batteries are under study in various academic and industrial laboratories. Even though these studies only began recently, the results are very encouraging and it is evident that the interest in this type of battery will consistently grow in the years to come.

Projected energy density values of the polymer electrolyte cells compare very favorably with those of existing systems, such as nickel/cadmium and lead/acid. The comparison is favorable also in terms of cost, especially in reference to established liquid electrolyte, rechargeable lithium batteries (49).

Furthermore, the polymer electrolyte, lithium batteries appear to be reliable, safe and relatively easy to assemble.

In the case of use of 'conventional' PEO-based electrolytes, an operation temperature of about $100^{o}C$ is generally required to assure low internal IR drops and fast kinetics. This may represent one of the drawbacks of the polymer electrolyte batteries. In fact, while a high temperature of operation would not be a serious problem for batteries directed to high rate applications, it may become a major concern for power sources designed for the electronic consumer market.

It is reasonable to forecast that modifications in the nature of the polymer host will soon provide electrolytes having improved low temperature conductivity, immediately allowing a wider temperature range of operation for the Li/polymer batteries. Furthermore, optimization in the cell structure should easily lead to considerable improvements in the polymer/polymer interfaces for consistently lower contact resistances.

Under these conditions and if properly designed, the Li solid-state, lithium polymer system may play a key role in advanced electrochemical technology. In fact due to the specific characteristics of the polymer components, ultrathin cells can be envisioned for applications, which by suitable selection of the electrolyte and the electrode materials, may range from flat plastic batteries for smart credit cards to flexible optical displays for smart windows.

ACKNOWLEDGEMENTS

The financial support of the Comitato Nazionale per la Ricerca e per lo Sviluppo della Energia Nucleare e delle Energie Alternative (ENEA), Contratto ENEA Dipartimento di Chimica, is acknowledged.

REFERENCES

1) C.A. Vincent, Progress in Solid State Chemistry, 17, 145 (1987).

2) F. Bonino and B. Scrosati, 'Polymeric electrolytes', in 'Materials for Solid-State Batteries' B.V. Chowdari and S. Radha Eds., World Scientific Publishing Co., Singapore, 1986, page 41.

3) B. Scrosati, Progress in Solid State Chemistry, 18, 1 (1988).

4) F. Bonino and B. Scrosati, 'Polymeric electrolytes', in 'Materials for Solid-State Batteries', B.V. Chowdari and S. Radha Eds., World Scientific Pu. Co., Singapore, 1986, page 41.

5) C.A. Vincent and J.R. MacCallum Eds. 'Polymer Electrolyte Review 1', Elsevier Applied Science Publ., London, 1987.

6) 'Handbook of Conducting Polymers', T.A. Skothein Ed., Marcel Decker Inc., New York, 1986.

7) D.E. Furton, J.M. Parker and P.V. Wright, Polymer, 14, 589 (1973).

8) M.B. Armand, J.M. Chabagno and M. Duclot, 2nd Intl. Conf. Solid Electrolyte, St. Andrews, Scotland, 1978.

9) C.D. Robitaille and D. Fauteux, J. Electrochem. Soc. 133, 315 (1986).

10) R. Huq and G.C. Farrington, Solid State Ionics, in press.

11) F. Bonino, S. Pantaloni, S. Passerini and B. Scrosati, J. Electrochem. Soc., in press, 1988.

12) S. Pantaloni, S. Passerini, F. Croce, B. Scrosati, A. Roggero and M. Andrei, Electrochim. Acta, submitted.

13) B.C. Tofield, R.M. Dell and A. Hooper, AERE Harwell R.11 261, 1984.

14) J.R. MacDonald Ed., "Impedance Spectroscopy", J. Wiley and Sons, N.Y. 1987.

15) L.L. Yang, H. Yang, R. Hug and G.C. Farrington, J. Electrochem. Soc., in press.

16) M.B. Armand, J.M. Chabagno, M. Duclot in "Fast Ion Transport in Solids", P. Vashista, J.N. Mundy and G. Shenoy Eds., North Hollands, N.Y. 1979, page 131.

17) P.R. Sorensen and T. Jacobsen, Solid State Ionics, 9/10, 1147, (1983).

18) P.R. Sorensen and T. Jacobsen, Electrochim. Acta, 27, 1671 (1972).

19) M. Watanabe, M. Togo, K. Sanui, N. Ogata, T. Kobayashi and Z. Ohtaki, Macromolecules, 17, 2908 (1984).

20) M. Leveque, J.F. Le Nest, A. Gandini and H. Cheradame, J. Power Sources, 14, 27 (1987).

21) P. Ferloni, G. Chiodelli, A. Magistris and M. Sanesi, Solid State Ionics, 18/19, 265 (1986).

22) J.E. Weston and B.C.H. Steele, Solid State Ionics 7, 75 (1987).

23) A. Bouridiah, F. Dolard, P. Deroo and M.B. Armand, J. Appl. Electrochem., 17, 625 (1987).

24) J. Evans and C.A. Vincent, Polymer, in press.

25) P.G. Bruce and C.A. Vincent, J. Electroanal Chem., 225, 1 (1987).

26) C.A. Vincent and J.R. MacCallum Eds., "Polymer Electrolyte Review 1", Elsevier Applied Science Publishers, London, 1987.

27) B. Scrosati, British Polym. Journal, in press.

28) A. Hooper and B.C. Tofield, J. Power Sources, 11, 33 (1984).

29) M. Gauthier et al., J. Electrochem. Soc., 132, 1333 (1985).

30) M.Z.A. Munshi and B.B. Owens, Solid State Ionics, 26, 41 (1988).

31) F. Bonino, M. Ottaviani, B. Scrosati and G. Pistoia, J. Electrochem. Soc., 135, 12 (1988).

32) D. Fauteux, Solid State Ionics, 17, 133 (1985).

33) K.M. Abraham and S.P. Brummer, in "Lithium Batteries", J.P. Gabano Ed., Academic Press, London 1983.

34) M.S. Wittingham, Prog. Solid State Chem., 12, 41 (1978).

35) K. West, B. Zachau-Christiansen, T. Jacobsen, J. Power Sources, 20, 165 (1987).

36) S. Panero, M. Pasquali and G. Pistoia, J. Electrochem. Soc., 130, 1225 (1983).

37) G. Pistoia, S. Panero, M. Tocci, R.V. Moshtev and V. Manev, Solid State Ionics, 13 311 (1984).

38) A. Selvaggi, B. Scrosati and Wang Gang, J. Power Sources, in press.

39) G. Vassort, M. Gauthier, P.E. Harvey, F. Brochu and M. Armand, Electrochem. Soc. Honolulu Meeting, 87-2, October 1987.

40) J.L. Bredas and G. B. Street, Acc. Chem. Res., 18, 309 (1985).

41) S. Panero, P. Prosperi and B. Scrosati, Electrochim Acta, 32, 1465 (1987).

42) P. Buttol, M. Mastragostino, S. Panero, and B. Scrosati, Electrochim. Acta, 31, 783 (1986).

43) S. Panero, P. Prosperi, B. Klaptse and B. Scrosati, Electrochim. Acta, 31, 159 (1986).

44) S. Panero, P. Prosperi, F. Bonino, B. Scrosati, A. Corradini and M. Mastragostino, Electrochim Acta, 32, 1007 (1987).   -

45) B. Scrosati, S. Panero, P. Prosperi, A. Corradini and M. Mastragostino, J. Power Sources, 19, 27 (1987).

46) D.D. Perlmutter and B. Scrosati, Solid State Ionics, in press (1988).

47) N. Mermillod, J. Tangeny and F. Petiot, J. Electrochem. Soc., 133, 1073 (1986).

48) S. Panero, P. Prosperi, B. Scrosati, Electrochim. Acta, 32, 1461 (1987).

49) B.B. Owens, EPRI AP-5218 Report, June 1987.

THE USE OF EXAFS IN THE STUDY OF POLYMERIC ELECTROLYTES

Roger J. Latham

School of Chemistry
Leicester Polytechnic
Leicester LE1 9BH, U.K.

INTRODUCTION

In conventional X-ray absorption experiments the absorption coefficient for any particular element varies smoothly with the energy/wavelength of the incident photon beam. When the beam energy is sufficiently high enough to overcome the binding energy of a core electron, a photoelectron will be ejected. The abrupt change in absorption of the incident beam is referred to as an absorption edge, and these features are superimposed on the smooth curve; thus, a K edge refers to the minimum energy required to eject a 1s electron and the closely spaced L edges refer to ejection of 2s and 2p electrons.

Extended X-ray Absorption Fine Structure, EXAFS, refers to the oscillations observed in the X-ray absorption for a range of several hundred eV beyond the absorption edge. This phenomenon occurs because the outgoing ejected photoelectrons interact with nearest-neighbor atoms in the material, which can act as secondary sources of scattering. The final state wavefunction of the outgoing electron can be modified by back scattering from these neighbor atoms. The fact that the outgoing and back scattered wavefunctions can interfere constructively or destructively gives rise to the small sinusoidal oscillations known as EXAFS. It can therefore be expected that the phase of these oscillations must depend on the distance from the absorbing to neighbor atom, and that the amplitude will depend on the number of nearest neighbors at a given distance. Shorter bonds are characterized by more widely spaced EXAFS oscillations and the amplitude increases with the number of nearest neighbors.

EXAFS is therefore a technique which can give local structural information about distances and types of nearest neighbors to atoms or ions of a chosen element within a given material [1]. It is a particularly useful technique in that the material can be either in the crystalline or amorphous state.

EXAFS experiments are becoming more widely applied with the intense spectrum of continuous X-rays that is produced by synchrotron radiation. The number of laboratories is at the moment limited. The only facility available in the U.K. is at SERC Daresbury Laboratories.

EXPERIMENTAL DATA

A necessary feature of EXAFS facilities is the need for 'user friendly' computer hardware and software in order both to acquire and deconvolute the data.

Having transferred the experimental data from the local station computer to the main files, the first stage is background subtraction from the raw data. After this the EXAFS is transformed into k space, where k is the wave vector of the incoming beam and is given in Angstroms by:

$$k = [0.263 \ (E - E_o)]^{1/2}$$

E is the energy of the incident beam and E_o is the energy corresponding to the adsorption edge of the target element. Fourier transform of the spectrum gives a radial distribution function from which the nature, number and distance of the neighbor atoms can be determined. Phase shift parameters relating to the target and nearest-neighbor atoms are required here and these must be tested by an EXAFS examination of a model compound. This must be a compound with the nearest neighbors to the target atom similar to that in the material under investigation, and a full structural determination must be available.

POLYMERIC ELECTROLYTES

EXAFS is particularly useful in the study of structure-conductivity relationships [2,3]. Care must be exercised in the interpretation of results for mixed morphology systems of the type encountered in polymeric

146

electrolytes in which there may be both crystalline and amorphous phases. In a polymeric electrolyte the presence of these phases may depend on factors such as thermal history and sample preparation [4,5]. Probably the most intensively studied polymeric electrolytes have been those based on polyethylene oxide [6]. The local environment of the mobile species is of particular interest and a number of issues may be addressed:

(a) What is the cation environment? What is the coordination with the backbone of the PEO?
(b) What is the local environment of the anions?
(c) Can EXAFS help to answer questions about the effect of moisture in film preparation?
(d) Does ion pairing take place in polymeric electrolytes?

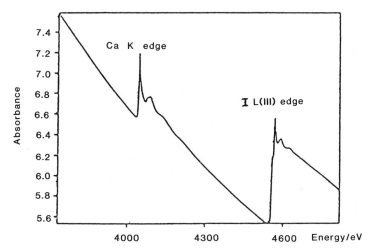

Figure 1. **Experimentally recorded photoelectron transmission spectra (from [7]).**

Figure 1 shows a typical experimentally recorded photon transmission spectrum for polymeric electrolyte film. Figure 2 shows the EXAFS spectrum together with the Fourier transform. Results from these initial EXAFS studies [7] of $(PEO)_4:CaI_2$ polymeric electrolyte films showed that in such systems at room temperature there was no evidence for ion pairing within the first two cation neighbor shells and that the ten nearest-neighbor species to the Ca^{2+} ions are in fact oxygen. Rigorous conditions of dryness were not employed in sample preparation and experimental handling, which made it difficult to assign these to the polymer backbone or to water for a hydrated salt. More recent results suggest that the

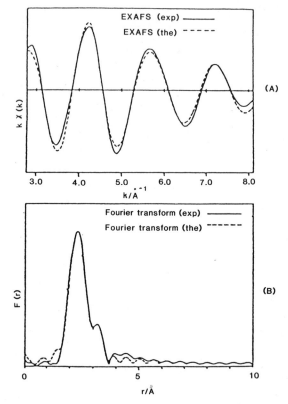

Figure 2. Experimental (——) and theoretical (---) EXAFS and Fourier transform spectra for Ca (from [7]).

calcium ions are still highly coordinated by oxygen even when, in contrast, meticulous attention has been paid to the preparation and handling of dry $(PEO)_n:CaI_2$ complexes. The number of nearest-neighbor oxygen species does seem to depend on the composition of the material. We have also studied $(PEO)_n:ZnX_2$ complexes [8] where X = I,Br. Results from these experiments show that the cation for the $ZnBr_2$ complexes is surrounded by six oxygen and two halide nearest neighbors, and the cation for the ZnI_2 complexes is surrounded by four oxygen and two iodine nearest neighbors. In both these cases the number and type of nearest neighbors does not change with polymer electrolyte composition.

In summary, EXAFS offers the possibility of studying polymeric electrolyte samples in which:

(a) the temperature above and below the glass transition temperature, T_g, and those at various other stages in heating/cooling cycles have been used.

(b) the environment of the action may be changed by adopting a range of partnering anions.

(c) 'in situ' methods to investigate (i) particular morphological areas of a film, and (ii) electrolyte behavior at various stages of battery discharge have been developed.

REFERENCES

1. J. Stohr, EXAFS and Surface EXAFS: Principles, Analysis and Applications, in "Emission and Scattering Techniques", NATO ASI Series C. P. Day, ed., Reidel, p. 213 (1981).
2. C.R.A. Catlow, A.V. Chadwick, G.N. Greaves, L.M. Moroney and M.R. Worboys, An EXAFS Study of the Structure of Rubidium Polyethylene Oxide Salt Complexes, Solid State Ionics, 9/10: 1107 (1983).
3. A.V. Chadwick and M.R. Worboys, NMR, EXAFS and Radiotracer Techniques in the Characterization of Polymer Electrolytes, in "Polymer Electrolyte Reviews", J.R. MacCallum and C.A. Vincent, eds., Elsevier Applied Science, London (1987).
4. R.J. Neat, A. Hooper, M.D. Glasse and R.G. Linford, Thermal History and Polymer Electrolyte Structure, Solid State Ionics, 18/19: 1088 (1986).
5. R.J. Neat, M.D. Glasse, R.G. Linford and A. Hooper, A Structural Model for the Interpretation of Composition Dependent Conductivity in Polymeric Solid Electrolytes, in: "Fast Ion and Mixed Conductors", F.W. Poulsen et al., ed., Riso National Labs, Roskilde, p. 341 (1985).
6. C.A. Vincent, Polymer Electrolytes, Prog. Solid State Chem., 17:145 (1987).
7. K.C. Andrews, M. Cole, R.J. Latham, R.G. Linford, H.M. Williams and B.E. Dobson, EXAFS Studies of Divalent Polymeric Electrolytes: An Investigation of PEO_4:CaI_2 at Room Temperature, Solid State Ionics, in press.
8. M. Cole, M.D. Glasse, R.G. Linford and M. Sheldon, to be submitted to J. Electrochem. Soc.

AMORPHOUS ROOM TEMPERATURE POLYMER SOLID ELECTROLYTES

W. Wieczorek, K. Such and J. Przyluski

Institute of Solid State Technology, Warsaw University of
Technology, ul. Noakowskiego 3, 00-664 Warszawa, Poland

INTRODUCTION

Polymer electrolytes are ionic conductors of considerable technical
interest due to the wide range of their possible applications in various
devices. These materials exhibit many interesting properties, such as
compatibility with alkali metal electrodes, a wide stability window, and
easy method of preparation in a thin film configuration; however,
relatively low ambient temperature ionic conductivity is their big
disadvantage. From the considerations of Berthier et al. (1), it is
evident that conductivity in polymer electrolytes is connected with the
amorphous phase of the studied samples. Following this assumption, many
methods leading to production of polymer electrolytes, in an amorphous form
stable at room temperature, were developed. In our laboratory new concepts
of producing highly amorphous polymer foils, such as preparation of polymer
blends or grafted copolymers, utilization of PEO (poly(ethylene oxide))
copolymers of high molecular weight, and synthesizing of PEO-based polymer
solid electrolytes with ceramic additives were introduced. All these
methods enabled us to prepare PEO-based solid ionic conductors whose
conductivity was 5×10^{-5} S/cm at room temperature and was two orders of
magnitude higher than that obtained in the standard procedure.

EXPERIMENTAL

PEO-PPO copolymers were prepared by a copolymerization reaction using
aluminum organic catalysts. The molecular weight of the obtained samples

was in the range of 10^4 - 10^5 g/mol and increased by raising the EO (ethylene oxide) concentration in the starting reaction mixture.

Grafted copolymers were synthesized by polymerization of methyl methacrylate in a highly viscous PEO solution. Various organic solvents such as chlorobenzene, acetonitrile, and 1,2-dichlorobenzene were utilized for completing the reaction.

PEO-based polymer solid electrolytes with ceramic additives were produced by the procedure described by us in our earlier papers (2,3,4). Polymer electrolyte foils were obtained by a standard casting method using NaI (reagent grade) as a dopant and acetonitrile as a solvent. The solvent was distilled twice from molecular sieve before use. All the experiments were carried out under vacuum. For impedance measurements, samples were placed between two gold plated electrodes and the measuring holder was put into an evacuated glass vessel. Their temperature was changed from 20°C to 100°C during measurements. The frequency range was 5Hz to 500 kHz. For structure investigations, X-ray diffraction experiments and SEM observations were applied.

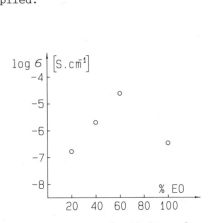

Figure 1. **Room temperature conductivity for PEO-PPO-10% NaI system.**

RESULTS AND DISCUSSION

Figure 1 shows room temperature conductivity of PEO-PPO copolymers doped with NaI versus EO content in the reaction mixture. The maximum conductivity was found in the sample containing 60% EO. Conductivity values were at least two orders of magnitude higher than those measured for pure PEO or PPO (poly(propylene oxide))-based electrolytes. From X-ray diffraction it is clear that PEO-PPO copolymers are amorphous. There are no diffraction peaks characteristic of pure crystalline PEO in the diffractograms. The static character of the polymer chain was evident from NMR ^{13}C experiments. EO units were randomly distributed along the chain and hence a statistic copolymer was obtained rather than a block copolymer. The mechanical properties of the sample studied were not good enough. It seems that this problem may be solved by increasing the molecular weight of the polymer.

Results of the conductivity study of PEO-PMMA (polymethyl methacrylate) grafted copolymers are depicted in Figure 2. As can be seen, the role of the applied solvent seems to be crucial. In our opinion the temperature of the polymerization reaction and viscosity of the reaction environment are also important parameters. The highest values of conductivity were found when 1,2-dichlorobenzene was used as a solvent in the polymerization reaction. It has the highest boiling temperature and viscosity among all the studied solvents. During the reaction, a highly inhomogeneous mixture consisting of pure PEO, pure PMMA and a grafted phase was formed. It seems that the presence of the grafted phase is responsible for the conductivity increase; however, it is not easy to separate these phases from the others. The conductivity increased to the range of 10^{-5} S/cm at room temperature in the case of the samples prepared in 1,2-dichlorobenzene solution. The crystallinity of pure PEO-PMMA copolymer is equal to about 40% as was determined by X-ray diffraction experiments. Changes in the structure in comparison to pure PEO were confirmed by SEM observations. The sizes of crystalline PEO spherulites were decreased from about 400μm for pure PEO to about 50μm for PEO-PMMA grafted copolymer.

A new method of producing highly conductive room temperature polymer solid electrolyte was developed in our group (2,3). It was based on addition of an inert ceramic powder to a polymer matrix. Similar investigations on mixed phase solid electrolyte based on the PEO-NaI-NASICON system were also carried out by us. Significant changes in the phase structure of polymer electrolyte were found in all investigations mentioned above. A stable amorphous phase is formed after addition of ceramic particles to the polymer (2,3). Ionic conductivity of polymer foils containing various amounts of ceramic powder as a function of

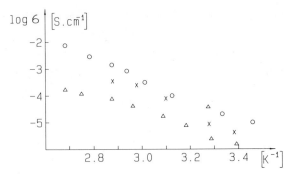

Figure 2. Conductivity vs. reciprocal temperature for PEO-PMMA grafted copolymers with 10% NaI additives, obtained in various solvents: o - 1,2-dichlorobenzene, x - chlorobenzene, △ - acetonitrile.

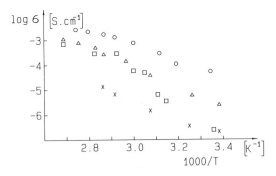

Figure 3. Conductivity vs reciprocal temperature for $(PEO)_{10}NaI$ system with $\alpha-Al_2O_3$ additives (0.3μm - grain sizes)
□ - 0% $\alpha-Al_2O_3$, △ - 20% $\alpha-Al_2O_3$
o - 10% $\alpha-Al_2O_3$, x - 30% $\alpha-Al_2O_3$

temperature is shown in Fig. 3. A maximum conductivity equal to 5×10^{-5} S/cm at room temperature was found for a sample containing 10% w/w α-alumina (grain size 0.3 μm, Merck).

The influence of ceramic concentration on the conductivity of the polymer electrolytes is also depicted in Fig. 3. For a concentration of the ceramic powder greater than 20%, conductivity decreases rapidly which is probably due to the formation of highly resistive ceramic layers in the samples. From earlier investigations it was clear that the grain size of alumina is a crucial parameter. For a particle size less than 2 μm, conductivity of the polymer electrolyte increases in comparison to that measured for samples not containing ceramics; but for larger alumina grains, conductivity decreases rapidly.

CONCLUSIONS

Some concepts of producing highly conductive room temperature polymer solid electrolyte in a stable amorphous form are reviewed. An increase of the conductivity to a value of 5×10^{-5} S/cm at room temperature, which is at least two orders of magnitude higher than for samples prepared by the standard procedure, is shown. Further investigations are in progress.

ACKNOWLEDGEMENTS

The authors wish to thank Prof. Z. Florianczyk and Dr. J. Plocharski for their valuable help. This work was financially supported by the Polish Ministry of Education under the C.P.B.P. 01.15 research programme.

REFERENCES

(1) C. Berthier, W. Gorecki, M. Minier, M.B. Armand, J.M. Chabagno and P. Rigaud, Solid State Ionics, 11 (1983) 91.
(2) W. Wieczorek, K. Such, H. Wycislik, J. Plocharski, to be published in Solid State Ionics.
(3) J. Przyluski and W. Wieczorek, Proc. of the II Asian Conference on Solid State Ionics, Singapore, 18-23 July 1988.
(4) J. Plocharski and W. Wieczorek, to be published in Solid State Ionics.

DESIGNING A SOLID ELECTROLYTE

I. QUALITY CRITERIA AND APPLICATIONS

John B. Goodenough

Center for Materials Science and Engineering, ETC 5.160
The University of Texas at Austin, Austin, TX 78712-1084 USA

INTRODUCTION

The design of a solid electrolyte or of a reversible battery electrode provides an excellent illustration of the interdisciplinary process labeled "materials engineering." The several aspects of this process are diagrammed in Fig. 1.

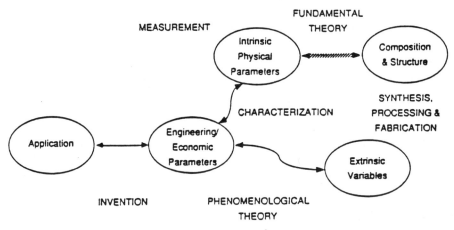

Fig. 1   Multidisciplinary diagram of materials engineering.

Materials engineering begins with the invention of a particular application and the specifications, in terms of engineering/economic requirements, of the properties an enabling material must possess. The second step is a phenomenological analysis that expresses the engineering parameters in terms of intrinsic material properties and the extrinsic variables that can be manipulated in the processing and fabrication of the overall device. The economic parameters generally depend more on device efficiency and fabrication costs than on materials costs, and these two factors depend sensitively on device design as well as on the intrinsic material properties. Others in this school have the task of speaking to the problems of device design; my task is to address the third step in the process -- the optimization of intrinsic material properties.

Ionic conductors of technical interest can be classified into four categories: ion exchangers, solid electrolytes, electrochemical stores, and chemical stores. Of interest for this school are the electrochemical applications. Therefore in my first lectures I address the

Solid State Microbatteries
Edited by J. R. Akridge and M. Balkanski
Plenum Press, New York, 1990

problem of designing a solid electrolyte; in my last lecture, I address the problem of designing a secondary-battery electrode.

ELECTROLYTE QUALITY CRITERIA

1. **Definition**. An electrolyte is an ionic conductor and an electronic insulator.

2. **Use**. Electrolytes are used in electrochemical cells. An electrochemical cell is designed to have d.c. electronic current flowing in the external circuit balanced by only ionic current in the internal circuit. The elctrolyte serves as an internal electronic insulator and ionic conductor between the cell electrodes. With liquid electrolytes it is necessary to introduce a separator -- commonly a glass fiber impregnated with the liquid electrolyte -- to keep the electrodes physically separated; a solid electrolyte also acts as the separator between the two reactants/electrodes of the cell.

3. **Quality Criteria**. Given the use of a solid electrolyte, the following engineering/ economic quality criteria for the material can be specified immediately:

- Unity transport number. The transport number of an electrolyte is defined as:

$$t_i \equiv \sigma_i/\sigma \tag{1}$$

where σ_i is the conductivity of the working ionic species and

$$\sigma = \sigma_i + \sigma_e + \Sigma_j\sigma_j \tag{2}$$

is the total conductivity, including any electronic conductivity σ_e and the sum of the conductivities of all other mobile ionic species. Any electronic current across the electrolyte dissipates the cell charge when on open circuit (self-discharge) and generates an internal voltage drop $\Delta V_e = I_e R_e$, where I_e and R_e are the internal electronic current and resistance of the cell. Therefore the ideal electrolyte passes no internal electronic current ($\sigma_e = 0$). Moreover, mobile counter ions encounter blocking electrodes and pile up at the electrode-electrolyte interface. Ionic pile-up creates a counter internal electric field that lowers the voltage V across the electrolyte of the cell. Therefore the ideal electrolyte conducts only the working ionic species ($\Sigma_j\sigma_j = 0$) and a

$$t_i \equiv \sigma_i/\sigma \approx 1 \tag{3}$$

at the operating temperature is an essential intrinsic quality parameter or electrolyte "figure of merit."

- High σ_i In an electrochemical cell, the operating voltage is necessarily low (< 5 V), so high power IV requires a high current I. A low internal I^2R loss requires, in turn, a small resistance

$$R = L/\sigma_i A \tag{4}$$

where L is the thickness of the electrolyte separator and A is its area.

- Easy fabrication into a large-area, dense, thin membrane. From equation (4) it follows that the electrodes of a high-power cell must be separated by a small distance L over a large area A. In addition, the electrolyte membrane must be made -- and must remain -- impervious to the reactants by retaining its mechanical and chemical integrity.

- Matched reaction window. A liquid electrolyte has a large energy gap E_g separating the highest occupied molecular orbital (HOMO) and the lowest unoccupied molecular orbital (LUMO). A solid electrolyte has a large energy gap E_g separating the top E_v of a filled valence band and the bottom E_c of an empty conduction band. In both cases, the energy gap

E_g corresponds to the thermodynamic "reaction window." Fig. 2 illustrates a matching of the thermodynamic reaction window for two types of electric-power cells: (a) liquid or gaseous reactants with a solid electrolyte and (b) solid, metallic reactants with a liquid (aqueous) electrolyte. Thermodynamic stability of the electrolyte requires that its lowest unoccupied state have a higher energy than the highest occupied state of the reactant, that its highest occupied state have a lower energy than the lowest unoccupied state of the oxidant. If either of these two conditions is violated, electrons may be transferred to or from the electrolyte to reduce or oxidize it. Therefore the requirement of thermodynamic stability restricts the e.m.f. of a power cell to an open-circuit voltage

$$V_{oc} < E_g \text{ (thermodynamic stability)} \qquad (5)$$

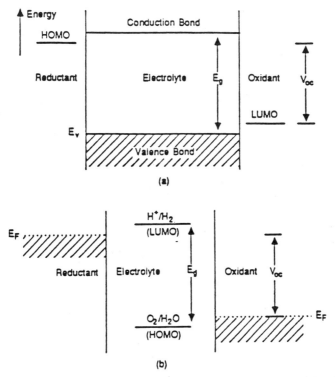

Fig. 2.   Matched reaction windows: (a) liquid or gas reactants with solid electrolyte, (b) solid, metallic reactants with an aqueous electrolyte.

However, in some power cells poor kinetics at an electrode/electrolyte interface allows a $V_{oc} > E_g$; this situation is classically illustrated by the 2-V lead-acid cell $Pb/H_2SO_4/PbO_2$, which has an aqueous electrolyte with a thermodynamic window of only 1.23 eV between the LUMO at the H^+/H_2 redox energy and the HOMO at the O_2/H_2O redox energy. Nevertheless, matching of the thermodynamic reaction window to the reactant donor and acceptor levels is the more prudent strategy in the design of an electrolyte for two given reactants.

It follows from this discussion that, unlike semiconductor technology, the selection of an E_g that is large enough to both maintain an intrinsic $\sigma_e \approx 0$ and allow a large V_{oc} is not in itself adequate; the positions of the band edges E_c and E_v -- or of the LUMO and HOMO -- of the electrolyte must also be matched to the positions of the reactant donor and acceptor levels or reactant-electrode work functions. Therefore we must anticipate a reiterative search for the best electrode and electrolyte materials before an optimum matching is achieved.

If the purpose of the electrochemical cell   is to electrolyze a liquid electrolyte, then a voltage $V > E_g$ is applied to two solid electrodes so as to induce electron transfer to the LUMO and from the HOMO of the electrolyte, but stability of the electrolyte in the absence of an applied voltage requires matching of the electrode work functions to the electrolyte HOMO and LUMO.

• Low interfacial ion-transfer resistance. In a power cell, ions are transported across the electrode/electrolyte interface; in an electrolysis cell, electrons are transferred across. Resistance to charge transfer across an interface contributes to the overvoltage (or cell polarization), where the overvoltage is defined as the change of voltage ($V-V_{oc}$) under a load current I. The overvoltage is a voltage loss in a power cell; it is a voltage increase in an electrolysis cell.

Fig. 3 illustrates a typical performance (polarization) curve for an electrochemical power cell. The voltage drop at low currents -- region *(i)* -- is due to the sum of the interfacial resistance and the internal IR voltage of the electrolyte, whereas the linear drop with increasing current in region *(ii)* is due to the internal IR voltage alone. The final additional voltage drop at high currents -- in the "diffusion-limited" region *(iii)* -- is due to depletion of acceptor sites or mobile ions at an interface of the cell. A higher σ_i on either side of the interface displaces region *(iii)* to higher currents.

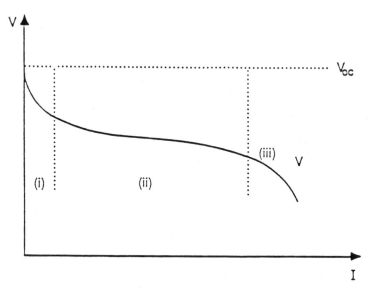

Fig. 3.   Typical Polarization Curve

A low interfacial resistance requires not only the lack of a blocking surface layer, but also the maintenance of intimate contact between the electrolyte and a reactant electrode or at a three-phase reactant, electrode, electrolyte interface. This requirement constrains the choice of electrolyte material, which may be a liquid, polymer, composite, glass, or polycrystalline ceramic. As indicated in Table I, the appropriateness of a given choice depends upon whether the reactants are solid, liquid, or gaseous.

TABLE I. Choice of Electrolyte Material

ELECTROLYTE	REACTANTS SOLID	LIQUID/GAS	T_{op}
Liquid*	Yes	Difficult	Aqueous:low
			Molten salt:low/med
Polymer(elastomer)†	Yes	Yes	Low
Composite: Liq/polymer	Yes	Yes	Low
polymer or liq./solid	Yes	Yes	Low
solid/solid	?	Yes	Low/high
Glass	?	Yes	Low/high
Ceramic polycrystal	?	Yes	Low/high

*Counter ion mobile   †Counter ion may be mobile

Solid reactants are commonly used in primary and secondary batteries. These solid reactants are generally metals and therefore serve as both electrode and reactant, which reduces the reaction zone to a single interface. However, a solid reactant changes its volume during a cell reaction; mass is transported either from or to it. Even insertion-compound electrodes, which undergo topotactic insertion/extraction solid-solution reactions, exhibit some change of volume with composition. In addition, they exhibit a thermal expansion that is not matched to that of the electrolyte with which they are to maintain contact. Therefore a solid reactant requires contact with an electrolyte material that can flow so as to maintain intimate contact on a change of volume of the reactant. This constraint tends to eliminate from consideration a rigid glass or polycrystalline ceramic as the electrolyte of contact with a solid reactant unless the cell design allows the interface to "breathe," as can be the case in a microbattery composed of alternative thin layers of electrode and electrolyte materials. On the other hand, liquid or elastomer electrolytes require no special interfacial designs for the maintenance of contact with a solid reactant electrode. Composite liquid or polymer/solid or liquid/polymer electrolytes contain an immobilized liquid or a polymer that is flexible; ionic transport occurs in the more flexible phase, which retains contact at the interface with a solid reactant electrode. Spillage is a problem with liquid electrolytes; therefore, an all-solid battery with elastomer or composite electrolytes offers the most straightforward strategy. An aqueous liquid electrolyte is commonly "immobilized" by introducing inert oxide particles (e.g. MgO), which transforms the liquid electrolyte to a liquid/solid composite.

Solid electrolytes are particularly well suited for use with liquid/gas reactants; glass or ceramic electrolytes become mandatory at higher temperatures. With liquid or gaseous reactants and a solid electrolyte, only the thermal-expansion mismatch between a current-carrying, porous electrode and the solid electrolyte is a concern; and intimate contact between electrode and electrolyte is not critical as electron transfer requires only bonding of the electrode to the surface of the electrolyte. What is critical in this case is the maintenance of electrical contact at a three-phase interface: electrode-reactant-electrolyte, and the flow inherent in the liquid/gas reactant ensures contact where the electrode and electrolyte adhere to one another sufficiently to insure electrical contact.

Use of a liquid electrolyte with gaseous or liquid reactants is possible, but awkward. In this case, porous electrodes must be designed for operation under a specified range of reactant pressures if a large-area three-phase interface is to be maintained and exhaust products are to be rapidly removed from this interface. Any changes in the pore size of the electrode as a result of corrosion or joule heating under load alters the optimum reactant pressure and degrades performance. Such an awkward engineering solution can be avoided if suitable solid electrolytes are identified and made available.

TYPICAL APPLICATIONS

1. **Atom-selective mass transfer**. Solid electrolytes may transport either cations or anions. Therefore either metal winning or a molecular pump can be envisaged. Fig. 4 illustrates schematically an oxygen pump utilizing a solid-membrane O^{2-}-ion electrolyte. Oxygen may be pumped from one side of the membrane to the other in a controlled manner by applying a potential between the porous electrodes on either side of the membrane. If a current I is passed through an external circuit between the electrodes, the ion flux J_i from one electrode to the other across the electrolyte is

$$J_i = I/nF \tag{6}$$

where n = 2 is the number of electrons transported by an O^{2-} ion and F is Faraday's constant. Oxygen pumps may be used, for example, to control the p_{O2} of a chamber.

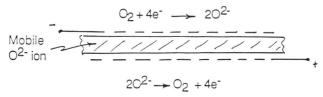

Fig. 4. Schematic of an oxygen pump.

2. **Open-circuit applications**. An oxygen pump may also be used to monitor the p_{O2} of a chamber by measuring the open-circuit voltage V_{oc} with a reference p_{O2} on the opposite side of the membrane. In this mode of operation, the oxygen electrolyte operates as an oxygen meter or sensor. Such sensors have found extensive industrial application. Some outstanding advantages of a solid-electrolyte system for this purpose include:

(a) a gas-specific, rapid, and continuous response of the e.m.f.

(b) a precise measurement that draws little current and is ideally suited as a transducer for recording and feedback control

(c) direct contact with the sytem to be measured, and

(d) stability at high temperatures.

On the other hand, poor oxygen mobility below 700 °C restricts the known O^{2-}-ion ceramic electrolytes to operating temperatures above 700 °C where an electronic conductivity tends to be troublesome -- see the discussion by W. Weppner.

The operating principle of an oxygen meter -- or any other concentration cell -- is the measurement of a difference in chemical potential across an electrolyte membrane that is caused by a difference in the equilibrium concentration of the mobile-ion species at the opposing electrolyte surfaces. If μ_{O2} and μ^o_{O2} are the monitored-gas and reference-gas O_2 chemical potentials on either side of the electrolyte membrane, then

$$-nFV_{oc} = \Delta G = \mu^o_{O2} - \mu_{O2} \tag{7}$$

where n=4 is the number of electrons involved in the reduction of O_2 to $2O^{2-}$. If use is made of the relation $\mu = \mu_0 + RT \ln a$, where $a \sim p_{O2}$ is the oxygen activity, then it follows from (7) that

$$V_{oc} = (RT/4F)\ln(p_{O2}/p^0_{O2}) \qquad (8)$$

Derivation of equation (8) presupposes a perfect solid electrolyte ($t_O = 1$). In practice, there is some oxygen leakage through the electrolyte due to any electronic conduction, if not due to pores in the ceramic, which makes $t_O < 1$ and modifies (8) to

$$V_{oc} = t_O(RT/4F) \ln(p_{O2}/p^0_{O2}) \qquad (8')$$

3. **Energy conversion and storage**. Solid electrolytes are of particular interest for electrochemical cells that convert chemical energy into electrical energy on discharge and, if rechargeable, electrical energy back into chemical energy in the charging cycle. These processes may be performed in either a secondary (storage) battery or with fuel/electrolysis cells.

(a) **Storage batteries**. In a storage battery, the chemical reactants are contained within the cells, the electrical energy is delivered to (or derived from) an external circuit. The classical secondary cell contains two reversible, solid-reactant electrodes and a liquid electrolyte; we symbolize it as

$$S^- /L /S^+ \qquad (9)$$

where S^- is the negative electrode (the anode) and S^+ is the positive electrode (the cathode). Representative examples are the 2.0 ± 0.15 V lead-acid cell Pb/ H_2SO_4/PbO_2 and the 1.3 V Cd-Ni cell Cd/KOH/NiOOH. In a Cd-Ni cell, for example, the reversible electrode reactions are

$$Cd + 2(OH)^- \ = \ [Cd(OH)_2 + L]_{comp} + 2e^- \qquad (10)$$

$$NiOOH + H_2O + e^- \ = \ Ni(OH)_2 + (OH)^- \qquad (11)$$

At the anode on discharge, Cd^{2+} ions leave the Cd-metal electrode, discharging two electrons to the external circuit; the Cd^{2+} ions react with the OH^- ions to form a $Cd(OH)_2$ + aqueous electrolyte (L) composite at the surface. The composite allows ions to be transferred through the liquid component, but a hydrogen-bonding network holds the composite to the surface of the electrode; the reaction is therefore reversible. Since the reaction is between two distinct phases, it occurs at a constant potential relative to an external reference (Gibb's phase rule).

The reaction at the cathode is somewhat different in character. Protons are inserted into the positive electrode without changing the NiO_2 sandwich layers of the structure, but a reorientation of the hydrogen bonding between sandwich layers distinguishes the NiOOH and $Ni(OH)_2$ phases, so this reaction also occurs at the constant potential; the overall voltage of the cell is therefore independent of the state of discharge. Nevertheless, this cathode reaction is representative of a large class of intercalation electrodes. If these solid electrodes are designated I for intercalation, then the symbolism of (9) may be rewritten as

$$S^- / L /I^+ \qquad (12)$$

for the Cd-Ni cell.

With the discovery of fast Na^+-ion conduction at 300 °C in the large-E_g, polycrystalline ceramic Na β-alumina, an alternative cell design was invented [1]; it utilizes liquid electrodes and a solid electrolyte

$$L^- /S /L^+ \qquad (13)$$

The ceramic electrolyte is stable at high temperatures, so the liquid electrodes can be molten metals and conductor-invaded salts having similar melting points. The conduction-band edge E_c of Na β-alumina lies above the Fermi energy E_F of molten sodium, so sodium can serve as the negative electrode L^-; it satisfies the matching requirement for the reaction window.

Selection of a suitable positive electrode proceeds as follows. Insertion of sodium ions into a liquid electrode must be charge-compensated by electrons supplied from the external circuit to a LUMO redox potential E_L of the electrode. Matching to the electrolyte window requires that E_L lie above the valence-band edge E_v of the electrolyte, but a large open-circuit voltage $V_{oc} = E_F(Na)-E_L$ would have it lie close to E_v. The molten-sodium anode has a fixed composition independent of the state of discharge of the cell. However, in order to have a V_{oc} that is independent of the state of discharge, the cathode reaction must also go to a fixed composition, which means that the sodium should be inserted into a two-phase liquid; only in this case does the Gibbs phase rule constrain V_{oc} to a value that is independent of the composition of the cathode. Moreover the cathode should be molten in the same temperature range as the anode. A sodium polysulfide two-phase mixture Na_2S_5 + elemental S meets these requirements; however, it is an electronic insulator. Therefore a carbon felt making contact with the positive part of the cell is introduced to supply electrons throughout the melt. A $V_{oc} = 2.08$ V is realized at 300 °C.

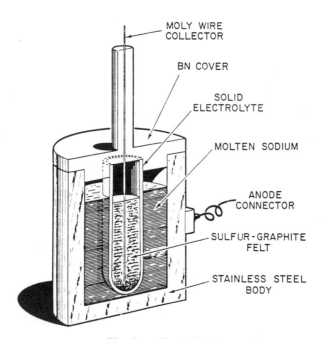

Fig. 5.   Early Na-S test cell.

Figure 5 illustrates an early design of this Na-S cell. Its potential advantage is a significantly higher energy and power density than the conventional aqueous-electrolyte batteries; its principal disadvantage is an operating temperature of over 300 °C. Moreover, fabrication in high yield of a chemically homogeneous, thin-walled ceramic test tube of large surface area proved to be a major problem in materials processing and testing. These

164

difficulties have led to a major parallel effort to develop lithium cells utilizing Li-insertion electrodes and an amorphous electrolyte

$$I^- / A / I^+ \qquad\qquad (14)$$

where A may be a low-melting-point salt, an elastomer, a composite, or a glass. A satisfactory combination for high-power discharge and reversibility has not yet been achieved.

(b) **Fuel/electrolysis cells.** A fuel cell converts chemical energy into electric power, an electrolysis cell converts electric power into stored chemical energy. Unlike the storage battery, which stores its reactants within the cell, the fuel cell receives gaseous or liquid reactants at a metal- electrode/electrolyte interface from an external store and vents its gaseous or liquid reaction product(s). The electrolysis cell performs the reverse process, but the chemical energy produced may be stored as well in a value-added product -- e.g. in electroplating or electrowinning--as in a fuel and an oxidant.

The most versatile oxidant is oxygen or air, and the hydrogen-air fuel cell has only pure water as its exhaust product:

$$2H_2 + O_2 \rightarrow 2H_2O \qquad\qquad (15)$$

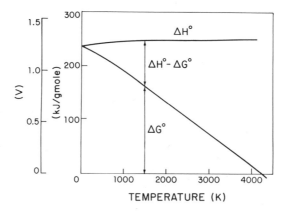

Fig. 6.   Enthalpy and Gibbs free energy versus temperature for the hydrogen-oxygen reaction (15).

Fig. 6 shows the Gibbs free energy and the enthalpy of this reaction as a function of temperature. At the limit of zero load current, the theoretical efficiency of the cell is

$$\eta_T = \Delta G / \Delta H_{298} = -nFV_{oc}/\Delta H_{298} \qquad\qquad (16)$$

where ΔH_{298} is the chemical energy stored at room temperature and n= 4 is the number of electrons transferred through the load per molecule of oxygen in reaction (15). Under load, the voltage is reduced from V_{oc} by the various losses discussed in connection with Fig. 3.

Moreover, it is clear from Fig. 6 that η_T decreases with increasing temperature. If a high-temperature electrolysis cell and a low-temperature fuel cell are run as a heat engine, the heat-engine efficiency at the limit of zero load current is

$$\eta = \frac{E_{out}}{E_{in}} = \frac{\Delta G_l - \Delta G_h}{T_h \Delta S} \approx \frac{T_h - T_l}{T_h} = \eta_c \qquad (17)$$

where η_c is the carnot efficiency and the subscripts h and l refer to high and low temperature. Clearly it is more efficient to run a fuel cell at lower temperatures. This consideration is important for the choice of electrolyte.

Given gaseous reactants, a solid electrolyte is the natural choice for a fuel cell. (See Table I.) Moreover, it is possible to use either an H^+ or an O^{2-} electrolyte. (See Fig. 7.) However, the conventional H^+-ion electrolytes are all hydrates or hydrated composites that lose water above 60 $^\circ$C unless pressurized; and at these lower temperatures, a catalyst is required to promote reaction at the reactant-electrode electrolyte interface. On the other hand, the conventional O^{2-}-ion electrolytes are poor ionic conductors below 700 $^\circ$C. Although considerable effort has been devoted to the development of a high-temperature fuel cell utilizing an O^{2-} ion electrolyte, the solid-polymer H^+-ion electrolytes offer the greater long-range promise. However, exploration of this important technology awaits the identification of catalysts stable in acid that are less expensive than platinum and, in the case of oxygen reduction, more active at room temperature. An acidic electrolyte is preferred for reaction (11) -- it is mandatory for a hydrocarbon fuel -- because CO and CO_2 react with OH^- ions to form carbonates that foul the cells. The solid polymer electrolytes are acidic.

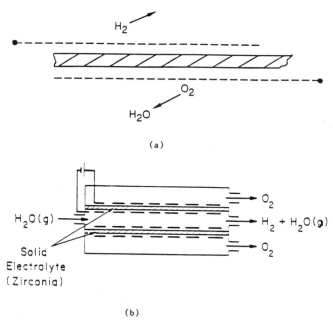

Fig. 7. Schematic use of (a) a solid polymer electrolyte in a hydrogen-oxygen fuel cell and (b) a ceramic O^{2-}-ion electrolyte in high-temperature electrolysis of steam.

4. Electronic components. The application of solid electrolytes to microelectronics has been limited, but some initiatives have been made. For example, placement of "blocking" electrodes on either side of a slab of electrolyte can provide a small capacitor with a remarkably high capacitance. (See Fig. 8(a).) At the interface of an electrolyte and a blocking electrode, electrons cannot pass into the electrolyte and ions cannot pass into the electrode. When a voltage is applied across such an interface, charges accumulate as in the charging of a capacitor whose plates are separated by distances of only a few atomic dimensions. The leakage current of these devices is low, but their breakdown voltage is also quite low.

A non-blocking electrode is one that permits a flow of ions across the electrode/electrolyte interface. The combination of a silver non-blocking electrode and a platinum blocking electrode at either end of a slab of Ag+-ion solid electrolyte has been used as a "timer." A layer of silver is plated onto the Pt electrode as illustrated in Fig. 8(b); this "sets" the timer. A constant current is then passed so as to erode the plated silver layers and deposit it on the silver counter electrode. After a certain time, depending upon the current and the amount of silver originally plated on the Pt electrode, the plated silver layer is depleted and the platinum electrode becomes blocking; at this time, the voltage across the device increases abruptly.

In these devices, the only requirement is a high ionic mobility at room temperature, and a Ag+- or a Cu+-ion electrolyte is quite acceptable.

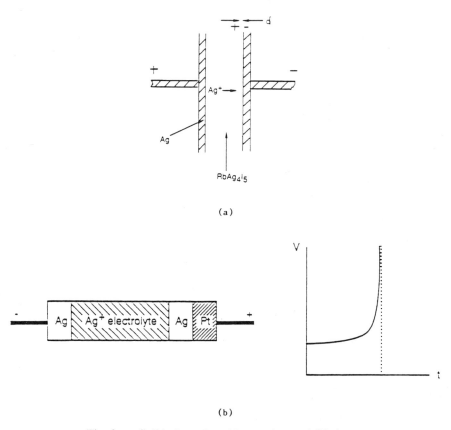

Fig. 8.   Solid-electrolyte (a) capacitor and (b) timer.

PHENOMENOLOGY

1. **Conductivity**. In a crystal, the conductivity defined by Ohm's law is generally a tensor. In amorphous or polycrystalline solids and in cubic crystals, as in liquids, the conductivity is isotropic; the current density induced by an applied electric field **E** is parallel to **E** and Ohm's law reduces to the scalar equation

$$j = \sigma E \tag{18}$$

However, in one-dimensional (1D) tunnel structures and two-dimensional (2D) layer structures, it is important to distinguish the conductivity $\sigma_{||}$ in the unique direction from the

167

conductivity $\sigma_\perp$ in the basal plane of a single crystal. In the phenomenological discussion to follow, an expression for the 3D isotropic conductivity is developed, but the adaptation required for a $\sigma_{||}$ or a $\sigma_\perp$ is indicated for 1D and 2D ionic conductors.

The current density imparted by a density n of carriers of charge q having a mean drift velocity v_d is $j = nqv_d$. Therefore from (18) and the definition of the drift mobility $\mu \equiv v_d/E$, the conductivity due to these charge carriers becomes

$$\sigma = nq\mu \tag{19}$$

Ionic transport is a thermally activated diffusion process, so any ionic mobility is related to the diffusivity by the Einstein relation

$$\mu_i = v_d/E = q_i D_i/kT \tag{20}$$

where kT is the Boltzmann temperature factor. This equation presupposes ionic diffusion within an array of energetically equivalent sites that are partially occupied. Since the jump time between sites is long compared to the period of an optical-mode lattice vibration, the mobile ion or vacancy carries its local deformation with it. The initial and final configurations are nevertheless energetically equivalent so long as the neighboring sites are energetically equivalent when all sites are empty or full.

Since the diffusion coefficient

$$D = D_0 \exp(-\Delta G_m/kT) \tag{21}$$

contains a motional free energy

$$\Delta G_m = \Delta H_m - T\Delta S_m \tag{22}$$

for an ion to jump to a neighboring, energetically equivalent site, equations (19) - (22) give a phenomenological expression

$$\sigma_i = (A/T)\exp(-E_A/kT) \tag{23}$$

$$E_A = \Delta H_m + \cdots \tag{24}$$

for the ionic conductivity that is found experimentally --except in the vicinity of a phase transition where temperature-dependent contributions to E_A may be found.

The upper limit for ionic conduction appears to be protonic conduction in a strong acid. A conductivity of about 0.5 $\Omega^{-1}cm^{-1}$ is found for 1 M HCl at room temperature, and fast ionic conduction in a solid is therefore defined as

$$\sigma_i \geq 0.1\ \Omega^{-1}cm^{-1} \quad \text{at } T_{op} \tag{25}$$

where T_{op} is the operating temperature. For a $T_{op} \approx 25\ ^oC$, it is necessary to have an $E_A < 0.2$ eV. Since it is difficult to attain a $\Delta H_m < 0.2$ eV in a solid, it is clear that a fast ionic conductor can be expected to have

$$E_A = \Delta H_m \quad \text{at } T_{op} \tag{26}$$

with no additional contributions and a $\Delta H_m < 0.2$ eV for fast ionic conduction at room temperature.

2. **Random walk.** For uncorrelated ionic motion (random walk), the velocity v_x for a field E_x applied along the x-axis is the product of half the x component of the jump vector **a** between energetically equivalent sites and the difference in the transition probability for an ion to jump in a forward or backward direction:

$$v_x = (a_x/2)\, z\, (1-c)\, (v_+ - v_-) \tag{27}$$

The factor $z(1-c)$ represents the fraction $(1-c)$ of z nearest-neighbor, energetically equivalent sites that are empty, c being the fraction of these sites that are occupied. The jump frequencies

$$v_{\pm} = v_0 \exp[-(\Delta G_m \mp \tfrac{1}{2}qE_x a_x)/kT] \tag{28}$$

contain, in addition to the attempt frequency v_0, the contribution from the electric field to the Gibbs free energy for an ion jump. In general, a $qE_x a_x <\!<2kT$ allows expansion of (28) and reduces (20) to

$$\mu_i \approx (q_i/kT)(a_x^2/2)z(1-c)v_0 \exp(-\Delta G_m/kT) \tag{29}$$

For an isotropic system, $a^2 = 3a_x^2$ and

$$\sigma_i \approx \gamma(Nq_i^2/kT)c(1-c)a^2 v_0 \exp(-\Delta H_m/kT) \tag{30}$$

$$\gamma = (z/6)f \exp(\Delta S_m/k) \tag{31}$$

where $N = n/c$ is the density of energetically equivalent sites. (For 1D conductivity, use $a^2 = a_x^2$; for 2D conductivity use $a^2 = 2a_x^2$). The factor f introduced into equation (31) represents a "correlation factor;" it includes the neglected motional correlations arising from long-range electrostatic interactions between the mobile ions as well as any geometric correlation.

In equation (30), an $N \sim a^{-3}$ makes $\sigma_i \sim a^{-1}$, so the ionic conductivity is inversely proportional to the jump distance a. The attempt frequency $v_0 \approx 10^{12}$ s^{-1} is given by the optical-mode vibrational frequencies. The correlation factor f can be important, but it enters as a secondary consideration for the designer of a new solid-electrolyte material. Of primary concern are the factor $c(1-c)$ and the motional enthalpy ΔH_m.

3. **Stoichiometric compounds.** At $T = 0$ K, stoichiometric compounds tend to have all their mobile ions ordered into subarrays of crystallographically equivalent sites that are either full ($c = 1$) or empty ($c = 0$), which makes the factor $c(1-c)$ vanish. The mobile-ion potential energies of the occupied sites are separated from the potential energies of the empty sites by an energy gap

$$\Delta G_g = \Delta H_g - T\Delta S_g \tag{32}$$

where ΔS_g is the change in configurational and vibrational entropy due to excitation of an ion from an occupied normal site to an empty interstitial site, Fig. 9. At finite temperatures, thermal excitations across the energy gap create densities $N(1-c)$ of vacancies in the normal sites and $N_I c_I$ of ions in the interstitial sites. As in the theory of intrinsic conduction in semiconductors, it follows that

$$N(1-c) = N_I c_I = \tfrac{1}{2}(NN_I)^{1/2} \exp(-\Delta G_g/2kT) \tag{33}$$

For a large ΔH_g, the product $c(1-c)$ remains small at any operational temperature T_{op}, and the experimental E_A of equation (23) becomes

$$E_A = \Delta H_m + \tfrac{1}{2}\Delta H_g \tag{34}$$

which is why most stoichiometric solids are poor ionic conductors. However, a few stoichiometric compounds having a small ΔH_m exhibit fast ionic conduction above an order-disorder phase transition at $T = T_t$; a relatively small ΔH_g at $T < T_t$ decreases with increasing temperature, vanishing at $T > T_t$. The phase transitions may be either smooth or first-order.

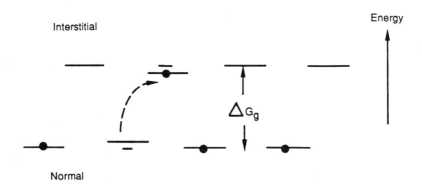

Fig. 9. Energy separation of normal-site and interstitial-site positions in an intrinsic ionic conductor.

Since the transitions are order/disorder in nature, they may be characterized by an order parameter

$$\xi = \Delta H_g/\Delta H_{go} \tag{35}$$

that varies from $\xi = 1$ at $T = 0$ K to $\xi = 0$ at $T = T_t$. (See Fig. 10.) A fundamental problem for the designer of a new, stoichiometric solid electrolyte is the identification of structural and compositional conditions that lead not only to a small ΔH_m, but also to a $T_t < T_{op}$. In the case of a smooth transition, which introduces fewer problems of materials mismatch on thermal cycling where T_t exceeds room temperature, the E_A of equation (23) becomes temperature-dependent over a finite temperature range $T' < T < T_t$. Failure to recognize this temperature dependence can lead to confusion in the interpretation of the pre-exponential factor A in these materials.

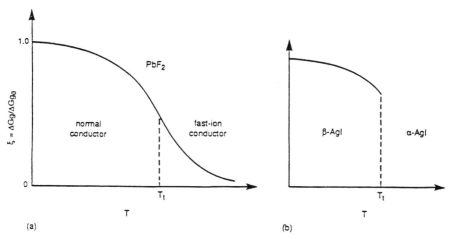

Fig. 10. Variation of the order parameter $\xi = \Delta G_g/\Delta G_{go}$ for (a) smooth and (b) first -order transitions.

4. **Doping**. An alternative strategy is to make $c(1-c) \neq 0$ by doping in analogy with creation of mobile electrons or holes in semiconductor technology. For a small dopant concentration, the mobile interstitial ion or normal-site vacancy is trapped at the dopant that creates it by an energy

$$\Delta G_t = \Delta H_t - T\Delta S_t \qquad (36)$$

and the phenomenological activation energy for ionic conduction in the doped material becomes

$$E_A = \Delta H_m + \lambda \Delta H_t \text{ with } \tfrac{1}{2} < \lambda < 1 \qquad (37)$$

As the dopant concentration increases to beyond the percolation threshold, the mobile ions may never need to overcome ΔH_t to find a continuous pathway through the structure, but the ΔH_m is generally increased due to random perturbations of the ionic potential. Moreover, at these higher concentrations, the long-range coulomb interactions between mobile particles may not only introduce motional correlations that reduce the factor f in equation (31), but also produce regions of ionic order in which a ΔH_g replaces ΔH_t. The growth of these ordered regions with time produces a deterioration of the electrolyte known as "aging."

Although a formal analogy exists between doping a semiconductor and an electrolyte, the magnitude of ΔH_t is generally an order of magnitude larger for the electrolyte. In a broad-band semiconductor, $\Delta H_m = 0$ and $\Delta H_t = E_D$ or E_A is reduced by a screening factor κ^{-2}, where κ is the dielectric constant. Since ionic conduction is diffusive, the electrostatic screening factor is significantly reduced, and a $\Delta H_t \approx 0.5 - 1.0$ eV is generally encountered as opposed to an $E_D < 0.1$ eV for a shallow electronic donor level in phosphor-doped silicon. Therefore any doping strategy for the design of a solid electrolyte must consider not only the structure/composition requirements for a small ΔH_m, but also how to reduce the effective trapping enthalpy ΔH_t.

5. **Geometric constraints on ΔH_m.** It is clear from these arguments that a primary consideration in the design of a solid electrolyte must be the minimization of ΔH_m. Structural geometry provides a simple approach to this minimization.

If partially filled, energetically equivalent sites are interconnected by common site faces, as is illustrated in Fig. 11, then ΔH_m may be minimized by satisfying the simple geometric constraint.

$$R_{face} \gtrsim R_{excl} \tag{38}$$

where R_{face} is the distance from the center of the common intersite face to the site position of an anion defining its perimeter. R_{excl} is the hard-sphere "exclusion" radius of the mobile ion; for hard-sphere ions it is the sum of the ionic radii of the mobile cation and a perimeter anion. If $R_{face} < R_{excl}$ holds, the mobile ion must push aside the perimeter anions in an intersite jump; the energy to do this enters ΔH_m. If $R_{face} \gg R_{excl}$ holds, the mobile ion is too small to remain in the center of its site; it becomes displaced to a new site position where it tends to be trapped. If the perimeter and/or mobile ions are polarizable (or quadrupolarizable), the exclusion radius is "softened," which reduces the effective R_{excl}. Therefore polarizable perimeter ions are generally advantageous to fast ionic conduction; and quadrupolarizable ions such as Ag^+ or Cu^+ are generally more mobile than alkali-metal ions of similar sizes. Hybridization of d and s orbitals at Cu^+ and Ag^+ ions allows changing the shape of the electronic core from spherical to ellipsoidal.

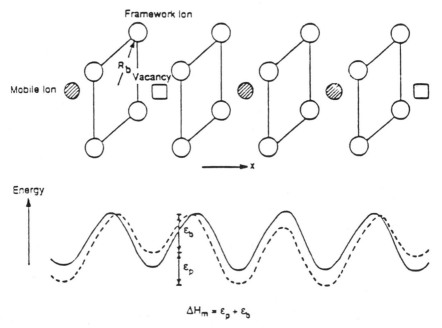

Fig. 11.　　Definition of "bottleneck" size $R_b \equiv R_{face}$ and variation of mobile-ion potential with position with (dotted line) and without (solid line) polaronic relaxation of host structures.

Fig. 11 also illustrates how a lattice relaxation about a mobile-ion vacancy can increase the motional enthalpy

$$\Delta H_m = \varepsilon_p + \varepsilon_b \tag{39}$$

by adding to the normal <u>barrier</u> energy ε_b between equally occupied sites a <u>polaronic</u> energy ε_p. This term enters because, in diffusional motion, the time for a charge carrier to jump to a near-neighbor empty site is

$$\tau_h > \omega_R^{-1} \tag{40}$$

where ω_R^{-1} is the period of the optical-mode vibration that is relaxed. The energy ε_p is reduced where $c(1-c)$ is a large fraction since relaxations at neighboring sites interfere with one another so long as the mobile ions remain disordered and their motions are uncorrelated.

Finally, if an occupied array of energetically equivalent sites is interconnected by sites of an interpenetrating set of equivalent sites accessible to the mobile ions, but of different energy, then a doping that introduces excess mobile ions is to be preferred over a doping that introduces vacancies on the occupied subarray. As illustrated in Fig. 12, this principle follows from the fact that electrostatic repulsions between neighboring ions in the two distinguishable sets of sites reduces the difference in the mobile-ion poential energies of the two sets of sites; occupancy of only one type of site does not.

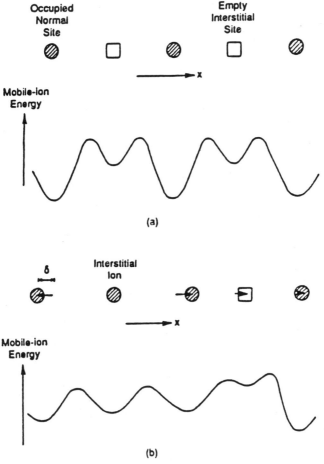

Fig. 12.   Flattening of the mobile-ion potential by occupancy of an interpenetrating subarray.

SUMMARY

From these simple considerations, the following principles emerge to guide the designer of a new solid electrolyte:

• <u>Large electronic E_g.</u> An ionic transport number $t_i \approx 1$ requires $\sigma_e \approx 0$ at T_{op}. Therefore an intrinsic electronic conductor with a large E_g becomes mandatory since an electronic mobility $\mu_e \gg \mu_i$ means that a small concentration of electronic charge carriers can make σ_e competitive with σ_i. Moreover a large E_g is required where a large open-circuit voltage V_{oc} is to be maintained across the electrolyte since, for thermodynamic stability, an $E_g > eV_{oc}$ is required.

• <u>Matched electrode work functions and electrolyte window.</u> For energy conversion and storage, a large $V_{oc} \lesssim E_g/e$ requires matching the electrode work functions to the electrolyte window, as illustrated in Fig. 2, if thermodynamic stability is to be maintained.

• <u>Easy fabrication into a dense membrane.</u> A solid electrolyte is employed as an electronic separator as well as an ionic conductor; and in power applications a large area is required to provide the required currents.

• <u>Compatibility with maintenance of contact with the reactants.</u> Solid electrolytes work well with gaseous or liquid reactants; they may lose contact with solid reactants that change their volume with state of discharge of the electrochemical cell unless specifically designed not to.

• <u>Minimization of E_A.</u> For room-temperature operation ($T_{op} \approx 25 \,^oC$), an $E_A \lesssim 0.2$ eV is required for "fast" ionic conduction.

In any case, the activation energy contains the motional enthalpy ΔH_m. Selecting a crystallographic structure that minimizes ΔH_m is mobile-ion specific. For mobile cations, the geometric constraint $R_{face} \gtrsim R_{excl}$ should be satisfied; polarizable anions and/or quadrupolarizable mobile cations have the beneficial effect of "softening" R_{excl}.

Stoichiometric compounds that are either undoped or doped only with isovalent substitutions generally have an activation energy $E_A = \Delta H_m + \frac{1}{2} \Delta H_g$ for ionic conductivity. However, in a few cases $\Delta H_g = \Delta H_g(T)$ decreases to zero at an order-disorder transition temperature T_t that is below the melting point T_m; at $T_t < T < T_m$ the solid contains a set of energetically equivalent (or quasiequivalent) sites that are only partially occupied. The problem is to identify solids having a $T_t < T_{op}$ and, for T_t above room temperature, a smooth transition at T_t.

Aliovalent doping of stoichiometric compounds having a large ΔH_g can introduce <u>extrinsic</u> ionic conduction with an $E_A = \Delta H_m + \lambda \Delta H_t$, $\frac{1}{2} \le \lambda \le 1$, for small doping. If a doping concentration in excess of the percolation limit can be achieved without the coulomb interactions between mobile ions introducing strong correlations or ordering, either short- or long-range, then the contribution from ΔH_t may be minimal. However, the percolation pathways represent a reduced cross-sectional area for ionic motion, which reduces the "effective area" of the electrolyte. Aliovalent doping can be an effective strategy where it stabilizes a solid phase having a small ΔH_m or immobilizes a liquid phase having a small ΔH_m. Where a parent phase contains energetically equivalent sites separated by an interpenetrating, empty array of sites sharing faces that satisfy $R_{face} \gtrsim R_{exc}$, there doping can reduce ΔH_m as well as introduce extrinsic charge carriers.

Support of this work by the Robert A. Welch Foundation, Houston, Texas and the Texas Advanced Research Program, Proposal #4257 is gratefully acknowledged.

REFERENCES

1.   N. Weber and J. T. Kummer, Proc. 21st Ann. Power Sources Conf. 37 (1967).

DESIGNING A SOLID ELECTROLYTE

II. STRATEGIES AND ILLUSTRATIONS

John B. Goodenough

Center for Materials Science and Engineering, ETC 5.160
University of Texas at Austin, Austin, TX 78712-1084

STOICHIOMETRIC COMPOUNDS

1. **Copper and silver iodides**. The iodide ion is large and polarizable; the Cu^+ and Ag^+ ions are quadrupolarizable because they have spherical d^{10} cores that may be transformed to ellipsoidal shape by energetically accessible, intraatomic d-s hybridization. Consequently the effective R_{excl} is particularly softened, and even the simple salts exhibit transitions to fast ionic conduction at a T_t below the melting point T_m.

At low temperatures (T < 369 °C) γ-CuI has the cubic zincblende structure of Fig. 1(a). High-temperature (T > 407 °C) α-CuI has the same cubic-close-packed array of I^- ions, but the Cu^+ ions occupy octahedral and tetrahedral sites with a nearly statistical distribution [1]. The intervention of an intermediate β-CuI having an hexagonal-close-packed array of I^- ions does not appear to be critical to the general argument of an order $\rightleftharpoons$ disorder $\alpha \rightleftharpoons \gamma$ transition. The γ-phase is a normal ionic conductor with an $E_A = \Delta H_m + \frac{1}{2} \Delta H_g \approx 1$ eV; the α phase is a fast ionic conductor with an $E_A = \Delta H_m \approx 0.2$ eV.

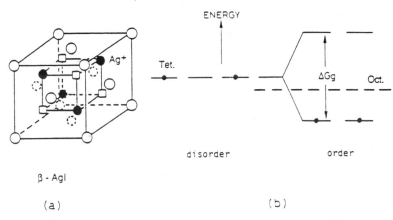

Fig. 1.   CuI: (a) zincblende structure and (b) Cu^+ -ion energy levels.

Solid State Microbatteries
Edited by J. R. Akridge and M. Balkanski
Plenum Press, New York, 1990

In the γ phase, Cu^+-ion ordering on the tetrahedral sites. (See Fig. 1(a)), has made inequivalent the energies of the occupied and empty sites, Fig. 1(b). A high-pressure transformation of γ-CuI to the rocksalt structure shows that the potential energy of the octahedral sites lies only a relatively small energy ΔG_g above that of the tetrahedral sites. Thermal excitations of Cu^+ ions to octahedral sites reduces the splitting between the occupied and empty sites. Expansion of this energy to terms linear in the number of thermally excited ions gives an enthalpy contribution

$$\Delta H_g \approx \Delta H_{go} - c_I \varepsilon \tag{1}$$

where c_I is the concentration of octahedral-site ions and ε is a finite energy. The term $c_I \varepsilon$ introduces a positive feedback that, at $T > T_t$, collapses the energy difference between tetrahedral-site subarrays and apparently eliminates the energy difference between octahedral and tetrahedral sites.

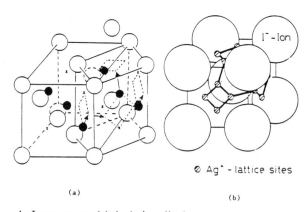

I^- - Ion

$\oslash$ Ag$^+$ - lattice sites

(a)

(b)

Fig. 2.   AgI structures: (a) Ag$^+$-ion displacements $T < T_t$ and (b) $T > T_t$.

A similar, but somewhat different situation prevails in AgI where Ag$^+$ ions have a stronger tetrahedral-site preference energy. The low-temperature phase of AgI consists of a mixture of zincblende and sphalerite structures; it corresponds to a mixture of the α and β CuI phases. For pedagogical purposes, only the zincblende phase is considered. As indicated in Fig 2(a), the Ag$^+$ ions of the zincblende structure are thermally excited into pseudotetrahedral sites where they induce a first-order bainite transition of the I$^-$-ion array at $T \approx$ 147 ^oC to the body-centered-cubic structure of Fig 2(b). Electrostatic interactions between Ag$^+$ ions prevent two of them from simultaneously occupying the same cube face of Fig. 2(b) even though each cube face contains four equivalent face-shared tetrahedral sites. Nevertheless, only two out of three faces can be occupied, which satisfies the condition $c(1-c) \neq$ 0. In addition, the shared tetrahedral-site faces satisfy the requirement $R_{face} \gtrsim R_{excl}$ with an extremely soft R_{excl}. Therefore at high temperatures the structure satisfies all the criteria for fast ionic condition, and

$$E_A = \Delta H_m = 0.05 \text{ eV} \tag{2}$$

is the smallest known activation energy for a polycrystalline solid electrolyte.

In order to lower and smooth T_t, isovalent substitutions for silver were investigated; RbAg$_4$I$_5$ was found to give fast Ag$^+$-ion conduction at room temperature, but at the price of a little larger ΔH_m. (See Fig. 3.)

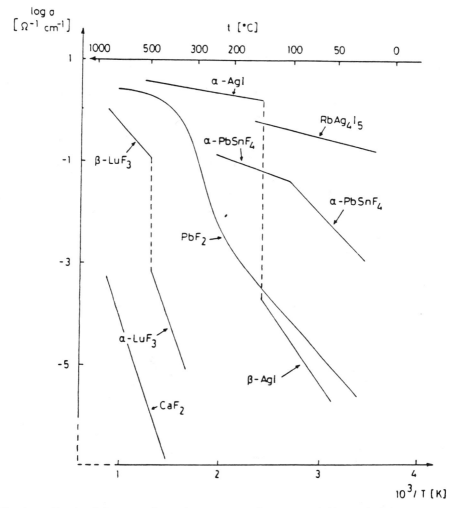

Fig. 3.   Conductivity vs. reciprocal temperature for some stoichiometric ion conductors.

2. **F⁻ -ion conduction in difluorides.** Several simple difluorides crystallize with the fluorite structure of Fig. 4(a). Of these, Pb^{2+} is the most polarizable cation; and PbF$_2$ shows the best F⁻ -ion conductivity, see Fig. 3. A smooth transition occurs from normal to fast ionic conduction at a $T_t \approx 438\ °C$[2] without any transformation of the face-centered-cubic array of Pb^{2+} ions. At T = 0 K, all the tetrahedral sites are occupied by F⁻ ions, so c = 1 for these sites. Attempts to account for the fast F⁻ -ion conductivity at higher temperatures have considered excitations to the empty octahedral sites; the results have not been satisfactory. On the contrary, high-temperature x-ray studies [3] have given direct evidence for a mobile-ion density that is distributed over the tetrahedral and edge (displaced toward octahedral sites) positions with little density in the octahedral sites themselves. Moreover, LuF$_3$ becomes a fast F⁻ -ion conductor below its melting point without any

transformation of the face-centered-cubic Lu^{3+} -ion array even though both tetrahedral and octahedral sites are filled [4]. It is therefore necessary to identify another type of low-energy excitation that can introduce a $c(1-c) \neq 0$.

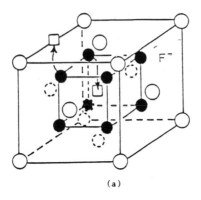

(a)

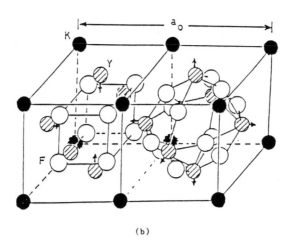

(b)

Fig. 4.   (a) Fluorite and (b) KY_3F_{10} structures.

        Since all the tetrahedral sites are occupied and only excitations to saddlepoint sites are evident from the x-ray data, it is necessary to consider the possibility of a correlated eight-atom -cluster rotation about a principal axis to the edges of an octahedral site; such a corre-lated motion would have a low activation energy were ΔH_m small enough. Evidence that such a cluster rotation is energetically feasible comes from the KY_3F_{10} structure of Fig. 4(b), which shows an occupancy of tetrahedral sites only in every other quadrant and an occu-pancy of all the octahedral-edge -- or saddlepoint -- positions in the remaining quadrants. Thus indirect evidence strongly supports a model in which eight-atom-cluster excitations to saddlepoint positions creates low-energy edge positions for the capture of ions from a neigh-boring quadrant. Such an ion capture would stabilize the cluster in its excited (rotated) state and generate a mobile F^--ion vacancy in the surrounding matrix. At highest temperatures, the mobile anions become disordered over tetrahedral and saddlepoint positions, but never occupying different types of sites within the same quadrant.

Isovalent substitution of Sn for Pb in the compound $PbSnF_4$ lowers the melting point and hence also T_t for the smooth transition of PbF_2. (See Fig. 3.)

3. O^{2-}-ion conduction. The oxide ion, like the fluoride ion, is a first-row element that readily adapts to 2, 3, 4, or 6-fold coordination. However, it is divalent; the higher valence complicates the search for a good O^{2-}-ion electrolyte in three ways:

(a) Oxide-ion bonding has a stronger covalent component; cations with polarizable cores tend to be statically polarized at lower temperatures in oxides.

(b) Stronger long-range electrostatic interactions between the mobile ions enhance correlated motion and order between oxide ions.

(c) Higher cation valences in equivalent structures introduce stronger crystalline fields and make it more difficult to find cations with polarizable cores.

As a consequence of these factors, dioxides with the fluorite structure of Fig. 4(a) exhibit interesting O^{2-}-ion conduction only if random oxide-ion vacancies or interstitials are introduced. Most attempts to do this are based on a doping strategy, which has its own set of limitations. On the other hand, vacancies can also be introduced into stoichiometric compounds; they are ordered at low temperatures, but they may become disordered at high temperatures. Two examples are illustrative, Bi_2O_3 and $Ba_2In_2O_5$.

The Bi^{3+}-ion has a polarizable $6s^2$ core, but at lower temperatures the Bi-O bonding is strong enough to create a static polarization of the core. However, above 730 °C Bi_2O_3 is stabilized in the cubic γ phase, which has an oxygen-deficient fluorite structure. In order to stabilize this phase to lower temperatures, it is necessary to substitute nonpolarizable cations for Bi^{3+}. The isovalent substitution $Bi_{2-x}Y_xO_3$, for example, has yielded materials with interesting O^{2-}-ion conductivities at temperatures $T > T_t(x)$, T_t decreasing with x; but in the limited range of x available, electronic conduction remains a problem at the high temperatures $T > T_t$ [5].

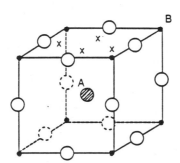

Fig. 5.   Ideal cubic ABO_3 perovskite structure; x markes pseudotetrahedral sites.

An alternative approach is to go to another cationic structure. Whereas the fluorite structure has a close-packed-cubic cation array, the cubic perovskite ABO_3 has two different cations ordered as in the CsCl structure. (See Fig 5.) The oxide ions ideally form 180° B-O-B bridges within the cation array, which completely occupies all the octahedral sites formed by the two B cations and four A cations. A variety of cooperative oxygen-ion displacements may occur, but the essential architecture for an ABO_3 perovskite constrains the oxygen sub-array to c = 1. However, a number of stoichiometric oxides have the same cation array, but with a total valence that leaves some oxygen vacancies.

For a solid electrolyte, it is preferable to create oxygen vacancies without introducing a mixed cation valency on either the A or the B sites. For example, the brownmillerite structure of $CaFeO_{2.5}$ contains oxygen vacancies ordered so as to place half the Fe^{3+} ions in octahedral and half in tetrahedral coordination, Fig. 5(a). However, the multiple-valence possibilities for ions make it an unsuitable host cation for an O^{2-}-ion solid electrolyte. On the other hand, $Ba_2In_2O_5$ has the same brownmillerite structure at room temperature, which invites investigation of whether it exhibits a transition above some $T_t < T_m$ to a disordered oxygen subarray in an oxygen-deficient perovskite structure. Fig. 6(b) shows that $Ba_2In_2O_5$ undergoes a first-order transition at a $T_t \approx 930$ °C from a normal ionic conductor with a brownmillerite structure to a fast ionic conductor with the disordered perovskite structure [6]. Unfortunately 930 °C is too high a temperature to be of practical interest, but this material illustrates a principle that may prove useful in the search for an improved O^{2-}-ion electrolyte.

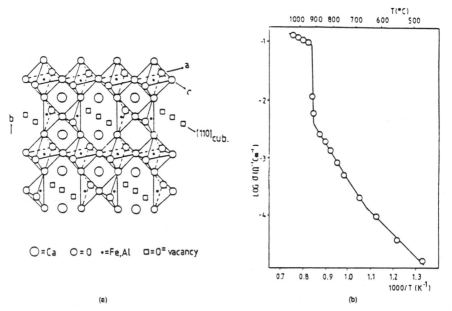

(a)　　　　　　　　　　　　　(b)

Fig. 6.　$Ba_2In_2O_5$: (a) brownmillerite structure $T < T_t$ and (b) conductivity vs.reciprocal temperature.

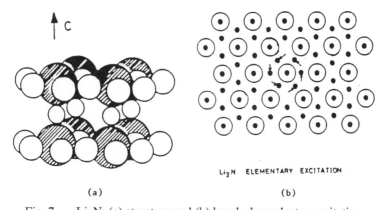

(a)　　　　　　　　　　　　(b)

Fig. 7.　Li_3N: (a) structure and (b) basal-plane cluster excitation.

4. Li⁺ -ion conduction. As illustrated in Fig. 7(a), Li_3N consists of -Li-N-Li-N- chains held together by Li^+ ions in triangular sites of close-packed N^{3-} -ion layers. At low temperatures, all the triangular sites are filled (c = 1). Nevertheless, at room temperature Li_3N exhibits good Li^+ -ion conduction within the Li_2N planes ($\sigma_\perp = 10^{-3}$ ohm^{-1} cm^{-1} >> $\sigma_{||}$)[7]. Unfortunately, the top of the N^{3-} : $2p^6$ valence band, E_v, has too high an energy for Li_3N to be useful in a Li power cell of high energy density. The high E_v is due to the large negative electron affinity of the N^{3-} ion.

The fact that the relatively small Li^+ ion should have a low ΔH_m for diffusion within a close-packed array of highly polarizable N^{3-} ions is not surprising. However, a good ionic conduction at room temperature where the Li-site occupancy is c ≈ 1 poses a problem similar to that encountered for F⁻ -ion conduction in PbF_2. Indeed, in this case also x-ray studies [8] have demonstrated that there are few Li^+ -ion vacancies in the basal planes and that the mobile Li^+ ions are constrained to triangular-site and saddlepoint positions in the basal planes. The x-ray data are completely analogous to those for PbF_2, which suggests that a cluster-rotation excitation is probably operative in this compound also. An elementary cluster rotation is illustrated in Fig. 7(b); it consists of six Li^+ ions neighboring a common N^{3-} ion rotated 30° about that anion into intersite edge (saddlepoint) positions. This cooperative excitation has a net motional enthalpy ΔH_m; from the x-ray data as well as the conductivity data, $E_A = \Delta H_m \approx 0.3$ eV [18]. Note that in a stoichiometric solid undergoing a cluster excitation, the polaronic contribution ε_p to ΔH_m (see equation (39) of Lecture I) vanishes; ε_b is reduced by a softening of R_{excl} due to the high polarizability of the large N^{3-} -ion.

Li^+ -ion conduction along the c-axis of Li_3N apparently proceeds by correlated jumps of two Li^+ ions [9]. A Li^+ ion of one layer displaces an interplanar Li^+ ion, which in turn moves into an adjacent layer where occupancy of edge positions by cluster excitation has created acceptor sites.

5. Framework structures. The Cu^+, Ag^+ and Li^+ crystalline electrolytes so far discussed fully occupy a subarray of interstices of a close-packed anion array at low temperatures (T < T_t) In each case, large and polarizable anions allowed the geometric constraint $R_{face} \geq R_{excl}$ to be "satisfied" by a softening of R_{excl}. However, three different mechanisms were identified for creating a c(c-1) ≠ 0 at temperatures T > T_t: (a) a phase transformation to a different anion subarray (AgI), (b) a small energy ΔG_g for excitation into interstitial sites that collapses at T_t (CuI), and (c) a cluster excitation stabilized by capture of extra mobile ions that generates mobile-ion vacancies in the surrounding matrix (Li_3N). If smaller anions are needed to obtain the desired electrolyte window, a problem encountered with Li_3N, then the anion-polarization softening is not large enough to make $R_{face} \approx R_{excl}$ in a close-packed anion array. Moreover, with the larger cations Na^+ and K^+, the geometric constraint is not satisfied even with the largest anions. Recognition of the need to satisfy this constraint has led to the exploration of framework structures as candidates for cationic solid electrolytes [10].

Any compound $A_x^+(M_mX_n)^{x-}$ that consists of a 3D host framework $(M_mX_n)^{x-}$ with an interconnected interstitial space is a framework structure. For example, the cubic tungsten bronzes Na_xWO_3 contain a cubic WO_3^{x-} framework of corner-shared WO_6 octahedra forming the cubic- perovskite BO_3 framework of Fig. 5. The Na^+ ions are randomly distributed in the body-center positions, which are the sites of the A cations in the ABO_3 cubic-perovskite structure. As a solid electrolyte, this WO_3 framework fails on two counts: (a) it is metallic, the formal $W^{6+/5+}$ mixed valence transforming via strong W-O-W interactions to a partially filled "5d" band, and (b) the cubic faces provide too small an interface, $R_{face} < R_{excl}$. Although the former could, in principle, be handled by making $Na_x(Ta_xW_{1-x})O_3$, the geometric constraint forces consideration of other framework structures where an $R_{face} \gtrsim R_{excl}$ is formed.

An oxide framework is preferred as its fabrication can be done in air and the $O^{2-}:2p^6$ valence band is stable enough to be chemically inert against most oxidizing agents. Sulfides and halides may also be acceptable, but most work to date has been done on oxides. The sum of the Na^+ and O^{2-} ionic radii is 2.4 Å; therefore, in an oxide an $R_{face} \gtrsim 2.4$ Å is desired for Na^+-ion electrolytes as the Na^+ ion is neither polarizable nor quadrupolarizable to any significant extent, and the polarizability of an O^{2-} ion is modest.

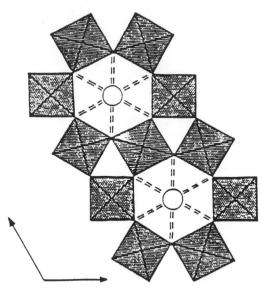

Fig. 8.   Structure of hexagonal $A_x WO_3$, $x \lesssim 0.33$. Framework of corner-shared octahedra is shaded; large alkali ions A occupy hexagonal tunnels parallel to the c-axis.

Three classes of framework structures must be distinguished: (a) tunnel or 1D structures like that of hexagonal tungsten bronze shown in Fig. 8, (b) 2D structures such as that of Na ß-alumina, Fig. 9, and (c) 3D structures such as that of high-pressure $KSbO_3$, Fig. 10, in which the mobile ions can move rapidly in three dimensions. Although the frameworks discussed in this lecture have "open" structures that permit fast diffusion of the larger Na^+ and K^+ ions, considerable attention will be given in my last lecture to a "spinel framework" containing close-packed anions. These latter are of interest for smaller mobile cations such as Li^+, especially if the anions are large and polarizable.

Although 1D structures may be used in microbatteries as electrolyte films having their unique axis parallel to the direction of desired Na^+-ion motion, they are not the optimal choice for a conventional polycrystalline ceramic membrane. Consequently, primary attention to date has been paid to the 2D and 3D materials.

The model 2D materials are the ß and ß" aluminas. Stoichiometric Na ß-alumina, $Na_2O \cdot 11Al_2O_3$, would contain $Al_{11}O_{16}$ spinel slabs with a [111] axis parallel to the c-axis; the slabs are connected by oxygen bridges between tetrahedral-site Al^{3+} ions of neighboring slabs; the bridging oxygen atoms and the Na^+ ions form intervening Na-O layers. There are three possible sites within the Na-O layers that are accessible to the Na^+-ion in space group $P6_3mmc$, the 2d, 2c, and 6h. These positions are now called, respectively, the Beevers-Ross (BR), the anti-Beevers-Ross (aBR), and the mid-oxygen (mO) positions. In

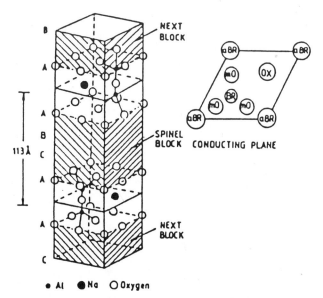

● Al   ● Na   ○ Oxygen

Fig. 9.   Na ß-alumina: (a) structure and (b) the Na -O plane.

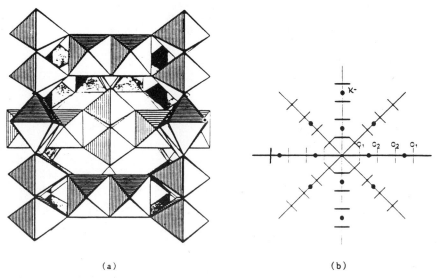

(a)   (b)

Fig. 10.   Cubic $KSbO_3$: (a) $(SbO_3)^-$ framework and (b) projection of <111> tunnels
showing ordering of K^+ ions in space group Pn3. $R_{face} \approx 2.40$ Å for O_2, 2.73
Å for O_1.

stoichiometric, isostructural $Na_2O \cdot 11Ga_2O_3$, the Na^+ ions are ordered into the BR sites at low temperatures, and a $c \approx 1$ makes the compound a poor Na^+ -ion conductor [11]. The ß" phase has a similar structure, but the spinel blocks are shifted relative to one another so as to make the Na -O planes mirror planes. In this structure the BR and aBR sites become equivalent, and fast Na^+ -ion conduction can occur at the stoichiometric composition. The BR and aBR sites of ß-alumina share a common rectangular interface perpendicular to the Na -O plane with two in-plane oxygen atoms at opposite corners of a face diagonal; the mO position is at the center of the interface. In Na ß-Al_2O_3, an R_{face} (=2.7 Å) > 2.4 Å satisfies the geometric constraint for a small ΔH_m for Na^+ -ion motion.

As normally prepared, sodium ß-alumina is not stoichiometric; it is self-doped with excess Na_2O. The ß"-Al_2O_3 is normally stabilized by substituting Mg^{2+} for Al^{3+} ions, which introduces extra Na^+ ions into the Na -O planes. These materials are practical Na^+ -ion electrolytes, but their discussion is deferred to the section on DOPING STRATEGIES. However, it is worth noting here that open framework structures can accommodate mobile neutral species as well as the mobile ion of interest, and Na ß-alumina is susceptible to hydrolysis in the presence of moist air.

The cubic high-pressure form of $KSbO_3$ has the body-centered $(Sb_{12}O_{36})^{12-}$ framework of Fig. 10(a)[12]. If the origin is taken at the cube face of Fig. 10(a), the framework contains empty tunnels along the four <111> axes; these tunnels intersect at the origin and body-center positions. The intersection sites are large octahedra sharing faces with the tunnel segments, and each tunnel segment consists of three face-shared octahedra that have been significantly flattened along the tunnel axis. On proceeding along a tunnel axis from an origin to a body-center position, successive interfaces are bounded by O_1, O_2, O_2, and O_1 anions as indicated in Fig. 10(b). The flattening of the octahedra makes the O_1-O_2 and O_2-O_2 ions peripheral to sites 16f and 8c, respectively, take the form of puckered hexagonal rings; consequently the shared O_1 or O_2 faces have an R_{face} that is comparable to R_{excl} for Na^+ in an oxide.

Annealed $KSbO_3$ samples -- first stabilized at atmospheric pressure by Spiegelberg [13] by incorporation of an unknown impurity -- have the primitive space group Pn3 with the mobile K^+ ions ordered; alternate tunnel segments contain two K^+ ions in sites 16f, the remaining half of the tunnel segments contain a single K^+ ion in site 8c. This order indicates that the K^+ -ion is more stable on the 8c site, but that coulomb repulsions between K^+ ions stabilize them in 16f sites in doubly occupied chain segments. Lack of occupancy of the intersection sites indicated that anions might be stabilized there, and $KSbO_3 \cdot \frac{1}{6} KF$ was prepared at room temperature [12]. The F^- ions occupy the intersection sites 2a and the extra K^+ ions are accommodated randomly in the tunnel segments. The tetrahedra formed by the O_1 faces and the F^- ions are large enough to permit K^+ ions to hop between channel segments. However, an $R_{face} \simeq 2.40$ Å for the smaller O_2 faces is more suited to Na^+ -ion than K^+ -ion conduction. The remaining problem, then, is to prepare a Na^+ -ion analogue. Since the K^+ -ions are mobile in the structure, this problem can be solved by a simple ion exchange at 330 ºC in a 20:1 bath of molten $NaNO_3$. The resulting metastable $NaSbO_3 \cdot \frac{1}{6}$ NaF exhibits good 3D Na^+ -ion conductivity even though the interstitial space is interconnected by different sites that are energetically only pseudoequivalent.

DOPING STRATEGIES

Four doping strategies can be distinguished: (a) isovalent doping to stabilize to lower temperatures a high-temperature phase, (b) introduction of interstitials to create a c(1-c) $\neq$ 0, (c) aliovalent doping to create a c(1-c) $\neq$ 0, and (d) introduction of colloidal particles to create a composite interface layer with c(1-c) $\neq$ 0.

1. **Isovalent doping.** Substitutions of one ion by another of the same valence does not change the concentration of mobile ions and does not introduce a strong electrostatic

contribution to any trapping energy. On the other hand, it tends to lower any order-disorder transition temperature T_t and to convert a first-order transition to second-order. Substitution of Y^{3+} for Bi^{3+} to stabilize the high-temperature cubic phase of Bi_2O_3 and substitution of Sn^{2+} for Pb^{2+} to lower the T_m and T_t of PbF_2 were discussed above.

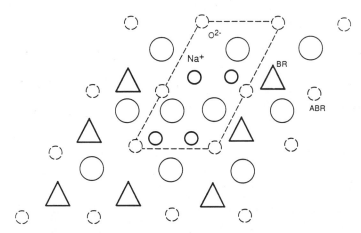

Fig. 11.   Na ß-alumina: positions of excess Na^+ and O^{2-} ions in NaO planes of $(Na_2O)_{1+x}11Al_2O_3$.

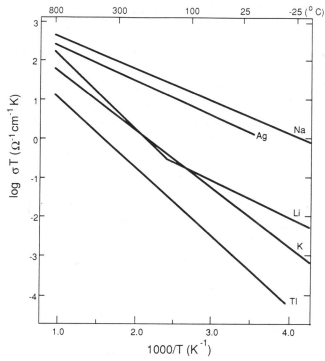

Fig. 12.   Ionic conductivities of several substituted ß-aluminas, after [15].

187

2. **Introduction of interstitials.** The introduction of interstitial host ions so as to create a $c(1-c) \neq 0$ for the mobile ions is generally not a feasible synthetic strategy. However, in the case of Na ß-alumina the stoichiometric phase is not stable. Three factors stabilize the introduction of excess Na_2O:(a) local charge neutrality is not preserved in the Al_2O_3 spinel slabs, (b) a thermal-expansion mismatch between the Na -O planes and the spinel slabs and (c) the availability of interstitial space in the Na -O planes. As normally prepared, Na ß-alumina has the chemical formula $(Na_2O)_{1+x} 11Al_2O_3$ with $x \approx 0.18$. The excess Na_2O is accommodated in the Na -O plane of Fig. 9 as shown in Fig. 11[14]. An interstitial O^{2-} ion perturbs the local potential so as to stabilize Na^+ ions in neighboring mid-oxygen positions relative to both the neighboring BR and aBR sites. The result is a distribution of the Na^+ ions over all sites available to it in a Na -O plane, which creates a $c(1-c) \neq 0$ on a set of Na^+ -ion sites made energetically pseudoequivalent. Consequently the 2D ionic conductivity of Na ß-alumina is high enough at 300 °C to be practical.

Fig. 12 compares the ionic conductivities for several monovalent cations in the ß-alumina framework [15]. These can all be prepared by simple ion exchange from Na ß-alumina, so the concentration of interstitial oxygen remains the same. An $R_{face} \approx 2.7$ Å at the mid-oxygen positions is particularly well suited to the size of the Na^+ ion; both the smaller Li^+ ion and the larger K^+ ion have lower mobilities in the structure. These data thus emphasize the need to match R_{face} to the R_{excl} of the mobile ion.

3. **Aliovalent doping.** Substitution of one host ion by another of different valence also changes the concentration of mobile ions and can therefore be used to create a $c(1-c) \neq 0$. This technique is used, for example, in the ß"-alumina phase and in the stabilized zirconias.

(a) **Na ß"-alumina.** Substitutions of Mg^{2+} for about 2/3 of the tetrahedral-site Al^{3+} ions converts the ß-alumina structure to the ß"-alumina structure and introduces extra Na^+ ions into the Na -O planes to charge-compensate. In this phase the BR and aBR sites are equivalent and only partially occupied by Na^+ ions, which makes it a better Na^+ -ion conductor than Na ß-alumina. Practical ceramic membranes generally contain a mixture of ß and ß" alumina in order to obtain an engineering compromise between high Na^+ -ion conductivity and mechanical strength.

(b) **Stabilized zirconia.** Pure ZrO_2 is monoclinic at ambient temperature; it transforms reversibly to a tetragonal structure above 1150 °C. For the fluorite structure to be stable, the ratio of cation to anion radius should be greater than 0.732; for ZrO_2 this ratio is only 0.724. Therefore substitutions of the larger Ca^{2+} and Y^{3+} ions are used both to stabilize the fluorite structure and to introduce oxygen vacancies in the systems $Zr_{1-x}Ca_xO_{2-x}$ and $Zr_{1-2y}Y_{2y}O_{2-y}$. About 12 mole % CaO is required to stabilize ZrO_2 in the cubic phase. Unfortunately this type of doping introduces a ΔH_t having both a coulombic and an elastic component, so the vacancies tend to be trapped at the Ca^{2+} ions. Therefore stabilized-zirconia membranes are only useful above about 700 °C. At these temperatures, the host cations can also have some mobility; but the temperature is low enough to promote ordering into Ca-rich and Ca-poor regions that permit minimization of the electrostatic coulomb repulsions between mobile anions by short-range ordering of the oxygen vacancies in the Ca-rich regions. Capture of the mobile anion vacancies within ordered regions lowers the anion conductivity. This capture process continues with increasing electrolyte use; it is known as "aging."

Finally, at temperatures in excess of 700 °C oxygen loss or gain introduces a troublesome electronic component into the conductivity. At high oxygen pressures, oxygen may dissolve in the lattice via the reaction

$$\tfrac{1}{2} O_2(g) + V_o^{\bullet\bullet} = O_o^x + 2h^\bullet \tag{3}$$

where $V_o^{\bullet\bullet}$ is a doubly ionized oxygen vacancy, O_o^x is an oxygen in a normal oxygen site, and $h^\bullet$ is a positive electron hole. Applying the law of mass action to (3) gives

188

$$K_1 = \frac{[O_o^x][h^\bullet]^2}{[V_o^{\bullet\bullet}]p_{O_2}^{1/2}} \tag{4}$$

Since $[O_o^x]$ and $[V_o^{\bullet\bullet}]$ remain large relative to any changes in a doped compound, they can be treated as constants, and equation (4) reduces to

$$[h^\bullet] = K_2 p_{O_2}^{1/4} \tag{5}$$

Thus at high oxygen pressures the concentration of positive holes varies as $p_{O_2}^{1/4}$.

At low oxygen pressure, the equilibrium reaction is

$$O_o^x = V_o^{\bullet\bullet} + 2e' + \tfrac{1}{2}O_2(g) \tag{6}$$

where e' is an electron, and from the law of mass action

$$[e'] = K_3 p_{O_2}^{-1/4} \tag{7}$$

Since the electronic species $h^\bullet$ and e' are much more mobile than the anion vacancies, the shape of the log σ vs p_{O_2} plot can be predicted. (See the lecture by W. Weppner.) Only at intermediate partial pressures of oxygen is the ionic transport dominant and, consequently, the conductivity p_{O_2}-independent. The range of p_{O_2} for which the ionic conductivity is dominant decreases with increasing temperature. Therefore, although the stabilized zirconias are used commercially above 700 °C, there is much room for improvement. A superior oxide-ion electrolyte operative at lower temperatures remains a challenging technical target.

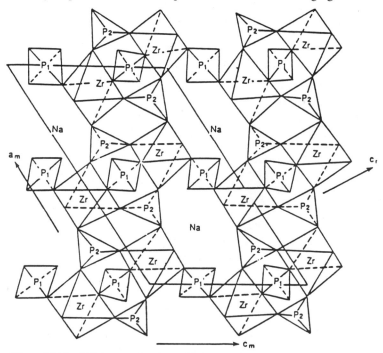

Fig. 13. $NaZr_2(PO_4)_3$: projection of hexagonal structure with unique axis c_r. Monoclinic axes a_m, c_m of NASICON are also indicated.

(c) **Framework oxides.** Aliovalent doping has proved more successful in some framework structures. For example, the compound $NaZr_2(PO_4)_3$ has the hexagonal $Zr_2(PO_4)_3$ framework illustrated in Fig. 13 [15]. It consists of PO_4 tetrahedra sharing corners with ZrO_6 octahedra. Along the c-axis, face-shared octahedra or triagonal-prismatic sites are ordered in the sequence

$$-Zr_o - V_t - Zr_o - M_1 - Zr_o -$$

where Zr_o is an octahedral-site Zr^{4+} ion, V_t is a vacant trigonal-prismatic site, and M_1 is an octahedral site available to Na^+ ions. These chains are interconnected by PO_4 tetrahedra to form an open framework containing three empty M_2 sites for every M_1 site that are also available to the Na^+ ions. The M_2 sites -- not shown in Fig. 13 -- form hexagonal-close-packed layers perpendicular to the c-axis and in the same plane as the O^{2-} ions in an M_1 site. The bottleneck between M_1 and M_2 sites consists of a puckered hexagonal ring bounded alternately by tetrahedral-site and octahedral-site edges. In $NaZr_2(PO_4)_3$ the M_1 sites are filled (c=1) and the M_2 sites are empty (c=0), so the compound is a poor Na^+-ion conductor even though $R_{face} \approx 2.5$ Å $> R_{excl} \approx 2.4$ Å satisfies the geometrical constraint for a small ΔH_m.

The compound $Na_4Zr_2(SiO_4)_3$ has the same framework [17]. In this compound both the M_1 and M_2 sites are filled (c=1); it is an even poorer Na^+-ion conductor than $NaZr_2(PO_4)_3$. However, the solid solutions $Na_{1+3x}Zr_2(P_{1-x}Si_xO_4)_3$ can be made [12], and compositions with x = 2/3 have an E_A = 0.24 eV with Na^+-ion conductivities competitive with those of Na ß"-alumina at temperatures T > 350 °C. On the other hand, substitution of Ta for Zr in $Na_{1+x}Zr_{2-x}Ta_x(PO_4)_3$ to obtain a c < 1 on the M_1 sites [18] does not prove to be a good strategy; the M_1 sites remain separated by an empty, interpenetrating array of M_2 sites of higher Na^+-ion potential energy. This situation illustrates the discussion made in connection with Fig. 11 of Lecture I.

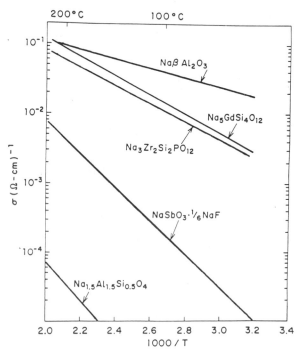

Fig. 14. Na^+-ion conductivity vs. reciprocal temperature for several framework structures.

Fig. 14 compares the conductivities of several of the best Na^+ -ion electrolytes that are known. All have framework structures. The Na ß-aluminas, stoichiometric $NaSbO_3 \cdot \frac{1}{6}$ NaF, and $Na_3Zr_2Si_2PO_{12}$ (NASICON) have been discussed. The remaining two compounds need only be commented on briefly.

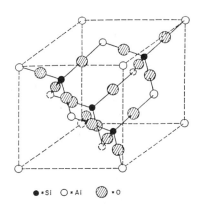

●・SI ○・Al ◐・O

Fig. 15.   The $(AlSiO_4)^-$ framework of carnegieite.

$Na_{1.5}Al_{1.5}Si_{0.5}O_4$ represents aliovalent doping via a change in the ratio of two cations appearing in an ordered host framework. Where the ordering energy is large, this strategy may result in multiple phases; however, it proved feasible synthetically in the case of the ordered $(AlSiO_4)^-$ framework of high-temperature carnegieite, $NaAlSiO_4$, shown in Fig. 15. The interstitial space is identical to the framework, but shifted by $a_0/2$ along a cube axis. Therefore only half the sites available to the Na^+ ions are occupied in the high-temperature phase. Unfortunately the framework collapses about the Na^+ ions at lower temperatures, trapping them in a filled, distinguishable subarray. K_2MgSiO_4 has an isostructural $MgSiO_4$ framework and remains cubic to room temperature; but in this case all the K^+ -ion positions are filled (c=1) in the cubic structure. In an ordered framework, two alkali-ion sites of the interstitial space are interfaced by an hexagonal ring where edges alternate between AlO_4 and SiO_4 tetrahedral edges. In order to retain the cubic phase to room temperature in the sodium compound, extra Na^+ ions are introduced by increasing the Al/Si ratio of the host framework [19]. Although the doped compound does not compete with the other Na^+ -ion electrolytes of Fig. 14, nevertheless it is a good Na^+ -ion conductor. In fact, by decreasing the Mg/Si ratio in $K_{2-2x}Mg_{1-x}Al_{1+x}O_4$, $x \approx 0.05$, it was possible to obtain a K^+ -ion conductor that is competitive with known K^+ -ion electrolytes even though the maximum value of x for this phase is small [19].

The compound $Na_5GdSi_4O_{12}$ has a complex framework; its discovery illustrates not a design strategy, but the serendipity of alert crystallographers [20-22]. The framework consists of cylindrical pillars parallel to a hexagonal c-axis that are bridged by Gd^{3+} corner-shared octahedra. The pillars themselves consist of corner-shared SiO_4 tetrahedra containing immobile Na^+ ions in their cores. The mobile Na^+ ions partially occupy the interstitial space of this complex framework. It is interesting to note that this structure was independently recognized to be a potential Na^+ -ion conductor by three different groups on the basis of a structural paper [23] that reported difficulty locating three Na^+ ions per formula unit; these proved to be mobile, and structural analysis of the conduction mechanism and its relation to the geometrical constraints has been performed [29].

(d) **Layered Oxides**. Aliovalent doping has also been used to obtain 2D solid electrolytes in layered -- as distinct from framework -- structures. For example, a number of $A^+M^{3+}O_2$ oxides form close-packed MO_2 layers of edge-shared octahedra that are held together by A^+ ions between the layers. Unlike the ß-aluminas, where the spinel slabs are held together by bridging oxygen in the Na -O planes, the A^+ -ion layer contains no oxygen. Consequently the spacing between MO_2 layers is free to adjust itself so as to optimize the A - O bond lengths. This degree of freedom can alleviate the geometrical requirement that $R_{face} \gtrsim R_{excl}$ if the concentration of A^+ -ions can be reduced to where the electrostatic A -O attraction is nearly balanced out by the electrostatic repulsion between oxide ions of adjacent MO_2 layers. In principle, the A^+ -ion concentration can be varied continuously by the substitutions $A_{1-x}M_{1-x}^{3+}M'_x^{4+}O_2$. Extensive studies on these layered compounds have shown that good ionic conductivities can be obtained, but the materials have not proven to be competitive with other solid electrolytes [25].

4. Composite electrolytes. The introduction of colloidal particles into a stoichiometric host matrix creates a composite with a large internal interface between the particles and the host. Depending upon the relative energies of the mobile-ion chemical potential at the particle surface and in the host matrix, mobile-ion interstitials or mobile-ion vacancies are created in a surface layer within the host. The width of the surface layer depends on the relative strengths of the electric field and the mobile-ion concentration gradient near the surface. This effect creates a $c(1-c) \neq 0$ in the interfacial layer within the host without introducing an important local trapping energy ΔH_t. However, the conduction pathway is made "tortuous," and the effective cross-sectional area of the electrolyte is reduced. The effect was first recognized and used in protonic conductors having an aqueous matrix [26], but the same principles apply to oxide particles dispersed in a solid matrix such as LiI. My third lecture will discuss protonic conduction and the principles associated with composite doping.

Support of this work by the Robert A. Welch Foundation, Houston, Texas and the Texas Advanced Research Program, Grant No. 4257 is gratefully acknowledged.

REFERENCES

1. J. B. Boyce, T. M. Hayes, and J. C. Mikkelson, Jr., Phys. Rev. **B23**, 2876 (1981).

2. W. Schröter and J. Nölting, J. Phys. (Paris) **41:C6**, 20 (1980).

3. M. Schultz, Proc. 2nd European Conf. Solid State Chemistry, Eindhoven Univ. of Technology, 7-9 June 1982.

4. M. O'Keefe and B. G. Hyde, J. Solid State Chem. **13**, 172 (1975).

5. T. Takahashi, H. Iwahara, and T. Esaka, J. Electrochem. Soc. **124**, 1563 (1977).

6. J. E. Ruiz-Diaz, D. Phil. Thesis, University of Oxford (1985).

7. U. Von Alpen, A. Rabenau, and G. M. Talat, Appl. Phys. Lett. **30**, 621 (1977).

8. H. Schulz and K. H. Thiemann, Act. Cryst. **A35**, 309 (1978); H. Schulz, Ann. Rev. Materials Science **12**, 351 (1982).

9. D. Brinkmann, M. Mali, and J. Roos, Phys. Rev. **B26**, 4810 (1982).

10. H. Y-P Hong, J. A. Kafalas, and J. B. Goodenough, J. Solid State Chem. **9**, 345 (1974).

11. G. V. Chandrashekar and L. M. Foster, J. Electrochem. Soc. **124**, 329 (1977).

12. J. B. Goodenough, H. Y-P Hong, and J. A. Kafalas, Mat. Res. Bull., **11**, 203 (1976).

13. P. Spiegelberg, Ark. Kemi **14A**, 1 (1940).

14. W. L. Roth, F. Reidinger, and S. La Placa, in Superionic Conductors, G. D. Mahan and W. L. Roth, eds. (Plenum Press, N. Y. 1976), p. 223.

15. R. A. Huggins, in Defects and Transport in Oxides, M. S. Seltzer and R. I. Jaffee, eds. (Plenum Press, N. Y. 1974), p. 549.

16. L. O. Hagman and P. Kierkegaard, Acta. Chem. Scand. **22**, 1822 (1968).

17. R. E. Zizova, A. A. Voronkov, N. G. Shumyatskaya, V. V. Ilyukhin, and N. V. Belov, Sov. Phys. Dokl. **17**, 618 (1973).

18. R. D. Shannon, B. E. Taylor, A. D. English, and T. Berzins, Electrochim. Acta. **22**, 783 (1977).

19. H. Y-P Hong, Adv. Chem. Ser. **163**, 179 (1977).

20. R. D. Shannon, H. Y. Chen, and T. Berzins, Mat. Res. Bull. **12**, 969 (1977).

21. H. V. Beyeler and T. Hibma, Solid State Comm. 27, 641 (1978).

22. H. Y-P. Hong, J. A. Kafalas, and M. Bayard, Mat. Res. Bull, **13**, 757 (1978).

23. B. A. Maksimov, Yu. A. Kharitonov, and N. V. Belov, Sov. Phys. Dokl. **18**, 763 (1974).

24. B. A. Maximov, I. V. Petrov, A. Rabenau, and H. Schultz, Solid State Ionics **5**, 311 (1982).

25. C. Delmas, A. Maazog, C. Fouassier, J. M. Réau, and P. Hagenmuller, in Fast Ionic Transport in Solids, Vashista, Mundy, and Shenoy, eds. (Elsevier, North Holland, 1979) p. 451.

26. W. A. England, M. G. Cross, A. Hammett, P. J. Wiseman, and J. B. Goodenough, Solid State Ionics **1**, 231 (1980).

DESIGNING A SOLID ELECTROLYTE

III. PROTON CONDUCTION AND COMPOSITES

John B. Goodenough
Center for Materials Science and Engineering, ETC 5.160
University of Texas at Austin, Austin, TX 78712-1084

PROTON CONDUCTORS

1. **Proton Movements**. Among the ions, the proton is unique in the character of its bonding and hence in the variety of movements available to it. In any discussion of the basic proton movements, it is necessary to distinguish at the outset the hydrogen bonding in hydrides from that encountered in ionic materials.

In a metal hydride, a hydrogen atom may have several equidistant nearest neighbors and diffuse over an energetically equivalent set of sites as do the mobile ions in other ionic conductors. Moreover, the hydrogen atom may donate most of its electronic charge to the host matrix or accept electron density from it or remain essentially neutral, depending upon the energies of the H/H^- and H^+/H redox couples relative to the Fermi energy of the host metal. Mixed proton/electron conductors are of no interest as solid electrolytes.

In an ionic material, the hydrogen atom forms molecular orbitals with anions having acceptor states significantly more stable than the H:1s energy level, so the protonic character is accentuated. As the smallest cation, the proton tends to coordinate at most two nearest anion neighbors. Back-bonding transfer of lone-pair electrons on the anion to the H:1s orbital would reduce the X-H bond length. If two coordinated anions are inequivalent, the proton is shifted toward the more polarizable anion, thus creating an asymmetric hydrogen bond.

An asymmetric hydrogen bond is also common where a proton coordinates two equivalent anions. Where π-bond repulsive forces between two coordinated anions prohibit a close X-H-X separation, competition between the two anions for the proton sets up a double-well potential for the equilibrium proton position. One well is made deeper than the other only by a motion of the proton from the center of the bond. Although a displacement toward one anion is energetically equivalent to a displacement toward the other, proton transfer from one well to the other requires a thermal excitation over a barrier enthalpy ΔH_w. The interwell jump frequency in such a case is $\nu = \nu_0 \exp(-\Delta H_w/kT)$, and such an asymmetric bond may be represented as, for example,

$$O\text{-}H \cdots O \rightleftharpoons O \cdots H\text{-}O$$

The barrier enthalpy decreases sharply with anion separation. For oxides, it vanishes for separations less than about 2.4 Å; at smaller separations the bond is symmetric:

$$O\text{-}H\text{-}O \quad \text{or} \quad O^{,H,}O$$

Solid State Microbatteries
Edited by J. R. Akridge and M. Balkanski
Plenum Press, New York, 1990

195

Symmetric bonds may be bent a few degrees from 180° as a result of higher order bonding and asymmetric π-bond repulsive forces.

Formation of asymmetric bonds associates the proton with a specific anion. The anion may remain fixed while the proton librates about the bond axis, rotates from one bond direction to another

$$
\begin{array}{ccc}
\text{O}\text{-}\text{O} & & \text{O} \\
\text{H} & \rightarrow & \\
\text{O}\quad\text{O} & & \text{O-H}\cdots\text{O}
\end{array}
$$

or tumbles randomly from one bond direction to another. The weaker the hydrogen bond, the more tightly the proton is bound to one anion and the lower is the temperature at which the motion changes from liberation to rotation to tumble and eventually to molecular translation.

Formation of symmetric hydrogen bonds tends to occur with F^{-} ions or where two equivalent oxide ions are strongly polarized to the opposite side by neighboring cations. The $O_2H_5^+$ dioxonium ion, for example, consists of two water molecules bonded by a symmetric hydrogen bond with an O-H-O bond angle bent about 6° from 180°. The $O_3H_7^+$ molecule contains two slightly bent, symmetric O-H-O bonds between a central OH unit and two terminal OH_2 groups. Protons within a symmetric bond are tightly bound; they can only move as a polyatomic cation.

The simplest polyatomic cation is the oxonium ion OH_3^+. Where the proton moves in association with a polyatomic cation, it rides <u>piggyback</u> whether the motion is libration, reorientation, tumbling rotation, or translation. Translational piggyback diffusion is commonly referred to as <u>vehicular</u> motion. In vehicular motion, the mobile molecule may carry a positive charge (e.g. NH_4^+, OH_3^+, $O_2H_5^+$, $O_3H_7^+$), a negative charge (e.g. NH^{2-}, OH^-), or be neutral (e.g. NH_3 or OH_2).

Cooperative displacements of protons within a network of asymmetric hydrogen bonds may give rise to a spontaneous polarization P_s of the solid. Ordering of the displacements occurs below a critical temperature in such ferroelectric materials; the critical temperature is called the Curie temperature by analogy with ferromagnetism. Reversal of P_s to $-P_s$ in a dc electric field gives rise to transient currents; it is a giant displacement current. The two ferroelectric states P_s and $-P_s$ may be represented schematically by the states (a) and (b).

$$\text{O-H}\cdots\text{O-H}\cdots\text{O-H}\cdots\text{O-H}\cdots\text{O-H}\cdots\text{O} \qquad\qquad \text{(a)}$$

$$\text{O}\cdots\text{H-O}\cdots\text{H-O}\cdots\text{H-O}\cdots\text{H-O}\cdots\text{H-O} \qquad\qquad \text{(b)}$$

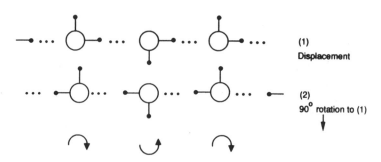

Fig. 1 Proton diffusion via Grotthus mechanism.

A bare proton may also diffuse through a hydrogen-bond system to give a dc current. It does so by the Grotthus mechanism, which consists of a combination of piggy-back reorientations and displacements as indicated in Fig. 1. The cooperative displacements give rise to a displacement current, but further movement is blocked until a cooperative rotation resets the chain so that a second displacement motion can occur.

In summary, five basic proton movements have been described:

(a) proton-electron diffusion (in hydrides and insertion compounds)

(b) proton-displacement (in asymmetric hydrogen bonds)

(c) piggy-back rotation (molecular libration, reorientation, or tumble)

(d) piggy-back diffusion (vehicular translation)

(e) proton diffusion via (b) + (c) (Grotthus mechanism)

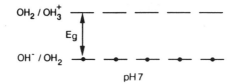

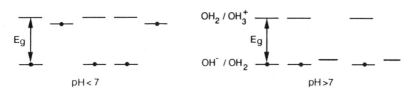

Fig. 2   Proton energy levels in an aqueous electrolyte. Filled circles mark occupied levels.

2. **Stoichiometric compounds**. In order to have proton translation, and hence a constant dc H^+-ion current, it is necessary to create the condition $c(1-c) \neq 0$ either for the proton or for the molecular species on which it rides piggy-back. Two stoichiometric compounds illustrate the problem, pure ice and the ferroelectric KH_2PO_4.

(a) **Pure ice**. Pure water has the proton energies of Fig. 2. At low temperature, where it becomes ice, all the molecules are OH_2; and excitation of a proton from one OH_2 molecule to another

$$2OH_2 \rightleftharpoons OH^- + OH_3^+ \tag{1}$$

requires an energy ΔG_g. The analogy with an intrinsic semiconductor is obvious, and the density of translational charge carriers is

$$[OH^-] = [OH_3^+] = \tfrac{1}{2} \exp(-\Delta G_g/2kT) \tag{2}$$

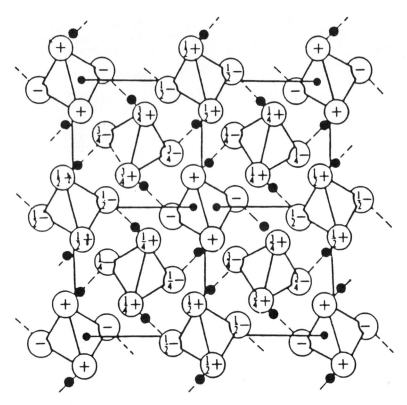

Fig. 3 A view on the a-b plane of ferroelectric KH_2PO_4.

which makes the activation energy for protonic conductivity

$$E_A = \Delta H_m + \tfrac{1}{2}\Delta Hg \qquad (3)$$

where the value of ΔH_m depends upon whether the proton vacancies OH⁻ or the interstitial protons at an OH_3^+ move by vehicular or Grotthus diffusion.

(b) **KH_2PO_4**. In the ferroelectric compound KH_2PO_4, half of the oxide ions are only weakly bound to a proton in a hydrogen bond and half are strongly bound, as is illustrated in Fig. 3. Moreover, the hydrogen bonds are ordered so as to coordinate each phosphorus atom with two O^{2-} and two OH⁻ ions. The asymmetric polarizability of the four oxide ions neighboring a P causes the P atom to be displaced from the center of symmetry of its tetrahedral site toward the two O^{2-} ions where it can form stronger covalent bonds. The cooperative displacements create a unique polar axis on which there is a net dipole moment. The dipoles at neighboring unique-axis strings of unit cells have their movements aligned parallel, so the crystal may have a net saturation polarization P_s; hence KH_2PO_4 is a ferroelectric. Above a Curie temperature T_c, a disordering of the hydrogen displacements within the hydrogen bonds destroys the saturation polarization P_s, but the directional character of the hydrogen bond is retained. So also is the ordering of 2OH⁻ and $2O^{2-}$ ions per $(H_2PO_4)^-$ unit; a large ΔG_g is required for the proton transfer

$$2(H_2PO_4)^- \rightarrow (HPO_4)^{2-} + (H_3PO_4) \qquad (4)$$

Therefore Grotthus diffusion does not occur to any significant degree even above T_c. Moreover, each $(H_2PO_4)^-$ unit is complete, so there are no anion vacancies that would allow a vehicular translation. As a consequence, KH_2PO_4 exhibits a transient displacement current on switching the sign of P_s below T_C in an applied electric field, but it cannot sustain very significant constant dc current. It remains a protonic as well as an electronic insulator.

3. **Doping**. As in a semiconductor, extrinsic protonic conduction is introduced into water by doping, i.e. by chemical alteration of the pH. A pH > 7 means an $[OH^-] \gg [OH^{3+}]$, which lowers the protonic electrochemical potential toward the OH_2 level in Fig. 2, and Grotthus proton diffusion of the type

$$OH_2 + OH^- \rightarrow OH^- + OH_2 \tag{5}$$

is the proton-vacancy analogue in ice of hole conduction in a p-type semiconductor. A pH < 7 means an $[OH^-] \ll [OH^{3+}]$, and Grotthus proton diffusion of the type

$$OH_3^+ + OH_2 \rightarrow OH_2 + OH_3^+ \tag{6}$$

is the proton analogue in ice of electron conduction in an n-type semiconductor. The concentration of mobile vacancies or protons -- and hence the value of c(1-c) -- depends on pH.

A vehicular mechanism competes successfully with the Grotthus mechanism in liquid water, but in this case also the mobile charge carriers are concentrations $[OH^-]$ and $[OH^{3+}]$. Whether a vehicular mechanism can compete successfully with the Grotthus mechanism in a solid phase depends upon the relative mobilities of the protonic and molecular species, which move via different mechanisms. Fast vehicle-ion conductivity requires access to a partially occupied set of energetically equivalent vehicle-molecule sites that form a continuously interconnected network.

As in the solids discussed in the previous lecture, doping may be accomplished by introducing interstitial water, which is analogous to introducing extra Na_2O into Na ß-alumina, or by aliovalent doping or by the introduction of a second colloidal phase to create a composite. Aliovalent doping with an anion such as Cl^- or SO_4^{2-}, which must be charge-compensated by H^+ or OH^{3+}, creates a pH < 7; with a cation such as Na^+, which must be charge-compensated by OH^- ions, it creates a pH > 7. These substitutions are analogous to the aliovalent doping of ZrO_2 to create O^{2-} -ion electrolytes in stabilized zirconias.

4. **Interstitial water**. In a 3D structure like KH_2PO_4, there is no room for the introduction of interstitial water, and the compound is not hygroscopic. However, intercalcation of interstitial water can be facile in framework and layered compounds, and a number of such oxides have been investigated for H^+ -ion transport. However, a word of caution is needed. Measurement of the protonic transport requires the making of pellets. Since sintering would involve a loss of the interstitial water, these pellets are made by cold-pressing wet samples. This procedure creates a "particle hydrate," to be discussed below, and the protonic conductivity can easily reflect the conductivity in an immobilized, intergranular aqueous matrix rather than the bulk conductivity of interest.

(a) **Layered hydrates**. Layered oxides that may contain variable amounts of water between the layers include, for example, various clays. An important family of layered hydrates are the lamellar acid salts $[M(IV) (XO_4)_2]_n^{2n-}$, where $M(IV)$ = Ti, Zr, Hf, Ge, Sn, Pb, Ce, or Th and X = P or As [1]. The best characterized are those with the α-layered structure illustrated in Fig. 4 for α-$[Zr(PO_4)_2]H_2 \cdot H_2O$. In this structure the Zr^{4+} ions form hexagonal arrays of edge-shared octahedra that corner share to (PO_4) tetrahedra above and below the Zr plane [2]. Successive layers are stacked so as to place each terminal O^{2-} ion of one layer directly below or above the Zr^{4+} ion of an adjacent layer. In the hydrate α-$[Zr(PO_4)_2]H_2 \cdot H_2O$, the water molecules are accommodated between the terminal oxygen atoms of adjacent layers so as to form a puckered close-packed array of oxygen atoms, and the protons form hydrogen bonds between the terminal oxygen and the water oxygen of those puckered oxygen arrays. Although the water is lost above 110 °C, interlayer diffusion of OH_2 or OH_3^+ is slow so long as there is insufficient water loss to create vacancies on the

OH$_2$ sites. Moreover, although there are twice as many O-O separations as there are protons to create hydrogen bonding, the Grotthus diffusion mechanism is also slow -- perhaps due to a protonic preference to reside as OH^{3+} at the interstitial water rather than at the more acidic terminal oxygen. Room temperature conductivities of about 3×10^{-4} ohm^{-1} cm^{-1} are found [3]. In order to obtain an interesting protonic conductivity, it is necessary to add additional water between the layers, and polyhydrated materials have room temperature conductivities approaching 10^{-3} ohm^{-1} cm^{-1} [4]. Modest H$^+$ -ion conductivities have also been observed at surfaces covered by a layer of adsorbed water.

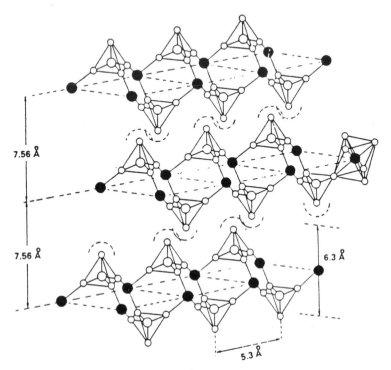

Fig. 4   Arrangement of octahedrally coordinated Zr(IV) layers and phosphate groups in α-[Zr(PO$_4$)$_2$]·H$_2$O, after ref. [1].

The uranyl phosphate hydrates represents a group of layered hydrates in which good H$^+$ -ion conductivity has been reported [5], and considerable effort has been devoted to unravelling the conduction mechanism in H$_3$OUO$_2$PO$_4$·3H$_2$O, commonly referred to as HUP. The structure contains layers of (O = U = O)$^{2+}$ and (PO$_4$)$^{3-}$ ions with the UO$_2$ axis normal to the layer; the four water molecules and associated proton per formula unit form planar networks between the (UO$_2$PO$_4$)$^-$ layers. HUP precipitates when uranyl nitrate solution is mixed with 2 to 3 M phosphoric acid; it is not stable in contact with solutions having pH > 2 to 3. The compounds rapidly lose water and conductivity above 50 °C; anhydrous HUO$_2$PO$_4$ is formed above 157 °C. The interstitial water and associated interlayer planar network are essential for bulk protonic conductivity.

The interlayer planar network consists of squares of hydrogen-bonded water molecules, H(1) bonds, connected to one another at opposite corners by H(3) bonds that are at least dynamically symmetric in an O$_2$H$_5$$^+$ unit. In addition, each water H(2) bonds with a layer oxide of a (PO$_4$)$^{3-}$ group. Neutron powder diffraction [6] has established that the H(1) positions are only three-fourths occupied. As illustrated in Fig. 5, the proton-vacancy transfer is assumed to occur via rotations about an O-H(2)$\cdots$O axis. Vacancy transfer across the dimeric unit bridging two squares is fast relative to the time of a diffraction measurement. In this compound, loss of one proton per formula unit from the planar network to the $(UO_2PO_4)^-$ array creates a hydrogen vacancy that permits a proton-vacancy diffusion via a Grotthus mechanism. Protonic -- and not vehicular -- diffusion has been established by ^1H and ^{31}P NMR [7].

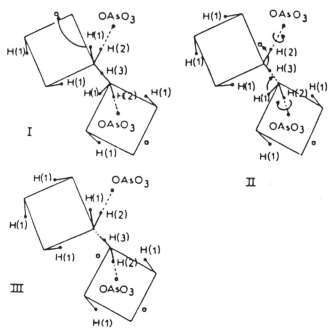

Fig. 5   Grotthus diffusion mechanism in interlayer water of HUP as determined for an isostructural, deuterated arsenate analogue, after ref. [6].

Although two-dimensional proton diffusion occurs within the layers, this mechanism does not account for the protonic conductivity of HUP as reported in Fig. 6. HUP is normally prepared as a particle hydrate [5], and the observed conductivity varies with the concentration of excess water [8]. The principal contribution to the conductivity of a wet, cold-pressed pellet comes from the intergranular aqueous matrix. As external water is lost from the aqueous matrix, it is replenished by the diffusion of internal water to the external space. Irreproducibility of the conductivity measurements of HUP from one laboratory to another is due to the fact that four parameters must be taken into account: temperature, time, water vapor pressure, and initial water content (both intragranular and intergranular).

(b) Framework hydrates. Framework hydrates are not difficult to prepare as powders. For example, antimonic acid $HSbO_3 \cdot nH_2O$ can be prepared by several routes, and the structure of the product is dependent on the route taken. Removal of K^+ from $KSb(OH)_6$ on an ion-exchange column or the hydrolysis of $SbCl_5$ each give an antimonic acid with the $(Sb_2O_6)^{2-}$ framework, Fig. 7, of the cubic pyrochlore $A_2B_2O_6O'$. Three water molecules may occupy the interstitial space occupied by the A_2O' atoms in this structure. The acidic

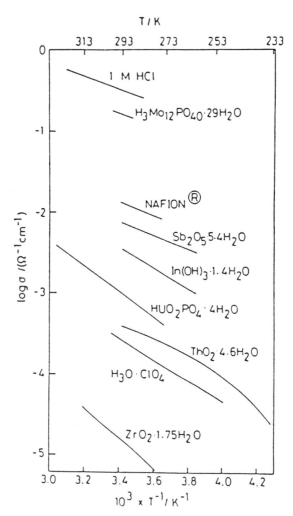

Fig. 6   Conductivity versus reciprocal temperature for several particle hydrates compared to that of 1 M HCl.

character of the framework biases the reaction

$$(H_3O)_2Sb_2O_6 \cdot nH_2O \; = \; (H_2O)H_2Sb_2O_6nH_2O \quad (n \leq 1)$$

strongly to the left, and proton conduction within the framework may take place by either a vehicular motion or a Grotthus mechanism. Similarly, ion exchange of the layered ilmenite $KSbO_3$ or the high-pressure cubic $KSbO_3$ structure (see illustration in Lecture II) gives antimonic acids retaining these structures with a water content that varies with the number of available sites in the structure. However, powdered framework hydrates are not useful as solid electrolytes; they must be fabricated into membranes. Cold-pressing of wet particles gives a particle hydrate in which the principal conduction pathway is intergranular. (See next section.) The conductivity in Fig. 6 is for a particle hydrate.

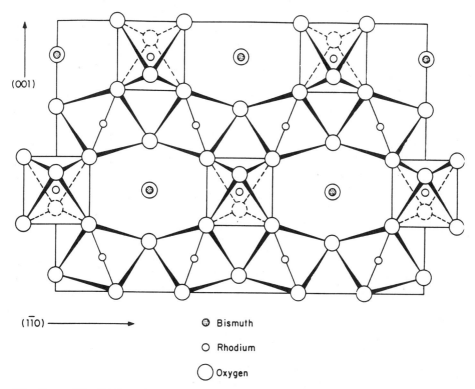

(001)

($\overline{1}$10) ──────────▶

⊚ Bismuth

○ Rhodium

◯ Oxygen

Fig. 7   The (110) projection of the cubic pyrochlore $Bi_2Rh_2O_7$ showing the (B_2O_6) framework of corner-shared octahedra.

Formation of a ceramic membrane from a framework oxide is also possible. Ceramic Na ß"-alumina is made commercially, for example. The problem is to retain the ceramic structure after ion exchange of H_3O^+ for Na^+ ions. A motivation for doing this is the retention of water within the framework to higher temperatures as well as the mechanical strength of a ceramic versus an immobilized liquid for the particle hydrates. However, protonic ion exchange on ceramic ß"-alumina tubes and plates nearly 1-mm thick presents a formidable problem. Internal stresses generated by changes in framework dimensions disrupt the ceramic unless care is exercised to tailor the lattice parameter of the parent structure. Successful ion-exchange of ceramic membranes has been accomplished [9] with a precursor ceramic $NaK_{0.67}Mg_{0.67}Al_{10.33}O_{17}$. The ceramic piece is shaped and sintered with this composition. Part of the Na^+ is then exchanged with additional K^+ in a NaCl/KCl melt so as to achieve an optimum expansion of the lattice before carrying out H_3O^+ exchange in aqueous acid solution. In this process, mechanical integrity requires leaving some alkali ions in this ceramic, and these ions inhibit the protonic conduction. Although steam electrolysis at 100 ºC has been demonstrated with such a ceramic membrane, the protonic conductivity remains too low for fuel/electrolysis cells using these membranes to be competitive with other systems.

(c) High-temperature proton conductors. It has only recently been appreciated that protons may exist in oxygen-deficient oxides to temperatures above 100 °C [10]. The most intensively studied are the aliovalent-doped perovskites

$$SrCe_{1-2x}Yb_{2x}O_{3-x} + yH_2O = SrCe_{1-2x}Yb_{2x}H_{2y}O_{3-x+y}$$

in which water occupies the oxygen vacancies created by doping ($y \leq x$) [11]. The strongly oxophilic character of the large, basic cations holds the structural water to high temperatures provided the concentration of water molecules is too low ($x \lesssim 0.05$) for a percolation pathway of escape and the preparation procedures achieve a $y \rightarrow x$. As $y \rightarrow x$, the oxygen-vacancy concentration vanishes, so competitive oxide-ion conduction also vanishes. Under these conditions, the only mobile ionic species can be the interstitial protons. However, these are trapped at the oxygen atoms neighboring the Y^{3+} ions, so only a small fraction are mobile at the temperatures where $y \rightarrow x$. Consequently the specific conductivity is only 10^{-2} ohm^{-1} cm^{-1} at 900 °C and 10^{-3} ohm^{-1} cm^{-1} at 600 °C. Nevertheless, this finding introduces a new concept for the exploration of alternate high-temperature proton conductors.

5. **Particle hydrates.** [12]. In the last lecture, the particle hydrates were mentioned in connection with composites, and it was stated that the introduction of small -- preferably colloidal size -- particles into a matrix represented an alternate way of doping the matrix without introducing a local trapping of the mobile species. In this section, this concept is explored more fully.

Particle hydrates consist of small particles, commonly oxides, imbedded in a hydrogen-bonded aqueous matrix. Because hydrogen bonds can be broken by application of pressure, these composite materials can be easily fabricated into dense sheets at room temperature by cold-pressing. This is important because fast protonic conduction is found in the presence of water, which is volatile at normal sintering temperatures.

Because the hydrogen-bonded aqueous matrix is confined to small dimensions between particles, the matrix remains solid above the melting point of water. However, proton motion within this matrix is similar to that in the liquid state. The design of a protonic solid electrolyte requires optimization of the matrix/particle cross-sectional area and mobile-proton concentration by variation of the particle size and composition.

The essential operative principal is quite straightforward. Metal ions at the surface of an oxide particle bind water in order to complete their normal oxygen coordination. The protons associated with this water distribute themselves over the oxide to create surface hydroxyl anions. If such a particle is imbedded in an aqueous matrix, its surface proton concentration comes into equilibrium with that of the matrix. If the oxide particle is "acidic," it pushes protons from its surface into a pH 7 aqueous matrix; if it is "basic," it attracts protons to the surface from a pH 7 aqueous matrix. (See Fig. 8.) Colloidal particles have a large surface-to-volume ratio, so they are more effective than larger particles at changing the pH of the aqueous matrix. Colloidal acidic particles carry a negative charge that is charge-balanced by mobile OH^{3+} ions (or "interstitial" protons) in the aqueous matrix. Basic particles are positively charged and are charge-compensated by mobile OH^- ions (or proton vacancies) in the matrix. Acidic particles have been more extensively studied. Clearly the more acidic and smaller is such a particle, the higher the mobile-proton concentration $[H^+]$ or $[OH^{3+}]$ for a given particle fraction. On the other hand, the mobilities of the protons (or proton vacancies) are greater the less structured is the water in which they move, and the structure of the aqueous matrix decreases with distance from the surface of the colloidal particle. Therefore the protonic mobility is higher the greater the fraction of water in the composite, and the protonic conductivity of a particle hydrate drops off with time if water is being lost. Technical applications are restricted to operating temperatures $T \lesssim 60$ °C unless operated under a high pressure of water vapor. Fig. 6 compares the protonic conductivities of a number of particle hydrates with that of 1 M HCl.

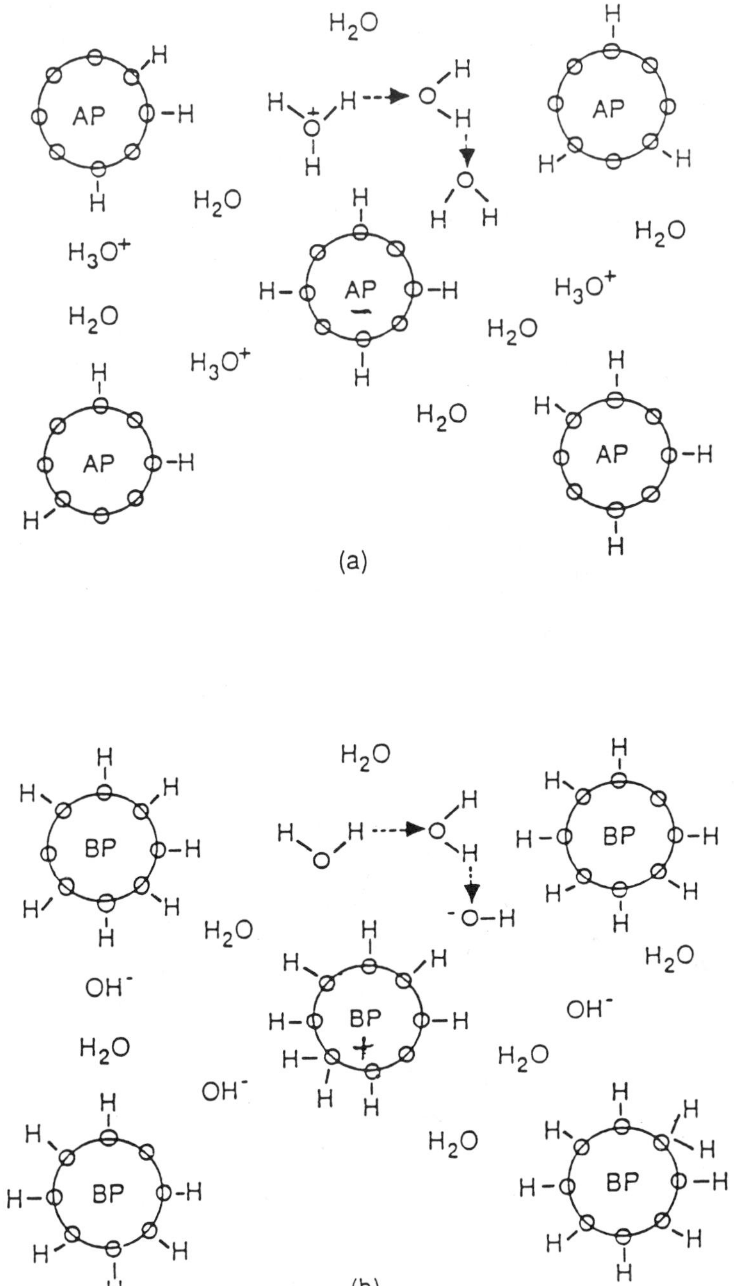

Fig. 8   Schematic representation of (a) an acidic-particle (AP) and (b) a basic-particle (BP) hydrate.

Most particle hydrates are amphoteric; they change the sign of their particle charge at a critical pH, the point of zero zeta potential (pzzp). These oxide-particle/aqueous-matrix composites are negative ion exchangers at pH < pzzp where the particles are positive; they are

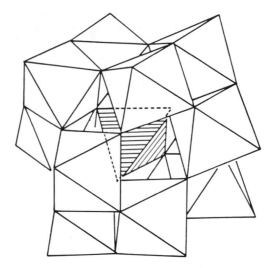

Fig. 9   Structure of the $(Mo_{12}PO_{40})^{3-}$ Keggin unit.

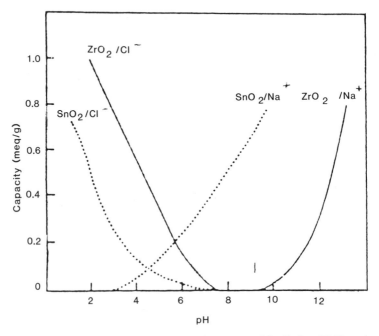

Fig. 10   Ion-exchange capacity versus pH for more acidic $SnO_2 \cdot nH_2O$ and more basic $ZrO_2 \cdot nH_2O$ particle hydrates, after C.B. Amphlett, "Inorganic Ion Exchangers," (Elsevier, Amsterdam 1964).

positive ion exchangers at pH > pzzp where the particles are negative. In order to have a high concentration of mobile ions with an aqueous matrix of initial pH 7, it is necessary to have a pzzp of much higher or lower pH.

The high protonic conductivity of solid $H_3Mo_{12}PO_4 \cdot 29H_2O$ (see Fig. 6) is due to the small size and strongly acidic (pzzp < 7) character of the polyanions $Mo_{12}PO_{40}$, which act as dispersed particles, as well as the large fraction of water. The $Mo_{12}PO_{40}$ polyanions, known as Keggin units, have the structure shown in Fig. 9. They consist of a central PO_4 group corner-sharing with triangular units of three edge-shared Mo(VI) octahedra. The four triangular units corner-share oxygen, but each octahedron has a terminal oxygen that forms a short Mo = O bond. No short O-H bond is formed at a terminal oxygen to compete with the Mo = O bond. The other surface oxygen atoms of the Keggin unit bridge two Mo(VI) ions, and their interaction with two Mo(VI) ions is much stronger than that of a water oxide ion to its two protons. Consequently the equilibrium

$$
\begin{array}{ccccc}
\text{H} & & & & \\
\text{O}^- & & & \text{O}^{2-} & \\
\text{Mo} \quad\quad \text{Mo} & + & H_2O & = & \text{Mo} \quad\quad \text{Mo} & + & OH_3^+
\end{array}
$$

is shifted strongly to the right, which makes the aqueous matrix acidic and a good proton conductor.

A more typical particle hydrate is $SnO_2 \cdot 2H_2O$, which contains amphoteric colloidal particles of SnO_2. If hydrated with pure water, $SnO_2 \cdot nH_2O$ is a weak acid, so it ion exchanges with cations. (See Fig. 10.)

Although $ZrO_2 \cdot 1.75H_2O$ is a particle hydrate similar to $SnO_2 \cdot 2H_2O$, its pzzp is too close to pH 7 (see Fig. 10) to compete with the better protonic conductors. Thoria is more basic than zirconia, and $ThO_2 \cdot 4.6H_2O$ exhibits a higher conductivity. (See Fig. 6.)

The particle-hydrate concept may be extrapolated to a simple salt like oxonium perchlorate, $OH^{3+} \cdot ClO_4^-$ in which the tetrahedral ClO_4^- anion acts as a particle in the equilibrium reaction

$$HClO_4 + H_2O = OH_3^+ + ClO_4^-$$

Solid oxonium perchlorate was identified early as a good protonic conductor [13]; and demonstration that the hydrogen self-diffusivity is about three orders of magnitude greater than the oxygen self-diffusity established that, in this compound, protonic diffusion dominates vehicular diffusion. However, the Grotthus mechanism requires some extra water to introduce proton vacancies into the OH^{3+} array. On the other hand, too great an excess of water disorders the ClO_4^- groups; at room temperature, $HClO_4 \cdot 2H_2O$ is a liquid containing dioxonium ions:

$$HClO_4 + 2H_2O = O_2H_5^+ + ClO_4^-$$

6. **Solid polymer electrolyes.** NAFION is representative of a class of perfluorinated polymer ion-exchange membranes [14]. The polymer contains polyfluoroethylene backbones covalently bonded to sulfonic-acid heads, which act as the ion-exchange groups. If pressed together with water, hydrophobic bonding holds the backbones mostly within non-aqueous regions; the aqueous matrix makes contact with the hydrophylic heads. This composite then acts as a particle hydrate; however, it is stronger because the non-aqueous aggregates are linked to one another by polymeric backbones (see Fig. 11) rather than by hydrogen bonding through the aqueous matrix.

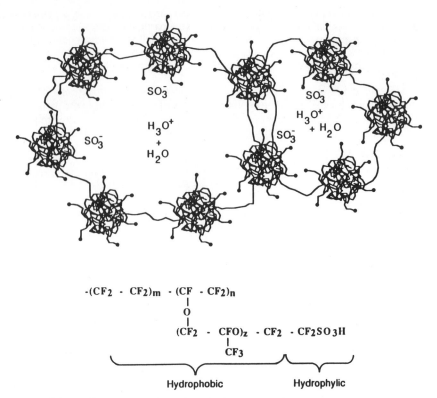

$$-(CF_2 - CF_2)_m - (CF - CF_2)_n$$

$$|$$
$$O$$
$$|$$
$$(CF_2 - CFO)_z - CF_2 - CF_2SO_3H$$
$$|$$
$$CF_3$$

Hydrophobic   **Hydrophylic**

Fig. 11   Schematic representation of NAFION polymeric backbones in aqueous matrix.

OXIDE/HALIDE COMPOSITES

The concepts developed for the particle hydrates can be extended further to embrace doping of stoichiometric halides with dispersed oxide particles. Since the initial observation [15] that the dispersion of small Al_2O_3 particles in LiI gives a dramatic enhancement of the low-temperature, extrinsic conductivity, a number of studies have been devoted to the phenomenon [16]. The following observations appear to be the most significant.

(1) The enhancement is greater the smaller the dispersoid particle size. Poulsen [17] has reported a linear increase in the conductivity of dry LiI/alumina composite electrolytes with increasing specific interface area.

(2) In the case of Al_2O_3 particles in AgI, dispersoid particles containing bound surface water were found to be more effective than predried particles.

(3) The enhancement exhibits a maximum between 10 and 40 mole percent dispersoid.

(4) The enhancement with respect to the pure phase is associated with the extrinsic conductivity occurring at lower temperatures.

(5) An enhancement is observed for anionic (e.g. F^- ions in PbF_2) conductors as well as cationic conductors of quite different mobile-ion electronegativity, e.g. Li^+ versus Ag^+ ions.

(6) A maximum in the enhancement as a function of sintering temperature appears to reflect changes in the particle distribution within the host.

(7) The phenomenon varies with the pzzp of the particle in a manner consistent with doping via a mechanism analogous to the situation in the particle hydrates [18], and the phenomenology of the enhancement mechanism can be modelled -- with an adjustable parameter -- in terms of an increase in the defect concentration responsible for extrinsic conduction in a narrow space-charge layer in the host at the particle-host interface.

The relevant reactions occurring at the interface of an oxide particle dispersed in a halide are compared in Table 1 with the corresponding surface reactions in the particle hydrates. Two types of halides are represented, the M^+ -ion conductors illustrated by MX (M = Li or Ag and X = Br or I) and the F^- -ion conductors illustrated by PbF_2.

Wet particles are those that have been exposed to the air at room temperature before dispersion in the salt; dry particles have been predried before dispersion. The surface of a wet particle contains bound water that completes the oxygen coordination of the surface cations; the associated protons are distributed over the surface. A dry particle has no bound water; a surface cation with deficient anion coordination may induce a reversible restructuring of the surface that, on contact with the salt, returns to normal with the capture of an anion from the salt. Any reversible surface restructuring is not indicated in Table 1; the surface-oxygen deficiency for full oxygen coordination at a surface cation is simply denoted as the neutral species $(V_O)_s$.

TABLE I. Surface Reactions

Oxide particle [a]	Host (x = Br or I)	Surface reaction	Net charge		Mobile ion
			particle	sc [b]	
Particle hydrates					
acidic	H_2O	$(H_2O)_s + H_2O \equiv (OH^-)_s + H_3O^+$ $(OH^-)_s + H_2O \rightarrow (O^{2-})_s + H_3O^+$	−	+	$H_i^\cdot$
basic	H_2O	$(OH^-)_s + H_2O \rightleftharpoons (H_2O)_s + OH^-$ $(O^{2-})_s + H_2O \equiv (OH^-)_s + OH^-$	+	−	V_H^-
Halides with dispersed oxides					
W	MX	$(H_2O)_s + 2MX \rightarrow (OM_2)_s + 2HX\uparrow$ $(H_2O)_s + MX \rightarrow (X^-)_s + H_2O\uparrow + M_i^\cdot$ $(OH^-)_s + MX \rightarrow (O^{2-})_s + HX\uparrow + M_i^\cdot$	0 − −	0 + +	$M_i^\cdot$ $M_i^\cdot$
D	MX	$(O^{2-})_s + MX \equiv (OM^-)_s + V_M^-$ $(V_O)_s + MX \rightarrow (X^-)_s + M_i^\cdot$	+ −	− +	V_M^- $M_i^\cdot$
W/acidic	PbF_2	$(H_2O)_s + \frac{1}{2}PbF_2 \rightarrow (OH^-)_s + HF\uparrow + V_F^\cdot$ $(OH^-)_s + \frac{1}{2}PbF_2 \rightarrow (O^{2-})_s + HF\uparrow + V_F^\cdot$	−	+	$V_F^\cdot$ $V_F^\cdot$
W/basic	PbF_2	$(H_2O)_s + \frac{1}{2}PbF_2 \rightarrow (H_2O)\uparrow + (F^-)_s + V_F^\cdot$	−	+	$V_F^\cdot$
D	PbF_2	$(V_O)_s + \frac{1}{2}PbF_2 \equiv (F^-)_s + V_F^\cdot$	−	+	$V_F^\cdot$

[a] W = wet, D = dry. [b] sc = charge layer.

The lithium salts LiI and LiBr must be distinguished from the silver salts AgI and AgBr. The Li^+ ions can displace surface protons from a wet particle to give the reaction

$$(H_2O)_s + 2\,LiX \rightarrow (OLi_2)_s + 2HX\uparrow \qquad (7)$$

which does not introduce a space-charge layer in the host. However, the oxide ions on both the wet and the dry particles attract Li^+ ions from the salt via the reactions

$$(O^{2-})_s + LiX \rightleftharpoons (OLi)_s^- + V_{Li}^- \tag{8}$$

until the electrochemical potential for Li^+ in the oxide becomes equal to that in the salt via the space-charge-layer capacitance created by the Li^+ -ion transfer. Therefore the enhanced extrinsic Li^+ -ion conductivity should be similar for wet and dry oxide particles dispersed in a LiX halide, and the Li^+ -ion conduction in the space-charge region should be via Li^+ -ion vacancies V_{Li}^-. Moreover, reaction (8) is shifted more strongly to the right for Li than for LiBr, so the enhancement should be greater for the iodide -- as is observed.

The Ag^+ ion, on the other hand, does not displace a surface proton; in the silver salts, reaction (7) is replaced by

$$(OH-)_s + AgX \rightarrow (O^{2-})_s + HX\uparrow + Ag_i^+ \tag{9}$$

where Ag_i^+ is an interstitial Ag^+ ion. The competitive reaction

$$(O^{2-})_s + AgX \rightleftharpoons (OAg^-)_s + V_{Ag}^- \tag{10}$$

which would be followed by

$$Ag_i^+ + V_{Ag}^- \rightleftharpoons Ag \tag{11}$$

is shifted strongly to the left, so the enhancement due to reaction (9) is not suppressed by reaction (11). For a dry particle, the reaction

$$(V_O)_s + AgX \rightleftharpoons (X^-)_s + Ag_i^+ \tag{12}$$

must also be shifted strongly to the left for Al_2O_3 particles in AgBr since little enhancement is observed in this case. The surface restructuring does not appear to be reversible in this case. Thus the chemical arguments outlined above tend to the prediction that the enhancement of Ag^+ -ion conductivity in AgBr for only wet Al_2O_3 particles is due to the dominance of reaction (9), which introduces interstitial Ag^+ ions, Ag_i^+, into the space-charge layer at the particle/halide interface.

In the case of PbF_2 containing a dispersed oxide, the reaction

$$(V_O)_s + \tfrac{1}{2}PbF_2 \rightleftharpoons (F^-)_s + V_F^+ \tag{13}$$

would be shifted more strongly to the right. Therefore a similar enhancement can be expected for wet and dry particles since the reaction

$$(OH^-)_s + \tfrac{1}{2}PbF_2 \rightarrow (O^{2-})_s + HF\uparrow + V_F^+ \tag{14}$$

would give a similar concentration $[V_F^+]$ of fluoride-ion vacancies in the space-charge region. Moreover, as with the silver salts, the reaction

$$(O^{2-})_s + PbF_2 \rightleftharpoons (OPb)_s + 2F_i^- \tag{15}$$

would be shifted strongly to the left, so the reaction

$$V_F^+ + F_i^- \rightleftharpoons F$$

is not available to annihilate the V_F^+ population.

Two factors are operative to shift reactions (13) and (14) to the right: (a) the more acidic the oxide, the stronger an F⁻ ion is bound to the particle surface as $(F^-)_s$ and the more strongly a proton is attracted from the surface to form HF; (b) the larger the surface cation of the oxide, the greater its normal anion coordination and hence the concentration of anion-vacancy sites $(V_O)_s$ and/or bound water molecules $(H_2O)_s$ at the surface. These predictions were tested by measuring the relative enhancement of F⁻ -ion conductivity in PbF_2 with CeO_2, ZrO_2, SiO_2, and Al_2O_3 particles all of 0.2 mm average size [18]. The respective isoelectric points for these oxides are 7, 4, 2 and 9; their cationic radii are 0.87, 0.59, 0.26, and 0.39 Å. According to the above argument, the maximum enhancement should occur for the maximum generation of V_F^+ species at the interfaces and hence for the oxides with the greatest cationic radius and smallest isoelectric point. A measured F⁻ -ion conductivity enhancement that varies as $ZrO_2 > SiO_2 > CeO_2 > Al_2O_3$ indicates that, in addition to the acidic character of the oxide, the ionic radius makes a noticeable contribution.

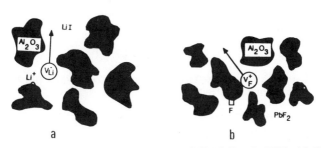

Fig. 12   Composite electrolytes: (a) LiI/Al_2O_3, (b) PbF_2/Al_2O_3.

Fig. 12 gives a schematic representation of the doping of a host LiI matrix with Li⁺ -ion vacancies, V_{Li}^-, and a host PbF_2 matrix with F⁻ -ion vacancies, V_F^+, by dispersed Al_2O_3 particles.

The extrinsic conductivities so far attained in these composites remain too low for high-power applications; but the LiI/Al_2O_3 electrolyte is used commercially in low-power primary batteries.

Support of this work by the Robert A. Welch Foundation, Houston, Texas and the Texas Advanced Research Program is gratefully acknowledged.

REFERENCES

1.   G. Alberti and U. Costantino, in "Intercalation Chemistry," M. S. Whittingham and A. J. Jacobson, eds., Ch. 5 (Academic Press, N. Y., 1982).

2.   A. Clearfield and G. D. Smith, Inorg. Chem. **8**, 431 (1969); J. M. Troup and A. Clearfield, Inorg. Chem. **16**, 3311 (1977).

3.   G. Alberti, M. Casciola, U. Costantino, G. Levi, and G. Riccardi, J. Inorg. Nucl. Chem. **40**, 533 (1978); M. Cascida, U. Costantino, and S. D'Amico, Solid State Ionics **22**, 127 (1986).

4.   M. Casciola, U. Costantino, and S. D'Amico, Solid State Ionics **22**, 17 (1986).

5.   M. G. Shilton and A. T. Howe, Mat. Res. Bull. **12**, 701 (1977).

6. A. N. Fitch, in "Solid State Protonic Conductors (I) for Fuel Cells and Sensors," J. Jensen and M. Kleitz, eds. p. 235 (Odense Univ. Press. 1982).

7. K. Metcalfe and T. K. Halstead, Solid State Ionics **26**, 209 (1988).

8. M. Kahil, M. Forestier, and I. Guitton, in "Solid State Protonic Conductors (III) for Fuel Cells and Sensors," J. B. Goodenough, J. Jensen, and A. Potier, eds., p. 258 (Odense Univ. Press, 1985).

9. M. Nagai and P. S. Nicholson, Mat. Res. Bull. **17**, 1131 (1982); K. Yamashita and P. S. Nicholson, Solid State Ionics **20**, 147 (1986).

10. T. Norby and P. Kofstad, Solid State Ionics **20**, 169 (1986).

11. H. Iwahara, T. Esaka, H. Uchida, and N. Maeda, Solid State Ionics **314**, 359 (19981);H. Iwahara, H. Uchida, and I. Yamasaki, Int. J. Hydrogen Energy **12**, 73 (1987).

12. W.A. England, M. G. Cross, A. Hammett, P. J. Wiseman, and J. B. Goodenough, Solid State Ionics **1**, 231 (1980).

13. D. Rousselet and A. Potier, J. Chim. Phys. **6**, 873 (1973).

14. D. Durand and M. Pinieri, in "Solid State Protonic Conductors (I) for Fuel Cells and Sensors," J. Jensen and M. Kleitz, eds. p. 141 (Odense Univ. Press, 1982).

15. C. C. Liang, J. Electrochem. Soc. **120**, 1289 (1973); C. C. Liang, A. V. Joshi, and N. E. Hamilton, J. Appl. Electrochem. **8**, 445 (1978).

16. F. W. Poulsen in "Proc. 6th Risø Symp. on Transport-Structure Relations in Fast Ion and Mixed Conductors, Risø National Laboratory, Raskilde, Denmark," F. W. Poulsen,N. H. Andersen, K. Clausen, S. Skramp, and O. T. Sørensen, eds. p. 67 (1985) gives a recent review.

17. F. W. Poulsen, J. Power Sources **20**, 317 (1987).

18. A. K. Shukla, R. Manoharan, and J. B. Goodenough, Solid State Ionics **26**, 5 (1988).

DESIGNING A SOLID ELECTROLYTE

IV. DESIGNING A REVERSIBLE SOLID ELECTRODE

John B. Goodenough

Center for Materials Science and Engineering, ETC 5.160
University of Texas at Austin, Austin, TX 78712-1084

PHENOMENOLOGY OF INSERTION COMPOUNDS

In the discussion of secondary batteries in Lecture I, two types of reversible electrodes were identified: elemental-metal anodes and insertion-compound cathodes. The elemental-metal anodes found application with aqueous electrolytes (e.g. the Cd-Ni alkali cell and Pb-PbO$_2$ acid cell) and with solid electrolytes (e.g. Na-S cell), but the number of possibilities is limited. Insertion compounds, on the other hand, may be used as anodes as well as cathodes; in fact a composite of two insertion compounds has particular promise for this application [1]. Therefore, my final lecture is devoted to a few observations about insertion compounds and their design as reversible solid electrodes.

An insertion compound is a <u>host</u> into/from which a <u>guest</u> species may be topotactically and reversibly inserted/extracted over a finite range of solid solution. Moreover, the guest species may be neutral, an electron donor, or an electron acceptor. These three possibilities were already anticipated in the discussion (Lecture III) of proton motion in metallic hydrides. The most common guest is water; e.g. the insertion of interlayer water into clays or the $\alpha-[Zr_2(PO_4)_3] \cdot nH_2O$ layered oxides discussed in Lecture III. A more dramatic illustration is the insertion of n-alkylamine in the Van der Waals gap of layered TaS$_2$ [2], which gives the structure of Fig. 1.

Hydrogen is an electron acceptor when a guest in an electropositive metallic host such as magnesium:

$$xH_2 + Mg = MgH_{2x}, \ 0 \leq x \leq 1 \tag{1}$$

More dramatic is the example of oxygen insertion into Ba$_2$YCu$_3$O$_6$ to convert it from an antiferromagnetic semiconductor to a superconductor with a $T_c \approx 90$ K.

$$Ba_2YCu_3O_6 + xO_2 \rightarrow Ba_2YCu_3O_{6+2x}, \ 0 \leq x < 0.5 \tag{2}$$

Of particular interest for secondary-battery electrodes is the case of an electron donor such as an alkali-metal atom. The first experimental investigation of this concept involved the electrochemical reaction

$$xLi^+ + xe^- + TiS_2 = Li_xTiS_2, \ 0 \leq x \leq 1 \tag{3}$$

Solid State Microbatteries
Edited by J. R. Akridge and M. Balkanski
Plenum Press, New York, 1990

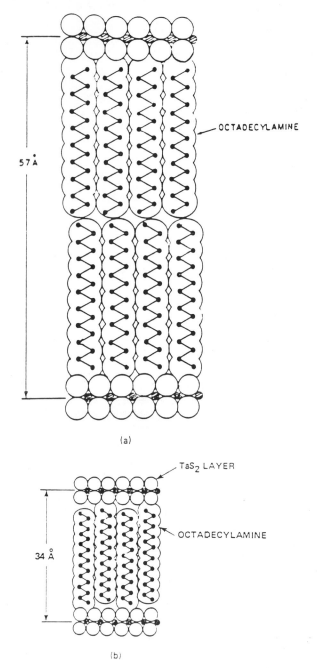

Fig. 1 Structure of TaS$_2$ inserted with 18-alkylamine: (a) bilayer and (b) interleaved anions, after M.S. Whittingham and A.J. Jacobson, eds., Intercalcation Chemistry (Academic Press, New York 1982).

in which Li$^+$ is inserted from the electrolyte across an electrolyte/electrode interface and electrons are supplied to the layered TiS$_2$ host by the external circuit. Lithium may also be inserted or extracted chemically by the respective reactions

$$xn\text{-}BuLi + TiS_2 \rightarrow Li_xTiS_2 + \tfrac{1}{2} x\, Bu_2 \qquad (4)$$

$$LiTiS_2 + \tfrac{1}{2}xI_2 \rightarrow Li_{1-x}TiS_2 + xLi\, I \qquad (5)$$

carried out in non-aqueous solvents.

The redox reactions may occur on the host cation, the host anion, or the guest species. For example, insertion of lithium into the oxospinel $Li[Mn_2]O_4$ reduces the host cation to give

$$Li_{1+x}\big[Mn^{4+}_{1-x}Mn^{3+}_{1+x}\big]O_4 \qquad (6)$$

whereas insertion into the thiospinel $Cu[Cr_2]S_4$ reduces the host anion to give

$$Li^+_xCu^+\big[Cr_2{}^{3+}\big]S_4{}^{(7+x)-} \qquad (7)$$

In the case of the Chervel phase $Li^+(Mo_6S_8)^-$, further insertion of lithium results in the formation of lithium trimers, thus reducing the guest species [8]

$$2xLi^+ + 2xe^- + Li^+(Mo_6S_8)^- = x(Li_3)^+(Mo_6S_8)^- + (1-x)Li^+(Mo_6S_8)^- \qquad (8)$$

Furthermore, the topotactic insertion/extraction reactions may occur by diffusion in one-dimensional (1D) channels, as in the hexagonal tungsten bronze (see Lecture II for this tunnel structure), or between linear chains of edge-shared tetrahedra (or octrahedra) as found in $KFeS_2$:

$$\text{1D: hexagonal } Na_xWO_3 \text{ or } Li_xKFeS_2 \qquad (9)$$

It may occur by 2D diffusion as in layered TiS_2 or the intergrowth structure of $Ba_2YCu_3O_6$:

$$\text{2D: layered } Li_xTiS_2 \text{ or intergrowth } Ba_2YCu_3O_{6+x} \qquad (10)$$

or it may occur by 3D diffusion as in the close-packed spinel $Li[Mn_2]O_4$, the hydride of Mg, or the open framework of hexagonal $Fe_2(SO_4)_3$, which has the framework structure of NASICON (Lecture II):

$$\text{3D: } Li_{1+x}[Mn_2]O_4 \text{ or } H_xMg \text{ or } Li_xFe_2(SO_4)_3 \qquad (11)$$

It may be noted that insertion compounds may also undergo ion-exchange reactions; however, solid electrolytes that undergo ion-exchange reactions are generally not insertion compounds as their energy gaps are too large to sustain a redox reaction of the host.

APPLICATIONS

It is worth noting that, in addition to their possible use as secondary-battery electrodes, insertion compounds are of interest for electrochromic devices, chemical storage, and low-temperature synthesis of phases not accessible by high-temperature techniques. We present examples of each:

$$\text{Battery Electrode: } xLi^+ + xe^- + \lambda\text{-}MnO_2 = Li_x[Mn_2]O_4 \qquad (12)$$

$$\text{Electrochromics: } xH^+ + xe^- + WO_3 = H_xWO_3 \qquad (13)$$

$$\text{Chemical Store: } xH_2 + Mg_2Ni = H_{2x}Mg_2Ni \qquad (14)$$

Low-T Synthesis:

$$\text{(a) } LiNiO_2 - 0.5(e^- + Li^+) \rightarrow Li_{0.5}NiO_2 \tag{15}$$

$$2Li_{0.5}NiO_2 \xrightarrow{300\ °C} Li[Ni_2]O_4$$

$$\text{(b) } Cu[Ti_2]S_4 + 0.465I_2 \rightarrow c\text{-}Cu_{0.07}[Ti_2]S_4 + 0.93CuI \tag{16}$$

$$\text{(c) } Cu[Zr_2]S_4 + 2Li \rightarrow Li_2[Zr_2]S_4 + Cu \tag{17}$$

$$Li_2[Zr_2]S_4 + Cu + \tfrac{3}{2}I_2 \rightarrow 2LiI + CuI + c\text{-}[Zr_2S_4]$$

Reactions (12), (16), and (17) are discussed below. In reaction (13), the electrode consists of a particle-hydrate composite of white WO_3 particles pressed in an aqueous electrolyte. The acidity of the WO_3 particles makes the aqueous matrix a proton conductor. Application of a potential to activate reaction (13) reduces the WO_3 particles, turning them blue-black. Reversal of the field extracts the inserted hydrogen, which erases the information stored in the blue-black color. This application does not require a large range of solid solution. Although electrochromics make an excellent "display," the reaction times are a little too slow to be competitive with the liquid-crystal displays.

Reaction (14) is of interest for hydrogen storage; the hydrogen is more densely packed in the hydride than it is in liquid H_2, and modest heating reverses the reaction.

Reaction (15) illustrates the synthesis of a compound, the spinel $Li[Ni_2]O_4$, that is unstable above 300 °C and could therefore not be obtained by normal high-temperature synthesis. In this case, oxidation of the $[Ni_2]O_4$ array beyond the formal valence states Ni^{3+} and O^{2-} cannot be sustained at higher temperatures.

ELECTRODE DESIGN

As illustrated schematically in Fig. 2, an insertion-compound electrode contains a low-resistance electronic conductor for collecting the current to/from the external circuit and small particles of the insertion compound pressed against it. The pressed particles form a porous mass of large surface area that provides electronic contact between all particles and the current collector. Particles of carbon are commonly pressed with the insertion-compound particles so as to enhance the electronic conduction between particles of the porous mass.

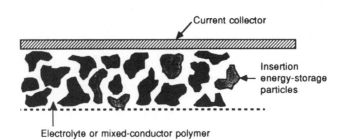

Fig. 2 Schematic of an electrode design.

Good contact between the large electrode surface and the electrolyte is generally established and maintained by pressing the electrode particles with a liquid or polymeric electrolyte that flows to adjust to changes in volume of the electrode particles during discharge or recharge. The electrode thickness must be optimized for capacity versus internal resistance associated with ionic conduction via a tortuous pathway through the porous mass. Also, electronic contact between particles must be maintained on extracting mobile ions from them, which reduces their volume. Any loss of electronic contact between particles during a discharge-recharge cycle causes a loss in the capacity of the battery.

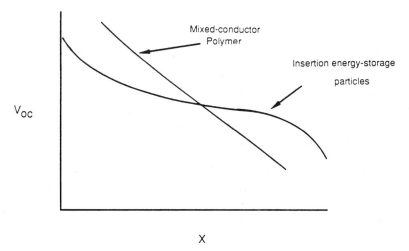

Fig. 3   Schematic of open-circuit voltage, V_{OC}, versus guest concentration x for two components of a composite electrode containing two insertion compounds.

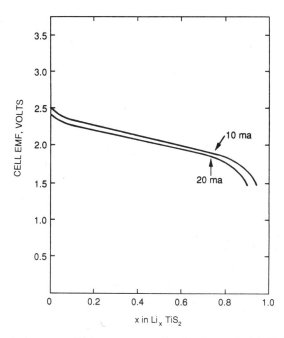

Fig. 4   Cell voltage versus lithium concentration for layered Li_xTiS_2 with a $LiClO_4$ electrolyte at two different currents, 10 and 20 mA/cm^2, after ref. [14].

To be useful as a battery electrode, the open-circuit voltage V_{oc} versus concentration x of the mobile guest should be nearly flat, Fig. 3. This curve is flat, i.e. V_{oc} is independent of x, over a two-phase region; but unless the difference in the two phases is only a small, diffusionless transformation of the host structure, movement of the two-phase interface is too slow for any high-power application. A few insertion compounds, e.g. Li_xTiS_2 of Fig. 4, remain reasonably flat throughout a single-phase solid-solution range, but more often the slope dV_{oc}/dx is unacceptably high.

The principle of a composite electrode made of two insertion compounds is also illustrated in Figs. 2 and 3. The particles of an insertion-compound electrode are pressed together with a polymeric insertion compound that is both an ionic and an electronic conductor, but with an unacceptably high dV_{oc}/dx slope to serve as an electrode material. The resulting composite is a good electronic conductor even where the storage-electrode particles lose contact because the polymeric matrix is a good electronic conductor. Moreover, ions from the electrolyte have access to all the storage-electrode particles via the polymer matrix because the matrix is also an ionic conductor. If the V_{oc} versus x for the polymer crosses that for the storage-electrode particles, equilibrium between the particles and matrix of the composite establishes a common V_{oc} for the electrode. Consequently the polymer matrix sustains a relatively small change in the mobile-ion concentration over a discharge-recharge cycle whereas the particles exhibit a large change. The V_{oc} of the composite is determined by the V_{oc} versus x curve of the particles. This approach to electrode design has another advantage; the surface of the composite electrode may be predominantly polymer matrix, which allows contact with a solid as well as a liquid electrolyte. Thus an all-solid battery can be envisioned.

ELECTRODE QUALITY FACTORS

These considerations, together with those discussed in Lecture I for a solid electrolyte, determine the quality factors by which an insertion compound is to be judged as a candidate electrode material. These factors may be summarized as follows:

(1) Insertion/extraction of the mobile guest species must be reversible.

(2) Chemical storage requires a large solid-solution range of the guest in the host; a small, diffusionless structural change of the host so as to accommodate the guest may result in a desirable two-phase region without altering the basic topotactic character of the insertion/extraction reaction.

(3) The Fermi energy of the electrode should be suitably matched to the electrolyte window so as to allow a maximum V_{oc}. (See Lecture I.)

(4) The V_{oc} versus guest concentration x should be flat, or nearly so, so as to retain a nearly constant output voltage with state of discharge.

(5) A high bulk ionic conductivity σ_i is required to minimize the internal IR drop.

(6) A high bulk electronic conductivity σ_e is also required to minimize the internal IR drop of the cell. Itinerant-electron (band) conduction is preferred, but small-polaron (hopping) electronic conduction is acceptable in low-power applications.
NOTE: a small polaron acts as a counter mobile ion.

(7) Low interfacial resistances for both ionic and electronic transport are also needed to minimize the internal IR drop. Moreover, the cell design and construction must be such that the required interfacial contacts are maintained through repeated discharge-recharge cycles in order to retain the capacity of the cell. This design is made simple by a minimization of the volume change of the insertion compound during a discharge-recharge cycle.

Matching of the Fermi energy to the electrolyte window makes it necessary to consider quite different classes of insertion-compound materials for the anode and for the cathode. For example, the cathode of a lithium battery should give as large a V_{OC} versus a metallic lithium counter electrode as is compatible with the available electrolyte window. With sulfides, it is possible to obtain a V_{OC} versus lithium of up to 2.5 V; in an oxide voltages of more than 4 V could be obtained with an electrolyte having a large enough window. On the other hand, the anode of a lithium cell should have as small a V_{OC} versus lithium as is compatible with the available electrolyte window. A logical candidate would be elemental lithium itself. Therefore elemental-metal or insertion-compound hosts have been used as anodes. But this discussion leads us to the question of strategies for the selection or identification of candidate insertion-compound hosts for anodes and cathodes.

STRATEGIES

A candidate electrode material contains a metallic -- or at least a highly electronically conducting -- host that can accommodate a large solid-solution range of a mobile guest atom that moves in the host with a high mobility. The problem has been most extensively explored for a Li^+ -ion guest, so my discussion of strategy and my illustrations are confined to this case. The structural considerations of Lectures II and III allow generalization to other guest species.

1. **Electronic conductivity.** As a guest species in a host framework, lithium is an electron donor to the host. The variable concentration of lithium over the solid-solution range of the topotactic reaction introduces a non-integral number of conduction electrons per equivalent host atom responsible for the conduction band -- or redox couple -- of the host. In this "mixed-valent" situation, the host is either metallic or a small-polaron conductor. What distinguishes these two possible cases is the relative strengths of the interatomic interactions between like atoms on energetically equivalent sites and the electron-lattice interactions. A measure of the strengths of the interatomic interactions is the bandwidth, which is given by tight-binding theory as

$$W \simeq 2zb \tag{18}$$

where z is the number of nearest neighbors and b is the nearest-neighbor resonance (electron energy transfer) integral defined by

$$b_{ij} \equiv (\psi_i, H'\psi_j) \simeq \varepsilon_{ij}(\psi_i, \psi_j) \tag{19}$$

In (19), H' is the perturbation of the atomic potential at $\underline{R}_i$ due to the presence of a like atom at $\underline{R}_j$, ε_{ij} is a one-electron energy, and (ψ_i, ψ_j) is an overlap integral. Qualitatively, the greater the overlap of the wavefunctions ψ_i and ψ_j on the neighboring atoms at lattice positions $\underline{R}_i$ and $\underline{R}_j$, the larger is the parameter b_{ij}.

For outer s and p electrons, which are primarily responsible for bonding in the solid, the parameter b is large, so the bandwidths are normally large (W > 5 eV). Electrons in partially occupied s and p bands are "itinerant" -- i.e. belong equally to all like atoms of the periodic array -- and give metallic conduction.

From the Uncertainty Principle, the time for an electron to tunnel from one atom to the next is

$$\tau \simeq \hbar/W \tag{20}$$

which, for s and p electrons, is short relative to the period ω_R^{-1} of the optical-mode lattice vibration that would trap the electron at a particular site.

The outer 4f electrons at rare-earth atoms are tightly bound to the atomic nucleus and are screened from neighboring atoms by closed $5s^2 5p^6$ shells. Therefore a small overlap of 4f wave functions on neighboring atoms makes for a narrow bandwidth (W $\approx$ 0.1 eV) and hence a bandwidth

$$W < \hbar\omega_R \tag{21}$$

which gives an electron tunneling time $\tau < \omega_R^{-1}$. In this situation, a mixed-valent configuration is better described by a redox couple in the solid, lattice relaxations having time to trap the mobile electronic species at an energetically different site by a local reorganization energy, as is illustrated in Fig. 5. The difference between the redox couple in a solid and a liquid is only that in the solid the atoms continue to reside in a periodic array satisfying the translational symmetry of the crystal. Unless the local lattice distortions are cooperative, giving rise to an ordering of the different valence states on crystallographically distinguishable sites, the translational symmetry requires that a particular valence state be mobile. However, tunnelling requires the introduction of a thermal energy that makes the energies on neighboring sites equal, so the electronic motion becomes diffusive. The electronic species remains dressed in its local deformation and is called a <u>small</u> <u>polaron</u>, whereas itinerant electrons have a drift mobility described by

$$\mu_e = e\tau_s/m^* \tag{22}$$

where τ_s is the mean-free-time between scattering events and m^* is the effective mass of the electron, the small polaron has a drift mobility described by the Einstein relation

$$\mu_p = eD/kT = (eD_o/kT) \exp (\Delta S_m/k) \exp (-\Delta H_m/kT) \tag{23}$$

where ΔH_m is the motional enthalpy associated with the thermal energy required to equalize the energies on neighboring sites before tunneling can occur. At room temperature the small-polaron mobilities are much lower than the itinerant-electron mobilities, so

$$\sigma_p = ne\mu_p \ll \sigma_e = ne\mu_e \tag{24}$$

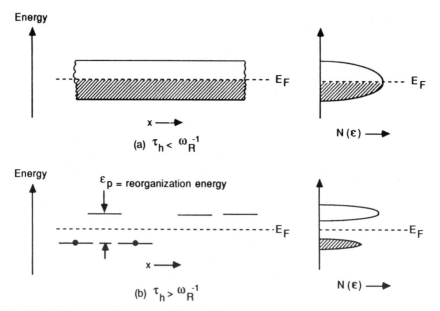

Fig. 5 (a) Itinerant-electron and (b) small-polaron energies versus energy-density of states N(E) for a mixed-valent system.

In the case of outer d electrons, the bandwidths are commonly $W \gtrsim \hbar\omega_R$, so mixed-valent systems tend to be metallic. However, first-row transition elements have narrower bands; and where the electronic configurations have strong Jahn-Teller couplings, a large electron-lattice coupling can result in an $\hbar\omega_R > W$. This is the case for the spinel $Li[Mn_2]O_4$, which contains high-spin Mn^{3+} and Mn^{4+} ions. The high-spin Mn^{3+} ions in octahedral sites have an outer-electron configuration $t_2^3e^1$, which corresponds to a twofold orbital degeneracy in the σ-antibonding e orbitals. Therefore the Mn^{3+} ions are strong Jahn-Teller ions, and $Li[Mn_2]O_4$ is a mixed-valent small-polaron conductor with $W < \hbar\omega_R$ for the $Mn^{4+/3+}$ redox couple.

Similarly structures with widely separated like transition-metal ions may have $W < \hbar\omega_R$. For example, the Fe atoms of the framework $Fe_2(MoO_4)_3$ structure are too far separated for $W > \hbar\omega_R$, and small-polaron behavior must be anticipated for the $Fe^{3+/2+}$ couple in $Li_xFe_2(MoO_4)_3$.

2. <u>Electron energies</u>. Elemental lithium is the natural choice for the anode of a lithium cell. Unfortunately recharging a lithium anode in the liquid electrolytes that have been investigated results in the formation of dendrites; these dendrites grow with successive discharge-recharge cycles until they make contact with the cathode and short out the cell.

Attempts to prepare host compounds having a conduction-band edge only a few tenths of an electron volt below the Fermi energy of metallic lithium have thus far met with failure. Therefore the anode materials of choice remain elemental lithium or a lithium alloy such as Li-Al or Li-Sn. The strategy for a host alloy would seem to be the identification of a metallic framework that can accept lithium as a hydride host accepts hydrogen. Moreover, the bands of the host framework should be so placed that the lithium is not too strong an electron donor to the host. How to do this without losing the weight advantage of elemental lithium remains a challenge.

The most promising cathode materials are oxides and sulfides. The design of the redox potential or Fermi energy in oxides and sulfides is therefore of importance. Fig. 6 illustrates the principles of such a design through a consideration of the simple binary compound MnO.

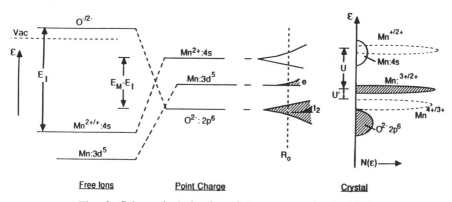

Fig. 6 Schematic derivation of electron energies for MnO.

The lefthand side is a schematic sketch of the energy levels of the free ions relative to the vacuum level. The O^{2-} ion is not stable in free space; the $O^{-/2-}$ electron affinity E_a is negative, so this level lies above the vacuum level. The energy $E_I = E_i - E_a$ is the difference in the second ionization potential E_i for manganese and E_a; it is the energy required to remove the remaining 4s electron at a Mn^+ ion and place it on an O^- ion a great distance away. Assembling the ions into the rocksalt structure introduces an electrostatic binding known as the Madelung energy E_M. The energy E_M stabilizes the O^{2-} ion and destabilizes all the cation orbitals, including both the empty 4s orbital and the high-spin $3d^5$ manifold. An $E_M - E_I > 0$ stabilizes the compound. Although covalent mixing reduces the effective charge on the ions, and hence the magnitude of E_M, this loss in energy is compensated by the quantum-mechanical repulsion between the admixed antibonding and bonding states. The translational symmetry of the crystal then broadens the s and p states into valence and conduction bands that are separated by a finite energy gap of ca 6 eV.

Placement of the Fermi energy in this gap and the potential of the Mn^{2+} ion for oxidation to Mn^{3+} depends on the placement of the $3d^5$ manifold relative to the band edges. If the Madelung energy also raises the Mn-$3d^5$ level above the O^{2-}:$2p^6$ level, then the 3d orbitals are antibonding with respect to the O-2p orbitals. In this case it is possible to remove an electron from the $3d^5$ manifold to make it $3d^4$, thus creating a formal valence Mn^{3+} even though covalent mixing introduces anionic character into the 3d wave functions.

Whether the $3d^5$ -manifold energy remains localized as a $Mn^{3+/2+}$ redox energy or becomes broadened into a band of itinerant-electron states depends upon the relative strengths of the intraatomic electron-electron interactions and the interatomic interactions, Mn-Mn and Mn-O-Mn. In MnO, the $3d^5$ electrons remain localized, and the Mn^{2+} ion carries a magnetic moment of ca $5\mu_B$.

Intraatomic electrostatic interactions between localized electrons of a given manifold introduces an energy gap between successive redox energies. This gap is particularly large on going from a half-filled to a more than half-filled manifold because of the loss of the intraatomic exchange stabilization of the minority-spin electron. Therefore the Mn^+ level lies more than 3 eV above the energy of the $3d^5$ manifold, which places it above the bottom of the conduction-band edge. Any attempt to create a Mn^+ ion by introducing an additional electron to the Mn^{2+} manifold puts the electrons in the 4s bands, which is a situation that is unstable relative to the formation of metallic manganese. On the other hand, the gap between the $Mn^{4+/3+}$ and $Mn^{3+/2+}$ redox energies is relatively small since the two σ-antibonding e orbitals per manganese atom are degenerate and addition of the fifth 3d electron to complete the 3d half-shell experiences the full intraatomic-exchange stabilization. This small gap has two important consequences: (a) it leaves the $Mn^{4+/3+}$ redox energy above the top of the O^{2-}-$2p^6$ band, which makes accessible the formal valence state Mn^{4+}, and (b) a disproportionation reaction of the type

$$2Mn^{3+} \rightarrow Mn^{2+} + Mn^{4+} \tag{25}$$

is energetically possible in Mn^{3+} oxides. Although reaction (25) is normally not stabilized in the bulk, it occurs commonly at the surface of manganese oxides in acid media, where the Mn^{2+} ions are subsequently dissolved.

A large crystal-field splitting of the octahedral-site 3d orbitals separates the $Mn^{4+/3+}$ couple from the $Mn^{5+/4+}$ couple by a large energy gap, so attempts to oxidize octahedral-site Mn^{4+} to Mn^{5+} can be expected to introduce holes into the O-2p bands.

Thus the qualitative location of the redox potentials of a transition-metal atom relative to the band edges can be deduced from the 3d-shell occupancies and the accessibility of the known redox potentials for the ions. This simple reasoning provides an excellent guide to the location of the Fermi energy relative to the lithium Fermi level, and hence to the V_{oc} versus lithium for a given host.

3. Ionic conductivity. Choice of a host structure for its high ionic conductivity is guided by the same principles used to choose fast ionic conduction in a solid electrolyte. However, the insertion electrode has an important advantage. The creation of a $c(1-c) \neq 0$ is accomplished by introducing mobile electrons rather than stationary ions as the charge-compensating dopant. If these electrons are itinerant, they introduce no trapping enthalpy ΔH_t, so the condition $E_A = \Delta H_m$ for fast ionic conduction is fulfilled.

ILLUSTRATIONS

1. Layered compounds. The publication of Fig. 4 by Whittingham [4] stimulated interest in insertion compounds as battery electrodes. However, the Li/TiS_2 battery proved a disappointment because of the non-reversibility of the lithium anode.

Layered TiS$_2$ has a close-packed-hexagonal S^{2-}-ion array; the Ti atoms occupy alternate basal-plane layers of octahedral sites. The S-Ti-S sandwich layers are held together by S-S Van der Waals bonding; the space between the layers is therefore referred to as the "Van der Waals" gap. One mole of lithium can be inserted reversibly into the empty octahedral sites in the Van der Waals gap to give Li$_x$TiS$_2$, $0 \leq x \leq 1$. Although a V$_{oc}$ $\approx$ 2 V relative to lithium is respectable, any attempt to modify the anode from elemental lithium can only lower this voltage. Therefore there is an interest in increasing the work function of the cathode material in order to allow for an anode of larger work function. This technical target immediately raises the question whether the desired cathode material can be realized with a similar layered sulfide, but with another transition-metal atom substituted for Ti. To answer this question, it is useful to go to the schematic energy versus density of one-electron states shown in Fig. 7.

As normally prepared, TiS$_2$ is not stoichiometric; a small Ti excess located in the Van der Waals gap makes it metallic. Stoichiometric TiS$_2$ is a small-bandgap semiconductor; the Fermi energy E$_F$ lies $\approx$0.2 eV above the top of the S-3p valence band. This small gap compared to the 3.0 eV gap of TiO$_2$ is due to the larger negative electron affinity of S^{2-} relative to O^{2-} and to the smaller Madelung stabilization of the sulfide. Lowering E$_F$ by more than ca 0.5 eV would place it deep enough inside the S-3p valence band for the trapping of sulfur-band holes in antibonding orbitals of a S-S bond via the formation of (S$_2$)$^{2-}$ polyanions. Therefore MnS$_2$, FeS$_2$, CoS$_2$, and NiS$_2$ have pyrite and/or marcasite structures rather than the layered structure of TiS$_2$.

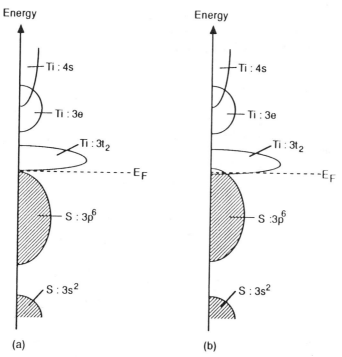

Fig. 7   Schematic densities of one-electron states in (a) layered and (b) cubic TiS$_2$.

This reasoning leads to a consideration of analogous oxides. However, Van Der Waals bonding between oxide ions is normally too weak to overcome the electrostatic repulsion between oxide layers, so TiO$_2$ crystallizes in the rutile structure. Layered oxides without a cation -- or water -- in the Van der Waals gap only occur where a cation displacement from the center of symmetry of the anion interstice creates an electric dipole, e.g. in V_2O_5 and MoO_3. On the other hand, a number of layered oxides $LiMO_2$ are known. They have the $NaFeO_2$ structure of Fig. 8, which is analogous to that of $LiTiS_2$ except for a change from hexagonal to cubic stacking of the close-packed anion layers. The pertinent question then is, How much lithium can be withdrawn before the electrostatic repulsion between M-O-M sandwich layers renders the structure unstable? The answer to this question proves to be dependent on the M cation. The most interesting results have been obtained with the system $Li_{1-x}CoO_2$ [5]. It appears that nearly all of the lithium could be extracted reversibly were it possible to find an electrolyte that isn't oxidized by the electrode at compositions $x > 0.5$. (See Fig. 9.) The stability of the O-Co-O sandwich layers at larger x is apparently due to a displacement of a Co^{4+} ion out of the center of symmetry of its octahedral site [6].

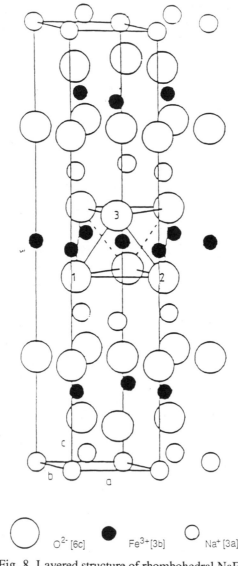

Fig. 8 Layered structure of rhombohedral $NaFeO_2$.

Once the stability and the higher V_{oc} versus lithium was established for $Li_{1-x}CoO_2$, the next concern was the magnitude of the Li^+-ion mobility relative to that in Li_xTiS_2. Insertion of lithium into TiS_2 requires work to prize apart the sandwich layers, which are held together by Van der Waals bonding. In $Li_{1-x}CoO_2$, any Van der Waals bonding is weaker than the electrostatic repulsive force between oxide-ion layers; therefore it is not a priori clear whether the smaller, harder O^{2-} ions necessarily result in a lower mobility of Li^+ ions in a layered oxide. In fact, measurement of the mobility gave a significantly higher diffusion coefficient ($\tilde{D} = 5{\times}10^{-12}m^2s^{-1}$ vs 1×10^{-12} m^2s^{-1}) for the oxide at room temperature [7,8].

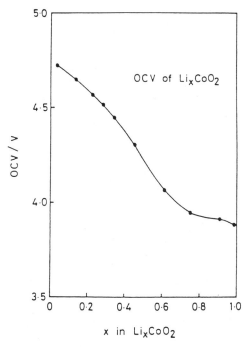

Fig. 9   Open-circuit voltage versus lithium concentration for the cell $Li/LiBF_4$ in propylene carbonate/Li_xCoO_2.

2. Spinels. The experiments on the layered oxides and sulfides not only confirmed the known stability of Li^+ ions in octahedral sites of an anion array, they also demonstrated that the Li^+ ions are mobile in a close-packed anion array having an interconnected space of edge-shared octahedra. It is therefore of interest to compare the room temperature Li^+-ion mobilities in 1D, 2D and 3D interstitial arrangements of edge-shared octahedra in oxides and chalcogenides.

Toward this end, experiments on room temperature lithiation of the ferrospinel Fe_3O_4, magnetite, provided evidence that the $[B_2]X_4$ subarray of the $A[B_2]X_4$ spinels may be considered a host framework having a 3D interstitial space within a close-packed anion array [9]. A cubic $A[B_2]X_4$ spinel (lattice parameter a_0 in Fig. 10) contains eight formula units and has its close-packed array of anions at the 32e positions of space group $Fd3m(O_h^7)$. The B cations occupy half the octahedral sites designated 16d, and the A cations the eighth of the tetrahedral sites designated 8a. The empty 16c octahedral sites form an interconnected 3D array of edge-shared sites identical to the 16d array, but shifted by $a_0/2$ along any principal axis. Each 16c site shares edges with six other 16c sites to form intersecting <110> chains; each also shares common faces with two 8a sites on opposite sides along a <111> axis.

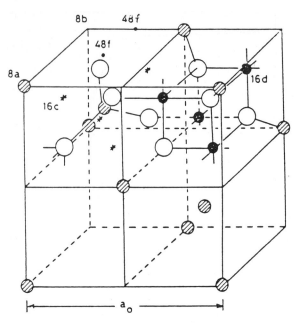

Fig. 10 Two quadrants of the $A[B_2]X_4$ cubic-spinel structure. A cations in positions 8a, B cations in 16d, anions X in 32e of space group Fd3m.

Where the M cations have a strong octahedral-site preference, the lithium spinels LiM_2O_4 are normal, which means that the Li^+ ions occupy the A sites and have access to the interstitial 16c sites. Known examples of normal LiM_2O_4 spinels include: $Li[Ti_2]O_4$, $Li[V_2]O_4$, and $Li[Mn_2]O_4$. All of these spinels have a mixed $M^{4+/3+}$ formal valence state and are therefore good electronic conductors. In fact, $Li[V_2]O_4$ is a narrow-band metal [11] whereas $Li[Mn_2]O_4$ is a small-polaron conductor [12]. Since the transition-metal array is clearly reducible/oxidizable in each of these systems, these spinels invite exploration of both insertion and extraction of Li^+ ions into and out of the 3D interstitial space.

In contrast, Na^+ ions are not stable in the octahedral/tetrahedral interstitial space of an oxide spinel; the Na^+ ion is too large. Larger ions require an open-framework structure. (See Lecture II.)

Initial experiments on these spinels were carried out on $Li[Mn_2]O_4$ [13]. The Mn^{3+} ion is high-spin $^5E_g(t_2^3e^1)$ with a twofold orbital degeneracy for the single hole of the half-shell. Although Mn^{3+} is therefore a strong Jahn-Teller ion, there is no cooperative Jahn-Teller distortion from cubic to tetragonal symmetry at room temperature. The guest Li^+ ions are ordered on the 8a sites of the interstitial space.

Electrochemical or chemical insertion of additional lithium in the system $Li_{1+x}[Mn^{3+}_{1+x}Mn^{4+}_{1-x}]O_4$ increases the concentration of octahedral-site Mn^{3+} ions, and for $x > 0.1$ the concentration is large enough to induce a cooperative Jahn-Teller distortion to tetragonal (c/a > 1) symmetry at room temperature. The transition is first-order, and V_{oc} vs composition x, Fig. 11, shows the curve is flat over the compositional range $0.1 \lesssim x \lesssim 0.8$, indicative of a two-phase region. X-ray diffraction confirmed the presence of a cubic and a tetragonal phase. The Jahn-Teller distortion is a displacive transition; it is therefore rapid and the cubic/tetragonal interface has a relatively small motional enthalpy. However, its mobility at room temperature is too low for a high-power application. On the other hand, a $V_{oc} \approx 3.0$ V versus lithium provides an interesting voltage, and this material is presently under development in Japan as a cathode of high energy density for low-power applications.

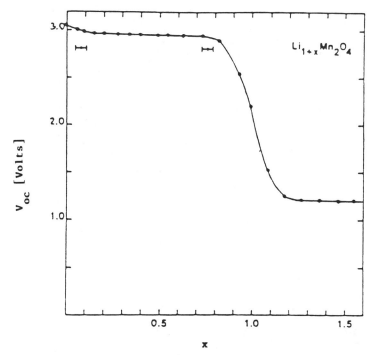

Fig. 11   Open-circuit voltage versus x for the cell Li/LiBF$_4$ in propylene carbonate/
Li$_{1+x}$[Mn$_2$]O$_4$. First plateau: cubic + tetragonal Li$_{1+x}$[Mn$_2$]O$_4$; second
plateau: tetragonal Li$_{2+y}$[Mn$_2$]O$_4$ + layered Li$_2$MnO$_2$.

Lithium may also be extracted from Li[Mn$_2$]O$_4$ either chemically [14] or
electrochemically [15]. The resulting host framework [Mn$_2$]O$_4$ is called λ-MnO$_2$.

Cubic Li[V$_2$]O$_4$ is most easily prepared by extracting half the lithium from the layered
compound LiVO$_2$ and then heating the layered Li$_{0.5}$VO$_2$ to T $\gtrsim$ 300 $^\circ$C [16]. Layered
LiVO$_2$, unlike the isostructural LiMO$_2$ compounds such as LiCoO$_2$ discussed above, exhibits
a remarkable first-order transition at a temperature T$_t$ $\simeq$ 460 K [17] that is due to the
formation of vanadium-atom trimers within the close-packed vanadium layers [18]. Room
temperature extraction of lithium from LiVO$_2$ is reversible only to the composition Li$_{0.7}$VO$_2$;
on further extraction some vanadium atoms become displaced from the vanadium to the
lithium layers. Interestingly, exposure of LiVO$_2$ to air at room temperature results in the
formation of Li$_2$O at the surface of the particles with a depletion of the Li content of the bulk
to a composition approaching Li$_{0.7}$VO$_2$ [16]. The room temperature mobility of lithium
allows absorbed oxygen on the surface of the samples to leach lithium from the bulk.
However, because this much lithium depletion is reversible, annealing in H$_2$ to remove the
surface oxygen restores the compound to its stoichiometric composition LiVO$_2$. On the other
hand, washing the compound removes the superficial Li$_2$O to leave only the bulk Li$_{0.7}$VO$_2$.

The composition Li[V$_2$]O$_4$ is metallic, whereas isostructural Mg[V$_2$]O$_4$ which
contains the single formal vanadium valence V^{3+} is an antiferromagnetic semiconductor
[19], and the system Li$_x$Mg$_{1-x}$[V$_2$]O$_4$ exhibits a transition with increasing x from small-
polaron to metallic behavior [20]. A similar transition from metallic to small-polaron
behavior occurs in the system Li$_{1+x}$[V$_2$]O$_4$, and the transition is marked by the change in the
V$_{oc}$ versus x curve shown in Fig. 12 [16]. A broadening of the x-ray-diffraction lines in the
compositional range 0.3 $\lesssim$ x $\lesssim$ 0.5 is suggestive of a subtle segregation into two cubic phases
of slightly different lattice parameter.

227

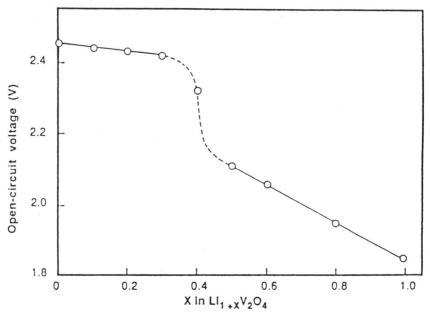

Fig. 12   Open-circuit voltage versus x for the cell Li/LiBF$_4$ in propylene carbonate/
Li$_{1+x}$[V$_2$]O$_4$.

Insertion of lithium into superconducting Li[Ti$_2$]O$_4$ was expected to leave the system metallic throughout the compositional range $0 \lesssim x \lesssim 1$ for Li$_{1+x}$[Ti$_2$]O$_4$. In fact, preliminary measurements indicate that there is no broad range of solid solution in this system, but a two-phase region between the spinel Li[Ti$_2$]O$_4$ and the ordered-rocksalt structure Li$_2$[Ti$_2$]O$_4$.

In summary, lithium insertion into the oxospinels Li[M$_2$]O$_4$ (M = Ti, V, Mn) have shown that reversible insertion/extraction reactions occur for $0 \lesssim x \lesssim 1$ in the system Li$_{1+x}$[M$_2$]O$_4$, but that the mobile electrons are itinerant in Li$_{1+x}$[Ti$_2$]O$_4$, small polarons in Li$_{1+x}$[Mn$_2$]O$_4$, and exhibit a transition from itinerant to small-polaron character in Li$_{1+x}$[V$_2$]O$_4$. The experiments also show evidence of a two-phase character within the compositional range 0<x<1 for each system and a Li$^+$ -ion mobility that is too slow for high-power applications. Removal of the c-axis degree of freedom on going from the 2D layered oxides to the 3D close-packed-framework oxides has lowered the Li$^+$ -ion mobility. This line of reasoning naturally leads to the cubic thiospinels having larger, more polarizable S^{2-} anions.

Preparation of the analogous Li[M$_2$]S$_4$ thiospinels is not straightforward for two reasons: first, stabilization of a Mn$^{4+/3+}$ couple is not possible in a sulfide host because the redox couple lies below the top of the S:3p^6 band; second, the defect layered compound Li$_{0.5}$MS$_2$ is more stable than the spinel phase. In order to obtain a metastable Li[M$_2$]S$_4$ phase, it is necessary to identify a low-temperature synthetic route. Since the tetrahedral-site preference of a Cu$^+$ ion is known to stabilize the Cu[M$_2$]S$_4$ phases with accessible M$^{4+/3+}$ couples as cubic spinels, the obvious synthetic route would appear to be the room temperature chemical removal of copper followed by insertion of lithium.

The copper can be oxidatively extracted from the normal thiospinel Cu[Ti$_2$]S$_4$ at room temperature with mild oxidizing agents [21,22]. However, not all of the copper is removed; the end product has an approximate compositon Cu$_{0.1}$[Ti$_2$]S$_4$. Removal of all of the copper requires either the presence of a small excess of titanium [22] or the substitution of a small amount of titanium by niobium [23]. The reason for this appears to be a small overlap of the titanium-3d and the sulfur-3p bands in the cubic phase.

Oxidative extraction of copper from $Cu[Ti_2]S_4$ is best achieved with a controlled excess of 0.2 M I_2/acetonitrile solutions at 45 °C [24]:

$$Cu[Ti_2]S_4 \; + \; (x/2)I_2 \; \rightarrow \; Cu_{1-x}[Ti_2]S_4 \; + \; x \; CuI$$

Iodine is preferable to the $FeCl_3$ or Br_2 solutions used by [21,22] because it is specific for the copper-extraction reaction and shows no tendency to oxidize sulfide ions. The CuI is readily removed since it is soluble in acetonitrile to the extent of 35 g/l at 35 °C. The limiting composition for oxidative extraction of copper is $Cu_{0.07}[Ti_2]S_4$.

Lithium insertion into $Cu_{0.07}[Ti_2]S_4$ gives a V_{oc} vs.x and a Li^+ -ion mobility that are nearly the same as those obtained for layered Li_xTiS_2 [22, 25]. In this case, loss of the c-axis degree of freedom has not impaired the Li^+ -ion mobility in the close-packed sulfide-ion array. In the $Li_xCu_{0.07}[Ti_2]S_4$ system, the inserted lithium atoms apparently occupy the 16c octahedral sites for all compositions $0 \leq x \leq 0.93$; they do not exhibit the preference for 8a sites found in the oxospinels.

The cubic $[Ti_2]S_4$ framework has an important advantage over the layered TiS_2; the lack of a c-axis degree of freedom makes its insertion reactions selective for smaller cations. Unwanted species from the electrolyte become inserted into the layered TiS_2 after repeated cycling.

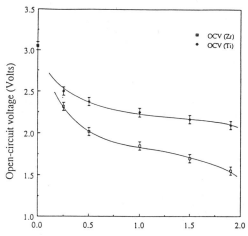

Fig. 13   Open-circuit voltages versus x for the cells (a) $Li/Li_xCu_{0.07}[Ti_2]S_4$ and (b) $Li/Li_xCu_{0.05}[Zr_2]S_4$ with $LiBF_4$ in propylene carbonate as electrolyte.

Attempts to perform an analogous study on $Cu[Zr_2]S_4$ thiospinel encountered difficulties immediately; oxidative extraction of copper from this compound is not straightforward. Preparation of $Cu_{0.05}[Zr_2]S_4$ was accomplished by first inserting lithium so as to extrude metallic copper and then removing the inserted lithium and extruded copper by oxidative reaction with I_2 [26]:

$$Cu[Zr_2]S_4 \; + 1.95n\text{-}BuLi \; \rightarrow \; Li_{1.95}Cu_{0.05}[Zr_2]S_4 \; + 0.95Cu$$

$$\text{Li}_{1.95}\text{Cu}_{0.05}[\text{Zr}_2]\text{S}_4 + 0.95\text{Cu} + 1.45 \text{ I}_2 \rightarrow \text{Cu}_{0.05}[\text{Zr}_2]\text{S}_4 + 1.95\text{LiI} + 0.95\text{CuI}$$

Comparison of the V_{oc} vs x curve for $\text{Li}_x\text{Cu}_{0.05}[\text{Zr}_2]\text{S}_4$ and $\text{Li}_x\text{Cu}_{0.07}[\text{Ti}_2]\text{S}_4$ is given in Fig. 13.

3. <u>Open frameworks</u>. Lithium insertion into oxides having an open framework provides my final illustration.

Several oxides of general formula $A_n\text{M}_2(\text{XO}_4)_3$ crystallize with $\text{M}_2(\text{XO}_4)_3$ frameworks consisting of XO_4 tetrahedra sharing all four corners with MO_6 octahedra and MO_6 octahedra sharing their corners with XO_4 tetrahedra. The character of the interstitial space of a framework depends upon the space-group symmetry of the framework, and the occupancy of the interstitial sites by A cations and the size of the A cation influence the symmetry of a framework.

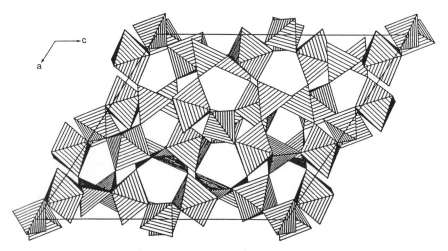

Fig. 14   Structure of monoclinic $\text{Fe}_2(\text{XO}_4)_3$ (X = Mo, W or S) drawn with STRUPLO program, R.X. Fisher, Am. Cryst. Assoc. Abstracts 12, Abstract PA9 (1984).

A few $\text{M}_2(\text{XO}_4)_3$ frameworks are stable in the total absence of A cations; these include the molybdates and tungstates of the smaller trivalent ions as well as sulfates of trivalent V, Cr and Fe [27]. These empty frameworks crystallize in two different structural types; one is the rhombahedral $\overline{\text{R3c}}$ framework of NASICON (Lecture II) and the other is a related orthorhombic (Pnca or Pbcn) structure that, on cooling, undergoes a diffusionless transformation to the monoclinic ($P2_1/a$) form of Fig. 14. The displacive phase transition commonly occurs above room temperature. All the sulfates adopt the rhombohedral structure; all the molybdates and tungstates adopt the monoclinic/orthorhombic structure. Only $\text{Fe}_2(\text{SO}_4)_3$ is known to crystallize in both modifications, rhombohedral and monoclinic at room temperature. Both modifications have an open structure with an interstitial space interconnected in 3D into which lithium can be inserted. The availability of a framework that not only supports fast Li^+-ion and Na^+-ion conductivity, but also contains transition-metal ions having an accessible redox potential has invited exploration of these materials as candidate electrodes for secondary batteries [27-29]. Initial investigations have used lithium as the insertion species, but sodium can also be inserted. A question to be resolved is whether the electronic conductivity is high enough for high-power applications; the Fe atoms are sufficiently far from one another in the structure that the mobile electrons form small polarons.

Lithium insertion into monoclinic $Fe_2(XO_4)_3$, with X = Mo, W, or S, proceeds by a two-phase mechanism; the monoclinic $Fe_2(XO_4)_3$ is transformed on lithium insertion into an orthorhombic $Li_2Fe_2(XO_4)_3$ compound. Consequently the V_{oc} vs. lithium concentration is flat, Fig. 15, but the insertion is facile and reversible because of the displacive character of the orthorhombic-to-monoclinic distortion. On the other hand, there is no phase change on lithium insertion into the rhombohedral form of $Fe_2(SO_4)_3$, so the V_{oc} vs. x of $Li_xFe_2(SO_4)_3$, $0 \leq x \leq 2$, decreases monotonically with x. Yet to be done is a comparison of the Li^+ -ion mobility in the rhombohedral versus the monoclinic/orthorhombic forms of $Li_xFe_2(SO_4)_3$.

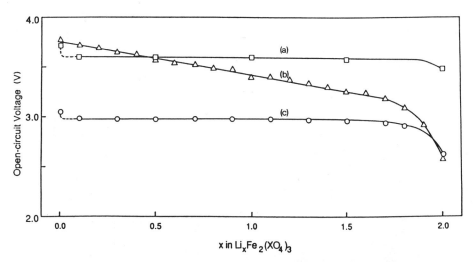

Fig. 15   Open-circuit voltage versus x for the cells $Li/LiBF_4$ in propylene carbonate/ $LiFe_2(XO_4)_3$ with (a) monoclinic X = S, (b) R$\overline{3}$c X = S, and (c) monoclinic X = Mo. The curve for monoclinic X = W is similar to that for X = Mo.

Support of this work by the Robert A. Welch Foundation, Houston, Texas and the Texas Advanced Research Program, Grant #4257, is gratefully acknowledged.

REFERENCES

1.   M. Maxfield, T.R. Jow, M.G. Senchak, and L.W. Shacklette, Proc. 4th Int.Meeting on Lithium Batteries, Vancouver, Canada, May 24-26, 1988.

2.   F.R. Gamble, J.H. Osiecki, M. Cais, R. Pisharody, F.J. DiSalvo, and T.H. Geballe, Science **174**, 493 (1971).

3.   R. Schöllhorn, 9th Conf. on Solid Compounds of Transition Elements, Univ. of Oxford 4-8 July 1988 (unpublished).

4.   M.S. Whittingham, J. Electrochem. Soc. **123**, 315 (1976).

5.   K. Mizushima, P.L. Jones, P.J. Wiseman, and J.B. Goodenough, Mat. Res. Bull. **15**, 783 (1980).

6.   M.G.S.R. Thomas, P.G. Bruce, and J.B. Goodenough, J. Electrochem. Soc. **132**, 1521 (1985).

7.   M.G.S.R. Thomas, P.G. Bruce, and J. B. Goodenough, Solid State Ionics **1**, 13 (1985).

8.   A. Honders, Ph.D. Thesis (Univ. of Utrecht, 1984).

9. M. M. Thackeray, W.I. F. David, and J. B. Goodenough, Mat. Res. Bull. **17**, 785 (1982).

10. D.C. Johnston, H. Prakash, W. H. Zachariasen, and R. Viswanathan, Mat. Res. Bull. **8**, 777 (1973).

11. D.B. Rogers, J. B. Goodenough, and A. Wold, J. Appl. Phys. **35**, 1069 (1964); D. B. Rogers, J. L. Gillson, and I.E. Grier, Solid State Commun. **5**, 263 (1967).

12. I.T. Sheflad and Ya. V. Pavlovtski, Inorg. Mater. **2**, 782 (1966).

13. J.B. Goodenough, M.M. Thackeray, W.I.F. David, and P.G. Bruce, Rév. chim. minérale **21**, 435 (1984).

14. J.C. Hunter, J. Solid State Chem. **39**, 142 (1981).

15. M.M. Thackeray, P.J. Johnson, L.A. dePicciotto, W.I.F. David, P.G. Bruce, and J.B. Goodenough, Mat. Res. Bull. **19**, 179 (1984).

16. A. Manthiram and J.B. Goodenough, Can. J. Phys. **65**, 1309 (1987).

17. P.F. Bongers, Ph.D. dissertation, Univ. of Leiden, Leiden, The Netherlands, 1957.

18. J.B. Goodenough, <u>Magnetism and the Chemical Bond</u>, (Interscience and John Wiley, New York, NY 1963) p. 269.

19. D.B. Rogers, R.J. Arnott, A. Wold, and J.B. Goodenough, J. Phys. Chem. Solids **24**, 347 (1963).

20. B. Reuter and J. Jaskowsky, Ber. Bunsen - Ges. Phys. Chem. **70**, 189 (1966).

21. R. Schöllhorn and A. Payer, Angew. Chem. Intl. Ed. Engl. **24**, 67 (1985).

22. S. Sinha and D. W. Murphy, Solid State Ionics **20**, 81 (1986).

23. A.C.W.P. James and J.B. Goodenough, unpublished.

24. A.C.W.P. James, D.Phil. Thesis, University of Oxford, Oxford, 1988.

25. A.C.W.P. James and J.B. Goodenough, Solid State Ionics **27**, 37 (1988).

26. A.C.W.P. James, N.J. Clayden, P.M. Banks, and J.B. Goodenough, Mat. Res. Bull. (in press).

27. A. Manthiram and J.B. Goodenough, J. Solid State Chem. **71**, 349 (1987) and references listed therein.

28. A. Nadiri, C. Delmas, R. Salmon, and P. Hagenmuller, Rév. chim. minér. **71**, 537 (1984).

29. W.M. Reiff, J.H. Zhang, and C.C. Torardi, J. Solid State Chem. **62**, 231 (1986).

TECHNOLOGY AND PHYSICS OF THIN FILM INSERTION COMPOUNDS

C.JULIEN and M.BALKANSKI

Laboratoire de Physique des Solides, associé au CNRS
Université Pierre et Marie Curie
4, place Jussieu 75252 Paris Cedex 05, France

The purpose of this paper is twofold : (i) to provide a brief summary on the thin film technology of layered compounds and particularly the chalcogenide films obtained using different methods including the MBE technique, (ii) to furnish physics of insertion films used as cathode in microbatteries.

It is of course not possible to cover all properties, but electrical and optical characteristics are discussed.

1. THIN FILM TECHNOLOGY OF LAYERED COMPOUNDS

1.1. Limitative parameters

The use of thin film technology for the deposition of layered materials used as positive electrodes in solid state micro-batteries may offer different main advantages :

i) thin films are well adapted for devices design using a advanced microelectronic technology.

ii) thinning of layers gives a lower electronic resistance in transversal direction in the case of poor semiconducting materials.

iii) thin film technology provide clean surface of the compound and may be improve the electrode-electrolyte interface contact reducing the interfacial resistance.

iv) film deposition of layered material may give a well defined crystallographic structure and by a very well orientation of the layers, i.e. the Van der Waals planes, increases considerably the ionic path into the lamellar. It is well known that in solid state batteries, the electrolyte and the electrode are sources of limitation on the current density. Current density limitation depends whether pulses or substained currents are required. For short pulses, the current density J in A/mm^2 may be given by the following expression :

$$J_{max} = 10^2 \, \sigma \, \Delta V_{max} \, / \, l \qquad (1)$$

where l is the layer thichness in microns, σ is the total conductivity in $(\Omega.cm)^{-1}$ and ΔV_{max} is the differential voltage (in Volts). ΔV_{max}, the allowed potential drop accross the electrode is an arbitrary choice if the requirement is for a short current pulse, with averaged sustained current, the potential loss will progressively increase from the initial value, but according to simple limiting current theory (Atlung et al., 1979) will reach a steady state provided the initial value was less than 4 RT/F (Volts).

Precise determination of the long term current density limitation

Solid State Microbatteries
Edited by J. R. Akridge and M. Balkanski
Plenum Press, New York, 1990

is complex and required data regarding the variations of chemical diffusion coefficient D and electrode potential E of the mobile species with concentration c. These data are not generally available for many materials, although average values of D and $\partial E/\partial c$ are often known. With a knowledge of the latter values, the following simple treatment gives the order of magnitude of current limitations due to the electrode. The potential loss due to diffusion in the electrode increases with time at constant current to a steady state value, corresponding to a parabolic concentration profile, whose gradient $\partial c/\partial x$ decreases linearly to zero accross the electrode. Then the potential loss is given by :

$$\Delta V = (\partial E / \partial c) \Delta c \tag{2}$$

From Faraday's and Fick's laws :

$$J = -2 FD (\partial c / \partial x)$$

$$J = 2FD (2 \Delta c / l) = 2 \sigma^* \Delta V / l \tag{3}$$

where $\sigma^* = 2FD (\partial E/\partial c)$ is the effective conductivity due to the minority carrier. Using this value for conductivity (eq. 1) can again be used to estimate the resistance, and limitation by the electrode of the average sustained current.

An admixture of an electrolyte or an electronic conductor is often used to boot the ionic or electronic conductivity of the electrode for use in secondary lithium batteries as the following average parameters.

$$D = 10^{-9} \text{ cm}^2 \text{ s}^{-1} \quad \text{and} \quad (dE / dc) = 50 \text{ V cm}^3 \text{ mole}^{-1} \tag{4}$$

The minority carrier conductivity, in this case due to the Li ion is $\sigma^* = 2.10^{-6} \ \Omega^{-1} \text{ cm}^{-1}$. Clearly, the addition of an electrolyte with $\sigma = 10^{-4} \ \Omega^{-1} \text{ cm}^{-1}$ should have a beneficial effect.

The limitation due to the interface between electrolyte and electrode must be considered. This is normally described simply as a charge transfer, i.e. an area-specific resistance or as an exchange current density. Few values have been documented for solid state interfaces, but resistance values below 1000 $\Omega.\text{mm}^{-2}$. A final comment is included concerning the charge density. This is calculated from the electrode composition charge as usual, leading to the formula :

$$Q_A = zFl / V_M \tag{5}$$

where V_M is the volume of electrode which accomodates one mole of the active ion.

Clearly the optimum thickness of the electrode is the maximum thickness allowed by space considerations and the current density requirements according to (eq.1). Thus the maximum charge density may be calculated from (eq. 5) and (eq. 1).

The relationship between current density, thickness and electrode area for a battery system which satisfies relevant power-weight and power-volume criteria is shown in the power-volume criterium indicates that in principle this requirement can be obtained by a variety of cell configurations (Steele, 1984).

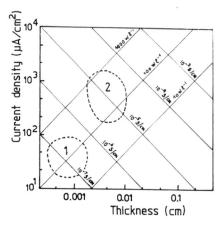

Fig.1 Relationship between current density and thickness for battery systems which satisfy relevant power-volume criteria.

The relationship between current density (mA/cm^2) and the maximum thickness of a cell (cm) is located by the diagonal line which gives the power-volume of 100 W/l. That means that it is possible for a cell which has a power density of 100 W/l operation with an open circuit voltage of 2 V and ohmic losses of 10%. Also displayed on the diagram are the minimum specified conductivity values required for IR losses less than 0.2 V as a function of current density and thickness. Clearly it is possible to have a relatively small electrode area high current density configuration (region II) which can be typified by the hybrid solid-state batteries incorporing composite electrodes. An alternative configuration is to have a very large electrode area requiring only small operating current densities (region I) which is approach adopted for battery incorporation thin film glass electrolytes (fig.1).

1.2. Evaporation techniques

Since passive thin films are widely used in microelectronic devices, a large variety of deposition techniques as electrolytic, sputtering, thermal evaporation, CVD, screen-painting ..., have been developed to obtain electronic conductive (Al, Au, Ni, Si...) or insulating layers of many different materials (SiO_2, Si_3N_4, Ta_2O_5...). During the last decade a new thin film growth technique, the molecular beam epitaxy (MBE) or the molecular beam deposition (MBD) has been set-forthfully to provide improved control over composition, thickness and doping profile in the direction of growth on an atomic scale (Ploog, 1980). Based on these unique features, MBD offers the feasibility of tailoring the electronic properties of the crystal to it desired function by growing high-quality epitaxial thin films with predetermined compositionnal profiles perpendicular to the growth surface. The MBD technique consists in the growth of semiconductor beams of the constituant elements with a crystalline susbtrate surface held at a suitable temperature under ultra-high vacuum (UHV) conditions.

For instance fig.2 shows the schematic MBD system in which it is easy to grow a entire solid state micro-battery, starting from the MBD chamber where the layered cathode material is formed and finishing into the MBD chamber which is the room for the lithium deposition.

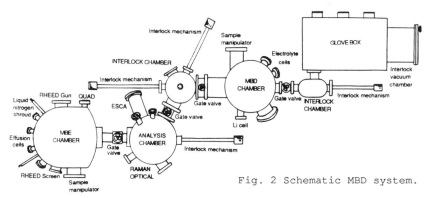

Fig. 2 Schematic MBD system.

It is crucial requirement of improved UHV conditions for successful growth of clean and well controlled layered thin film cathode. In addition, in the same conditions of clean atmosphere, surface studies like the reflection high energy electron diffraction (RHEED) analysis and electron spectroscopies as Auger electron spectroscopy (AES) or photoelectron spectroscopy (XPS) are desirable for most of the investigations of structural and electronic surface states of compound semiconductors. These three techniques take place in the system shown in the fig.2. Useful in-situ analysis can be performed during or after the thin film preparation : for instance, the growth mechanism can be monitored using a mass spectrometer (QUAD) which controls the constituant fluxes, the film structure can be investigated by means of the Raman spectroscopy, in this case the excited laser light and the scattered light are injected or collected through special UHV windows. Others, optical investigations may be envisaged as photoluminescence, visible or near-infrared reflectance, ellipsometry...

Several aspects distinguish MBE from other growth techniques besides the background vacuum ambient (Bachrach, 1980). The most distinguishing characteristics are : (i) a low growth rate, (ii) a low growth temperature (T ≤ 300°C for InSe), (iii) the ability of abrupt cessation or initiation of growth, (iv) a smoothing of the surface of the growing crystal during growth down to atomic steps, (v) the facility for in-situ analysis.

The intensities of the beams incident on the substrate crystal of temperature T_s are controlled by the temperatures of the effusion cells. Provided that the cell aperture is less than the mean free path of vapour molecules within the cell (that is Knudsen effusion), the flux F, in molecules / cm^2.s, is given by the expression :

$$F_i = (a \ P_i \ / \ \pi \ d^2)(2\pi \ m_i \ kT)^{-1/2}. \cos \varphi \qquad (6)$$

where P_i is the equilibrium pressure of species i at the cell temperature T, d is the distance to the substrate, m_i is the atomic mass of effusing species i, a is the area of the cell aperture in cm^2 and φ is the angle between the beam and the normal of the substrate. For typical growth rate of ~ 1 μm / h the fluxes used are approximately 10^{17}-10^{18} molecules / cm^2.s for group III elements and 10^{14}-10^{15} atoms / cm^2.s for group VI elements. By choosing appropriate cell which is generally a graphite or boron-nitride tubular crucible and substrate temperatures, growth films of the desired chemical composition can be obtained. According to the phase diagram of the material, there is only a single compound formation in the corresponding composition. The guidelines for achieving stoichiometric compounds by MBE have been found in the surface chemical dependence of the sticking coefficient of the volatile elements, the sticking parameter S being described the overall process of adsorption, diffusion and growth.

The discussion will describe the In-Se growth conditions in the next paragraph as an example of the MBD crystal growth technique of layered materials.

1.3. Physical properties and growth conditions of thin films

a.   Influence of the substrate-Nucleation and growth

Thin films are interently fragile and, for this reason are normaly retained on the substrate used as a base for deposition to give the film the strength needed for practical applications. The influence of the substrate on thin film cells is complicated by the fact that several films may be required and that each film will act as a substrate for the following film in addition to the original substrate. One of the problems is the adhesion which refers to the practical aspects of how a film sticks to the substrate during use. The peel energy will exceed the interfacial bonding energy by the amount of any work done in the peeling process less any stored energy in the film which contributes to the removal due to the internal stress developed during deposition of the film, which may change with increasing film thickness of the electrolyte cannot be varied over wide limits. Typically, the films less than 0.5 μm thick are prone to shorting but the smallest thickness is derisable in order to procedure high ionic conductance while cathode insertion films of thickness greater than 2-5 μm are difficult to fabricate.

Several methods for improving adhesion are known and have been applied to thin film electrochemical cells. The simplest is substrate cleaning. For instance, in the MBD growth technique, the silicon substrates are cleaned by using a cyclic oxidation and oxide-removal procedure to eliminate carbon and heavy metal contaminants. The final oxide layer which is formed in a $HCl:H_2O_2:H_2O$ (3:1:1) solution (Parker, 1985) is removed in-situ by a thermal treatment at substrate temperature of T_s = 780°C. Improved adhesion can sometimes be obtained by substrate heating, but this method can also lead to adverse effects such as grain growth and, in electrochemical cells can result in high resistance and or shorting (Liang et al., 1969).

The formation of a thin film on a substrate involves several steps. Initially the atoms of the source material impinge on the substrate ; these particles collide and rebound or may stick to the surface. In the case of sticking to the surface, they may still re-evaporated a short time later. However, adsorbed atoms can migrate across the surface of the substrate during the period they remain adsorbed and can form aggregated with others.

As grain growth takes place, the energy of a film changes because of two contributions : the increasing strain energy and the decreasing grain boundary energy. The first one occurs during the coalescence of grain boundaries while the second one as a reduction in boundary area occurs during grain growth, thus there can be an optimum grain size at which the film energy is a minimum (Changhari, 1972).

It is not surprising then that final film properties depend on the composition, partial pressure and temperature of residual gases as well as the temperature and composition of the crystal itself. In particular, it seems that layered materials and more specifically chalcogenide compounds are substrate temperature sensitive, it is that we will discuss next.

b. Substrate temperature dependence

The free surface energy of the film and also the condensation coefficient are mainly influenced by the substrate temperature. It should be kept in mind that at typical pressures of 10^{-6}-10^{-4} Torr the number of residual gas molecules striking the substrate surface is comparable with the number of vapor molecules on the surface. The higher substrate temperature allows to high molecules mobility in the surface while lower T_s permits the grain formation. This is of crucial importance in the layered thin film because further physical properties as the electrical conductivity which is primordial in the case of cathode battery application, must be higher as possible.

A good example can be given for In_2Se_3 thin film (fig. 3) where a comparison between the classical thermal (CTE) evaporation and the thermal flash evaporation (TFE) is shown. In both case the electrical conductivity changes by two orders of magnitude by increasing the substrate temperature. It appears that the TFE technique produces films with higher conductivity which reaches an optimum for T_s = 475 K. It seems that for substrate temperature above 550 K a certain decomposition of the film appears because the chalcogen atoms become more and more volatile.

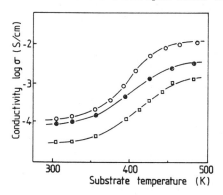

Fig.3 Variation of electrical conductivity against subtrate temperature for thermal (□) and flash (o),(o) evaporation. In_2Se_3 film thicknesses are (o),(□) 500nm and (●) 300nm onto pyrex substrates.

An important factor which must be taking into account during the film growth is the possible formation of different phases for the same compound. Fig. 4 shows the electrical resistance of two In_2Se_3 films as a function of the temperature. It is well-known that In_2Se_3 single crystal exhibits a crystalline phase transition $\alpha \rightarrow \beta$ at 473 K (Julien et al.,1985). During the thermal exploration of the film evaporated at 423 K, at the beginning the resistance decreases continuously with the increasing temperature and an abrupt increase is observed at 473 K because of the phase transition.

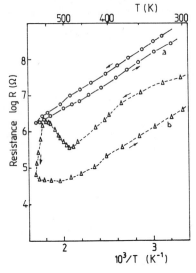

Fig.4 Electrical resistance of In$_2$Se$_3$ thin film as a function of the temperature :
a) film deposited at 423K,
b) film deposited at 500K
(from Eddrief et al.,1984).

While the resistance increases monotonically from the β-phase to room temperature, no phase transition appears and the thin film retains a β-phase resistivity. In thin films deposited onto a pyrex substrate at 500 K the room temperature is higher by two orders of magnitude and the sample remains in the β-phase. From the electrical conductivity the activation energy ΔE can be deduced and values of 0.53 eV, 0.63 eV and 0.69 eV for films deposited in the α-phase, for annealed films at 500 K and for films deposited in the β-phase respectively. Such values can be compared with the forbidden energy gap obtained from optical measurements and it is concluded that the enthalpy of the transition and the transition temperature for the α-->β transformation are dependent on the exact composition of the In$_2$Se$_3$ sample. In the selenium-rich sample the transition temperature is above 473 K while in the selenium-deficient sample the transition temperature is below 473 K.

c.   Thermal annealing treatment of thin films

       After investigations on the morphological and compositional properties of thin films, in the most case the degree of disorder and defects present in the amorphous structure changes due to heat treatment.
       The general features of the density of states of amorphous solids can be understood from the model proposed by Mott and Davis,(1971) and shown in the fig.5

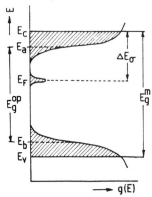

Fig.5 Density of states g(E) as a function of energy E. Regions with oblique lines are localized states.

The mobility gap is defined by E_g^m. The widths of the localized states near the mobility edge, E_c-E_a and E_b-E_v increase with the increase of disorder in the amorphous structure. The shift of the Fermi level E_F is expected by annealing the sample because the density of the gap states changes by annealing. $E_{ca} = E_c-E_a$ is the width of the band tail below the conduction band, ΔE_σ, the activation energy, E_g^{op} the optical energy gap.

The thermal activation energy ΔE_σ is determined by the inverse temperature dependence of the conductivity of the system, according the well-known relation $\sigma \propto \sigma_o \exp(-\Delta E_\sigma/kT)$. The mobility gaps E_g^m are determined experimentally as an energy of the peak value of photoconductivity spectrum. The intercepts of $(\alpha h\omega)^n$ versus $h\omega$ curves extrapolated to $\alpha = 0$ are taken as the value of the optical gap E_g^{op}, where α is the absorption coefficient, that is, the number of absorbed photon number per incident photon. From the annealing behaviours of ΔE_σ and E_g^{op} the effects of annealing on gap states are investigated. Fig.6 shows the annealing temperature dependence of $(\alpha h\omega)^2$ for InSe thin films deposited by the flash evaporation technique. The InSe films are heat treated under Ar atmosphere at different elevated temperatures from 493 K to 573 K. The values of optical gaps E_g^{op} of InSe films are found to increase with temperature of heat treatment, and this effect can be interpreted in terms of the density of state model of Mott and Davis.

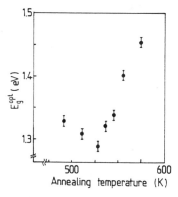

Fig.6 Annealing temperature dependence on the optical band gap of InSe films deposited by flash evaporation technique.
(Julien et al.,1985)

It is evidence that the thermal treatment reduces the disorder present in an amorphous film and that the transition from amorphous to polycrystalline or crystalline state which appears with the formation of grains forming more or less grain boundaries, produces dramatic changes in the physical properties such as optical absorption, electrical transport, etc... More deeper details will be given in the next paragraph, however an other interesting example is given by $MoSe_2$ thin films. The thermal annealing of such films increased the room temperature resistivities by more than one order of magnitude for the specimens obtained by r.f. magnetron sputtering at a low substrate temperature.

The effect of thermal annealing on the electrical resistivity is reported in fig. 7 (Bichsel and Levy, 1985) for a film grown at $T_s=-70°C$ (fig. 7a) and at $T_s=150°C$ (fig. 7b). In such material the post-deposition heat treatment of hot substrate temperature specimen does not lead to a significant change in their electrical properties as it can be done for the III-VI layered compounds. Hall effect measurements on $MoSe_2$ films annealed at 300°C for 20 hours in a vacuum of $6.6. 10^{-6}$ Pa were attempted but with little success. An upper limit for the Hall mobility can be set at 1 cm^2 $V^{-1}s^{-1}$. Similarly, the optical properties are weakly influenced by the process conditions. For the higher substrate temperature at $T_s=150°C$, the films show very similar absorption spectra with a difference which is less than the error in the thickness measurements ; while specimens sputtered at low temperature the absorption edge shifts slightly towards lower energies.

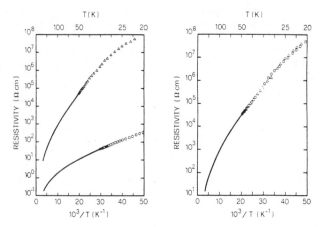

Fig.7 Effect of thermal annealing on the electrical resistivity of sputtered $MoSe_2$ films (O, as-sputteted films, Δ, films annealed at 300°C for 20 h in a vacuum of 6.6. x 10^{-6} Pa) : (a) : $T_s=-70°C$; (b) $T_s=150°C$.(from Bichsel and Levy, 1985).

d. Beam fluxes and growth rate in MBD technique

The film constituent atom arrival rates at the substrate may be calculated from the vapour pressure data. If we assume the vapour in the effusion cell of a MBD chamber is near equilibrium condition, the expression for the number of molecules per second striking the substrate of unit area is valided by (eq. 5). The typical growth rate used in MBD is 1 μm h^{-1} which in the case of MX = InSe requires the metal and chalcogen fluxes both to be 6.10^{14} atoms cm^{-2} s^{-1}. To obtain these fluxes, Knudsen-cell temperatures of T_{In} = 820°C and T_{Se} = 240°C are required, assuming that the sticking coefficient, of the group III and VI elements are unity. On the other hand, elemental sources of III-VI compounds sublimate via the following reactions :

$$M(s) \; ---- \; M(g) \tag{7}$$
$$X(s) \; ---- \; a_1 X(g) + a_2 X_2(g) + ... + a_n X_n \tag{8}$$

where vapour species over the group III elements are monomers and elemental S and Se generate a series of polyatomic molecules. Mass spectrometric studies have shown that Se vapor contains, in descending order of concentration Se_5, Se_2, Se_6, Se_7, Se_3 and Se_8 species (Honing and Kramer, 1969). Fig.8 shows the calculated molecular beam fluxes from Ga, In, Li, S, Se elemental sources.

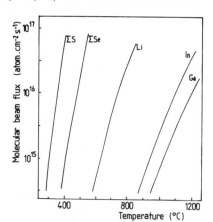

Fig.8   Calculated molecular beam fluxes from Ga, In, Li, Se and S element sources.

The growth rate of the deposited film depends both on the impinging molecular beam fluxes and on the substrate temperature. For

III-VI compounds the growth rate is almost constant in the temperature range below 400°C but it decreases at higher temperatures. These variations would be due to the variation of sticking coefficients of M and X. In order to analyse this dependence of the growth rate on the molecular beam flux, growth conditions and structure of polycrystalline In-Se thin films have been investigated by means of Raman scattering (RS) spectroscopy (Emery el al., 1988).

Fig.9 shows the Raman spectra of two films in the In-Se system which have been grown at T_s = 300°C with various pressure component ratio $R = P_{Se}/P_{In}$. The deposition rate was reduced to 0.3 μm.h^{-1} to obtain better crystallinity. At lower pressure component ratio of R = 2.2, the RS spectrum is identified to be that of InSe by means of the lines at 114, 176, 199 and 228 cm^{-1} (fig.9, curve a). The peaks between 170 and 235 cm^{-1} are very well defined because of a close resonance conditions in which the scattered photon energy is near the second excitonic band at 2.42 eV.

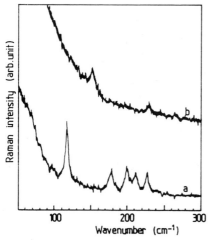

Fig.9 Raman scattering spectra of InSe and In$_2$Se$_3$ thin films recorded using the Ar$^+$ laser line at 514.5 nm. a)InSe film grown at T_s = 300°C with R = 2.2, 800 nm thick, b)γ-In$_2$Se$_3$ film grown at T_s = 300°C with R = 13, 1000 nm thick.

1.4. Preparation and physical properties of insertion thin films

In the previous section, some of the physical properties of thin films of layered materials were mentioned. It is found that the structure and morphology of vacuum-deposited thin films even depend on the evaporation geometry and conditions. In order to obtain electrochemical cells which are operated at relatively high current densities, crystallographic orientation should be expected and electrical properties of films must be optimized in order to reduce interface limitations. In examining the preferred orientations of films it is important to realize that there are three stages of film formation : initial, transition and final growth, and that factors influencing orientation are different for each stage. A discussion of these factors has been given by Davies and Williams (1985). If layered materials are to be used as a cathode film for thin film batteries, it should be used in such a way that good surface flatness and a preferred orientation as the crystallographic c-axis that is parallel to the substrate plane, is achieved.

In this paragraph, the physical properties of thin films of layered materials are reported and mainly, electrical transport phenomena which are of the crucial importance for microbattery applications.

a. TiS$_2$

TiS$_2$ has been well known as a desirable cathode material for secondary lithium cells (Whittingham, 1978). However, no practical process technology has yet been reported for fabricating oriented TiS$_2$ films, excepted the prior of Kanehori et al.(1983). TiS$_2$ having different crystallographic orientations were prepared by chemical vapor deposition (CVD) on substrate at temperature of 450°C. In the first example (film I), a gas composed of 0.1% of TiCl$_4$ and 3% of H$_2$S diluted with He was used. The total pressure was 6kPa. The strong orientation with the c-axis parallel to the substrate surface was found for a film grown under a 600 Pa total gas pressure using a gas composition of 0.6% of $TiCl_4$ and 3.6% of H_2S diluted with Ar. The deposition time of this film (II) was 0.5 hour.

Fig.10 shows the scanning electron micrographs of the surface and cross-section for these two TiS_2 films. The microstructures of these films consist of small, narrow plate-like crystals, each 0.5-3 μm wide and less than 0.1 μm thick, that intersect each other. Porosities of these films were found to correspond to 65% of theoretical density by gravimetric analysis.

|—— 1 μm |—— 2 μm   3.7 μm

Fig.10 Scanning electron micrographs of TiS_2 films
(from Kanehori et al., 1983).

The X-ray diffraction peak corresponding to the (110) pattern was only detected for the film of type II. This result indicates the strong orientation. In the more recent work, Kanehori et al.(1986) improved the fabrication of TiS_2 films by a plasma-enhanced CVD (PCVD) using a 1 W/cm^3 r.f-power density and 1 kPa of total gas pressure. Well oriented TiS_2 films were obtained of 25 μm thick. Such films appear to be a single TiS_2 phase. The influence of the concentration ration R of both gas H_2S to $TiCl_4$ on the composition of the film shows that, for a ratio of 7.4, PCVD-films were nearly stoichiometric : $Ti_{1.03}S_2$. Furthermore, the composition changes slightly from $Ti_{1.03}S_2$ to $Ti_{1.01}S_2$ when R changes from 4 to 12.

It is well-known that the stoichiometry plays an important role in the electrochemical behaviour of Li/ TiS_2 cells (Thompson, 1979). In the metal-rich material, absence of Li ordering is observed and a shift in the Fermi energy above that the stoichiometric TiS_2 occurs. In the dilute limit this shift in E_F can be estimated to be approximately 200 meV for x(Li) = 0.5.

TiS_2 thin films were also prepared by an activated reactive evaporation (ARE) technique (Zehnder et al., 1986). This method involves evaporation in the presence of a reactive gas, utilizing a plasma to activate or catalyze the reaction. To produce TiS_2 thin films, the Ti metal was evaporated in a plasma containing H_2S according to the chemical reaction Ti + 2H_2S ---> TiS_2 + 2H_2 by controling the Ti evaporation rate, the plasma voltage and current, the substrate temperature, H_2S pressure and H_2S flow rate. Thin films deposited at 400°C with a hydrogen sulphide low rate of 80 cm^3.s^{-1} were amorphous and their resistivity was excellent, ranging from 7 x 10^{-4} to 2 x 10^{-3} Ω.cm. The a- TiS_2 films, like bulk TiS_2 exhibit increasing resistivity with increasing temperature. This compares quite well with reported literature values for TiS_2 (Kukkonen et al., 1981).

b. MoS$_2$

MoS_2 which is a member of the group VI A transition metal dichalcogenides belongs to the layer-type structure with the metal atom having trigonal prismatic coordination within the sandwich layer (fig.11).

For many reasons, such as the appropriate energy state distribution and a wide window stability, this compound instead of TiS_2 is considered as one of the most interesting candidate for its use as positive insertion electrode in electrochemical generators (Haering et al., 1980) and lithium secondary batteries have been performed with powdered MoS_2 (Py and Haering, 1983).

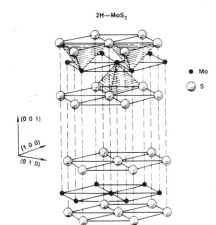

2H—MoS₂

(0 0 1)

(1 0 0)

(0 1 0)

● Mo
◐ S

Fig.11 Molybdenum disulfide (MoS_2) crystal structure.

MoS_2 thin films were prepared by means of radio-frequency (RF) sputtering. A number of substrate were coated, the investigations of thickness, structure and composition were carried out on layers deposited onto silicon, carbon, stainless steel substrates (Dimigen et al., 1985 ; Bichsel et al., 1986 ; Lince and Fleischauer, 1987).

The influence of sputtering conditions on the composition of MoS_x films was investigated by Dimigen et al. They found that a rather low argon pressure as well as a substrate bias potential causes a high sulphur deficiency in the sputtered layers compared with the target material ; this can be compensated by using an Ar-H_2S sputtering atmosphere. Fig.12 shows the stoichiometric ratio N_S/N_{Mo} of the number of sulphur atoms to the number of molybdenum atoms for films deposited onto various substrates as a function of the H_2S pressure during the reactive sputtering under various conditions. It is found that on steel substrate, MoS_x films have the MoS_2-2H structure with $0.95 \leq x \leq 2.1$ regardless of composition, the differences in composition may be due to sulphur vacancies and molybdenum interstitials. This explanation is feasible because the variation in stoichiometry over a wide range as a result of vacancies and interstitials has already been observed for other sulphides (Flahaut, 1978). The influence of the Ar pressure on the composition of the sputtered film is related to the actual H_2S pressure. When a pure molybdenum target is used, an increase in the Ar pressure (3.3 Pa instead of 1.3 Pa) results in an increase in the flux of molybdenum atoms arriving at the substrate surface.

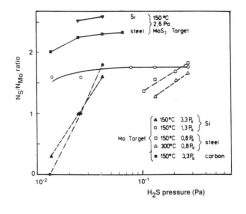

Fig.12 Atomic ratio of MoS_x films versus H_2S pressure during reactive sputtering under various conditions. A MoS_2 target was used as well as a molybdenum target (from Dimigen et al., 1985).

Orientation of sputtered MoS$_2$ films

Sputtered MoS$_2$ films, as observed, always have a tendency to grow onto substrates by forming three distinguished regions : (i) a ridge formation region, (ii) an equiaxed transition zone and (iii) a columnar-fiber-like structure (Spalvins, 1982). This configuration is well-shown in fig.13. It should be pointed out that parts of the columns remain on the substrate, which indicates that the adhesive force between the substrate and the roots of the columns is stronger than the cohesive force in the columns. The strength of the MoS$_2$ films certainly increases with the increase in the density as the structural zones change, with the highest density in the equiaxed zone, as illustred in the wear tracks. In the work of Spalvins, the mean microcrystallite size was about 0.008 µm and microcrystallites with the basal planes perpendicular to the substrate observed by TEM micrograph were displayed the corresponding diffractogram.

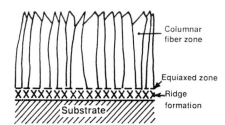

Columnar fiber zone

Equiaxed zone

Ridge formation

Substrate

Fig.13 Sputtered MoS$_2$ film with respect to morphological zones.

Bichsel et al.(1986) observed also that, using a precise control of deposition temperature, specific morphologies have been obtained which were independent of film thickness. At high deposition temperature the films exhibit a pronounced texture with the (002) planes of the MoS$_2$ crystallites perpendicular to the substrate. Recently, Lince and Fleischauer (1987) demonstrate that the MoS$_2$ films deposited on a 440°C stainless steel substrate have a morphology observed by SEM and X-rays which suggests the following process : the particles (crystallites) begin growing in three dimensions, although not isotropically, until they reach a certain thickness in the direction perpendicular to the (001) basal plane, i.e., ~ 300-400 Å ; the particles then grow only perpendicularly to the (hk0) planes. The films appear to grow in the "ridge-formation" (on platelike) region throughout the entire growth process, implying that the growth processes and morphologies of rf-sputtered MoS$_2$ films are sensitive to small changes in sputtering conditions as mentionned by Spalvins' sputtering conditions.

Electrical and optical properties of sputtered MoS$_2$ films

In the case of layered thin films, it should be stressed that films deposited at room temperature and above are polycrystalline with a grain size increasing with increasing substrate temperature. For such polycrystalline semiconductors, the interpretation of the electrical transport results is complicated by the presence of defects, such as dislocation or stacking faults, and of grain boundaries. The effect of grain boundaries is connected with the size of the grains and with the respective orientation of the boundaries. They are influenced by the process conditions and by the heat treatments.

The results of the electrical resistivity measurements obtained on as-sputtered MoS$_2$ films are shown in fig.14. The temperature dependence of the resistivity is given for four different specimens. All samples exhibit a semiconducting behaviour with a pronounced dependence of the transport properties on substrate temperature (T_s). The room temperature resistivity increases from $3.8.10^{-2}$ Ω.cm (T_s = -70°C) to 10.1 Ω.cm (T_s = 150°C) and the 20 K values vary by more than five orders of magnitude (from 6.3 Ω.cm to above 10^6Ω.cm, respectively). Similar influence of the deposition temperature on the electrical properties was observed for magnetron

244

sputtered MoSe$_2$ films (Bichsel and Levy, 1985). However, in the case of the selenide compound, slightly higher values of resistivity have been measured.

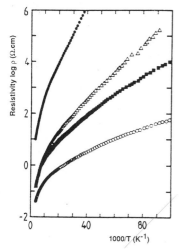

Fig.14 Electrical resistivity versus inverse temperature for as-deposited MoS$_2$ films for different substrate temperature T$_s$ and deposition rate v:
(o) Ts = 150°C, v =4.5 nm/mn (A),
(Δ) Ts = 25°C, v =1.5 nm/mn (B),
(▪) Ts = 25°C, v =3.5 nm/mn (C),
(o) Ts = -70°C, v =2.5 nm/mn (D).
(from Bichsel and Levy, 1986).

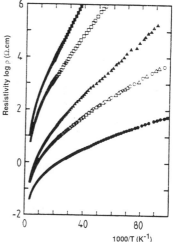

Fig.15 Effect of thermal annealing on the electrical resistivity of sputtered MoS$_2$ films: (▪) as-deposited film of type D; (Δ) type B; (o) type C; and (▪) annealed film (600°C for 10h. in vacuum of 10^{-4}Pa) of type D.
(from Bichsel and Levy,1986).

The difference of the resistivity characteristics between specimens is related to the structural state and demonstrates that the transport mechanisms in MoS$_2$ films are driven by the scattering at inter-crystalline boundaries. The band-bending at grain boundaries, which leads to the formation of potential barriers, is responsible for the high resistivity of samples sputter deposited at high temperature (T$_s$ = 150°C), Hall effect measurements would be useful to point out the influence of grain boundaries on the mobility. At 279 K, μ_H varied between 1-5 cm^{2}v^{-1}s^{-1} and the extreme values of carrier concentrations for the as-sputtered films were 6 x 10^{17} cm^{-3} for specimens at T$_s$ = 150°C and 3 x 10^{19} cm^{-3} for specimens obtained at lower temperature (T$_s$=-70°C).

The effect of post-deposition thermal annealing on the electrical resistivity of sputtered MoS$_2$ films is illustrated in fig.15. All the samples have been annealed at 600°C for 10 h. in a vacuum of 10^{-4} Pa. The resistivity of films obtained at low substrate temperature increases after thermal annealing while that of films sputter deposited at high temperatures decreases after the same treatment. This can be explained on the basis of the results of MoS$_2$ film crystallinity which showed that samples sputtered at room temperature and above are crystalline, whereas deposition at -70°C produces amorphous specimens. The heat treatment of samples obtained at T$_s$ = 150°C seems to decrease the density of structural defects in the films and thereby to reduce the number of trapping sites. A similar behaviour has been observed by Souder and Brodie (1972), but the resistivity of their sputtered films was much higher.

In the case of the sputtered MoS$_2$ films, the extrapolation of the absorption coefficient to zero gives an energy gap of 1.17 eV whereas photoemission data and band structure calculations for MoS$_2$ single crystals have predicted an indirect energy gap of about 1.1 to 1.2 eV and a direct gap of 1.6 eV. The lower obtained value and the functional dependence chosen for the extrapolation is justified, however the optical properties of MoS$_2$ films were weakly influenced by the sputtering process conditions and in spite of the broadening of the absorption edge and the less sharp structure at higher photon energies, the energy gaps obtained for polycrystalline MoS$_2$ films are near to the bulk single-crystal values.

c. MoSe$_2$

MoSe$_2$ is layered material which is semiconducting compound and exhibits interesting characteristics because of their potential electro-chemical applications. MoSe$_2$ is isostructural with MoS$_2$ and their sputtering and substrate conditions on the chemical composition was determined by varying the following parameters : argon pressure, power density, substrate temperature and substrate bias, and analysed using electron spectroscopy (XPS), (Bichsel et al., 1985). Typical ESCA-XPS survey spectrum is shown in fig.16. High resolution spectra were recorded to determine accurately the core level binding energy and the shape of the molybdenum and selenium 3d lines. The chemical shift of these lines in the spectra of the thin films were compared with those of the stoichiometric single-crystal reference and with the 3d binding energies for the pure molybdenum and selenium. The binding energies of the Mo $3d_{3/2}$, Mo $3d_{3/2}$ and Se 3d peaks (averages of the unresolved spin-orbit components) as well as the separation between the Mo $3d_{5/2}$ and Se 3d lines are 232.3, 229.4, 55.2 and 134.2 eV respectively for a film deposited at room temperature.

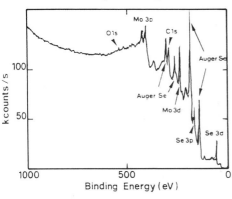

Fig.16 ESCA-XPS survey spectrum of a magnetron-sputtered MoSe$_2$ film (from Bichsel et al.,1985).

ESCA quantitative analysis

The quantitative analysis of the ESCA spectrum is made by calculating the peak area of the strong 3d lines of molybdenum and selenium after background substration (Shirley, 1972). The atomic concentration C_x of each element X(X = Mo or Se) is determined by :
$$C_x = (I_x / S_x) [I_{Mo} / S_{Mo} + I_{Se} / S_{Se}]^{-1} \qquad (9)$$
where I_x is the 3d peak area of element X and S_x the corresponding peak area sensitivity factor. This quantitative method gives accurate results when a stoichiometric standard is used to determine the peak area sensitivity factors. From (eq. 9), the ration of the atomic concentrations of selenium (C_{Se} and molybdenum (C_{Mo}) is given by :
$$C_{Se} / C_{Mo} = (I_{Se} / I_{Mo})(S_{Mo} / S_{Se}) \qquad (10)$$
For the stoichiometric standard $C_{Se} / C_{Mo} = 2$ and I_{Se} / I_{Mo} is calculated from the spectrum of the single crystal. Thus (eq.10) can be used to determine accurately the sensitivity ratio S_{Mo} / S_{Se} for the MoSe$_2$ single crystal compound. If the S_{Mo} / S_{Se} ratio is applied to the thin films, (eq.10) gives the atomic concentrations of Mo and Se in the films.

The results obtained by Bichsel et al., (1985) on MoSe$_2$ thin films show the clear influence of the sputter deposition parameters, in particular, the rise in argon pressure and substrate temperature resulted in a lower selenium concentration. The most dramatic change in chemical composition was observed for the films deposited with a substrate bias which strongly decreases their selenium content.

246

Physical properties of MoSe_2 sputtered films

The main sputtering parameter governing electrical and optical properties of the MoSe$_2$ films was found to be the substrate temperature T$_s$ (Bichsel and Levy, 1985). Lower electrical resistivity was obtained on film deposited at -70°C with no bias substrate voltage but the chemical deposition was found to be 35.9% of Mo and 64.0% of Se. The resistivity exhibits a marked temperature dependence following an exponential Arrhenius law $\rho = \rho_o \exp (\Delta E / kT)$ where the activation energy ΔE is strongly influenced by the process conditions, varying from 43 meV for cooled substrate to 71 meV for hot substrate. In the last case the high resistivity migh be attributed to the enhanced lamellar microstructure of the films which is responsible for the loose packing. The extreme values of the carrier concentrations were 9×10^{16} cm^{-3} for specimens at T$_s$ = 150°C and 7×10^{18} cm^{-3} for specimens at T$_s$ = -70°C.

In contrast with the electrical measurements, the optical properties of the MoSe$_2$ films were only weakly influenced by the process conditions, and the optical gap was determined to be 1.06 eV, compared with the values of 1.09-1.17 eV determined by Goldberg et al. (1975).

d. In-Se compounds

Amorphous In-Se films are currently of interest because they show a fairly large photovoltaic effect (Guesdon et al., 1987a) and attempts have already been made to utilize them as solar cells (Nang et al., 1977). Among the III-VI compounds, indium selenide and indium sesquiselenide are layered semiconductors which have been intercalated by Li and Ag foreign atoms (Balkanski et al., 1981 ; Julien and Hatzikraniotis, 1987 ; Julien et al., 1987).

Several authors have studied amorphous or polycrystalline InSe films, but only a few discuss their composition.

These compounds have been obtained in thin films by different techniques : thermal evaporation, flash (Guesdon et al., 1987b), co-evaporation (Ando and Katsui, 1981) and MBD (Emery et al., 1988).

Guesdon et al., have prepared InSe films by heating a stoichiometric mixture (In$_{50}$Se$_{50}$) of polycrystalline material. The films were deposited by slow evaporation or by flash evaporation on glass substrates heated at different temperatures ranging from 300 to 650 K in a vacuum of below 10^{-3} Pa. In the flash evaporation technique, the charge was powdered and evaporated from a boat temperature at about 1500 K.

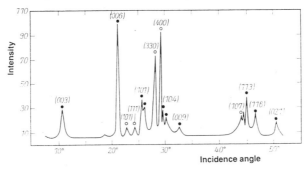

Fig.17 Typical X-ray diffraction spectrum at λ = 0.15405 nm of an In$_x$Se$_{1-x}$ thin film (with thermal treatment at 533 K during 24 h under Ar atmosphere).

The stoichiometric composition of In$_x$Se$_{1-x}$ film was investigated by means X-ray microprobe analysis and nuclear microanalysis which show that the film composition exhibits a large difference with the used starting InSe charge. The flash evaporation technique allows to obtain films close to InSe with reproducible conditions. The films are polycrystalline for substrate temperature higher than 433 K and the crystallite orientation is predominant with the a-axis parallel to the substrate plane. The crystallite size measurements performed by electron microscopy analysis show that for a substrate temperature at T_s = 463 K the crystallite size is around L = 20 nm while annealed films show a crystallite size up to L = 75 nm for annealing treatment at T_a = 533 K during 7 h. X-ray analysis carried out using the Debye-Scherrer diffraction method show that all diffractograms, the (001) lines are absent, this can be due to the fact that films were grown with the a-axis parallel to the substrate plane (fig.17).

In_2Se_3 has a more complex hexagonal layered structure than InSe, it is also a semiconductor with a direct band gap energy of 1.42 eV and an indirect band gap energy of 1.29 eV. Thin films of In_2Se_3 similarly to InSe, were grown onto different substrates, i. e. pyrex, NaCl or mica in the substrate temperature ranging from 50 to 250°C. The optimization of growth conditions was studied by means of X-ray microprobe analysis (Eddrief et al.,1984). The films grow stoichiometrically only up to T_s = 373 K (In = 0.398, Se = 0.602 ± 0.01). The films become more and more In deficient as the substrate temperature is below 323 K while, when grown at temperatures above 473 K, the films remain in the β-phase and the resistivity is higher. The qualitative composition investigated by means of electronic microscopy is represented in fig.18.

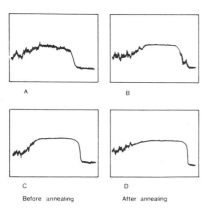

Fig.18 Profile analysis of In_2Se_3 thin film composition in depth from electronic microscopy. A and B are Se spectra before and after annealing, C and D are In spectra before and after annealing at 190°C during 24 h.

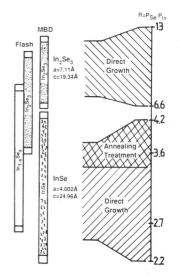

Fig.19 Diagram of the In-Se growth for different partial pressure of MBD effusion cell components.

The analysis of Se and In spectra of the films profile shows difference in depth before and after annealing the film at 463 K for 24 h. Attention is drawn to the fact that Se is highly mobile and an excess of Se was found in the region close to the substrate which corresponds to the beginning of the evaporation process due to the larger volatility of Se. Similar effects have also been observed in other III-VI compounds as GaSe or InSe.

The co-evaporation of In and Se is superior in that it avoids the problems occuring to the use of single-source materials. This is well resolute by the molecular beam deposition (MBD) in ultra-high vacuum conditions. The co-evaporation of In-Se compound exhibits the different phases which are obtained by varying the fluxes of each component. Fig.19 shows the diagram of growth for the different partial pressure ratios R = P_{Se} / P_{In} which is a good approximation of the flux ratio to the composition of deposited films on silicon wafers. The chemical analysis of

InSe films was carried out by X-ray photoelectron spectroscopy on films deposited at $T_s = 200°C$ using a flux ratio of $P_{Se} / P_{In} = 3.6$ and a deposition rate of about 2000 nm/h (fig.20). This spectrum is similar in line position and intensity to that of a γ-InSe bulk crystal grown from a non-stoichiometric melt using the Bridgman method, which is used as a reference sample. The energy separation between the In 4d and Se 3d states (35.6 eV) and their intensity ratio are identical to the reference. A chemical shift of about 0.9-1 eV is also observed between the In $3d_{3/2}$ state from the InSe film and an In film grown in the same chamber, in addition no metallic state was observed in the In 3d region.

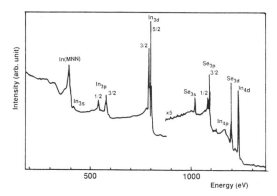

Fig.20 XPS spectrum of an InSe thin film taken using a Mgkα (1253 eV) source and recorded with a spectral resolution of 1 eV.

Optical properties of thin films

Optical measurements are very useful in order to determine the nature of thin film, they include absorption near the fundamental gap, Raman scattering or infrared spectroscopy. The absorption coefficient α can be calculated from the transmission value T using the following expression :

$$T = I / I_o = [(1 - R)^2 \exp(- \alpha d)][1 - R^2 \exp(- 2\alpha d)]^{-1} \qquad (11)$$

where I and I_o are the transmitted and incident light intensity respectively, R the film reflectivity and d the thickness. For high values of αd the second term in the denominator of (eq.11) can be neglected (e.g. if αd ~ 1 and R = 0.3 the error is less than 2%). Thus (eq.11) is simplified to :

$$T = (1 - R)^2 \exp(- \alpha d) \qquad (12)$$

The error made by taking (eq.12) instead of (eq.11) is everytime much less than the errors in thickness measurements which dominate the experimental accuracy. Since the measurements which are made on two samples of different thickness d_1 and d_2, give the absorption coefficient by the simplest formula :

$$\alpha(d_2 - d_1) = \ln(T_1 / T_2) \qquad (13)$$

Fig.21 shows the variation in the absorption coefficient as a function of the incident photon energy for In_xSe_{1-x} films for different substrate temperatures ranging from 300 to 473 K. The single crystal absorption coefficient parallel and perpendicular is also plotted for both polarizations (Guesdon et al., 1987a). The absorption coefficient α of the films decreases exponentially with decreasing photon energy hω according to the law α ~ A exp(hω/E_o) where E_o = 0.14 eV was obtained. Such an exponential tail for α < 10^4 cm^{-1} is characteristic of poly-crystalline semiconductors. This tail makes the occurate determination of the edge position difficult. The absorption coefficients $\alpha_|$ and $\alpha_{//}$ of the single crystal fall very rapidly while the decrease is soft in films. This can be explained by (i) the optical scattering of the defects and the grain boundaries (ii) the extrinsic absorption in the high concentration of defects states in the film grain boundaries (Bagley et al., 1979).

The structural characterization of MBD films is carried out by Raman scattering (RS) spectroscopy (Emery et al., 1988). Fig.22 shows typical RS spectra of InSe films with thickness d ≤ 1000 nm, when the pressure component ratio R = P_{Se} / P_{In} varies from 13 to 2.2. For values higher than R = 6.6 a γ-In$_2$Se$_3$ phase is identified by means of the lines at 153 and 232 cm^{-1} (Kambas et al., 1984), which are clearly visible in fig.22a and 22b. Decreasing R produces only the appearance of an InSe polycrystalline phase. At R = 4.3 this phase is still less pronounced and weak lines are observed in the RS spectrum (fig.22d-22e). The comparison between an as-growth and an annealed film allows the following remarks : first, annealing a film growth with R = 2.7 does not increase its crystallinity while a thermal annealing process clearly improves the crystallinity of InSe grown under a Se-rich condition.

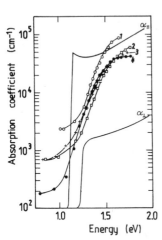

Fig.21 Absorption coefficient as a function of incident photon energy for polycrystalline In$_x$Se$_{1-x}$ deposited at different substrate temperatures (T$_s$ = 300 (1), 433(2), 463(3), and 473 K (4)) compared with α$_{//}$ and α| of single crystal.

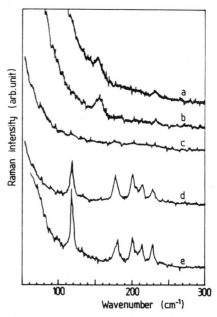

Fig.22 Raman spectra recorded at λ = 514.5 nm of InSe and In$_2$Se$_3$ thin films grown at T$_s$ = 300°C with various R = P$_{Se}$/P$_{In}$:
(a) R = 13, 1000 nm thick
(b) R = 6.6, 600 nm thick
(c) R = 4.2, 500 nm thick
(d) R = 2.7, 800 nm thick.

Electrical properties of In-Se films

In polycrystalline semiconductors the transport of carriers is driven by the scattering mechanisms at inter-crystalline boundaries rather than by intra-crystalline characteristics. Moreover, for compound materials the grain boundaries can be regions of non-stoichiometry which influence the transport properties. Although up to now no exhaustive interpretation of the electrical transport phenomena in polycrystalline compound semiconductors exists, various modelling techniques have been proposed to correlate the transport characteristics with the polycrystallinity of thin films (Kazmerski, 1980). The grain boundaries are considered to have an inherent space charge region due to interfaces states. Band-bending occurs which leads to the formation of potential barriers.

The basic model of potential barrier formation at a grain boundary, successful in several applications, is identical to a Schottky barrier. The boundary can be considered as two back-to-back Schottky barriers with a double depletion zone (fig.23).

Petritz,(1956) proposed a model for the conductance and carrier mobility in polycrystalline compound semiconductors based on thermoionic emission of carriers across a large number of grain boundaries with identical barrier heights. Under a weak potentiel difference V_b this thermoionic emission is given by :

$$j = e\ V_{th}\ N\ \exp(-\ \Phi_B\ /\ KT)\ \exp(-\ e\ \Delta V_b\ /\ kT) \qquad (14)$$

where e is the elementary charge, V_{th} is thermal velocity of carriers, Φ_b is the potential barrier height and N is the carrier concentration in the crystallite, i.e. very low resistivity region.

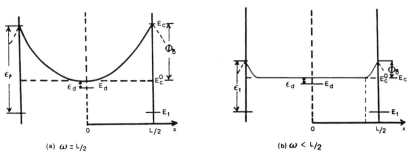

(a) $\omega = L/2$ (b) $\omega < L/2$

Fig.23 Conduction band profile, E_{CO} is the bottom of the conduction band in the grain centre.

     Later, Seto (1975) introduced a grain boundary trapping theory in which free carriers are captured by active trapping sites at the boundary. Potential barriers are formed due to the charge states at the grain boundaries. The band-bending is caused by charge trapping and the mechanism which dominates the transport of carriers between grains is the thermoionic emission across the boundary barriers. More recently, the conductivity phenomena observed in polycrystalline semiconductors have been modelled taking into account the contribution of both the thermoionic emission and tunneling current (Seager and Pike, 1982; Garcia-Cuenca and Morenza, 1985). These models seem to give accurate descriptions of the transport mechanisms which occur at the grain boundaries. However, it should be mentioned that the barrier heights at grain boundaries can vary widely, even in the same sample, probably due to differences in boundary states. Moreover, the grain boundary characteristics differ between large-grained and small-grained specimens.

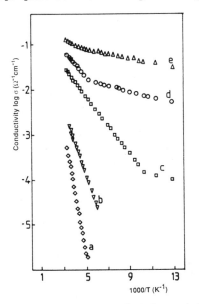

Fig.24 Electrical conductivity versus $10^3/T$ of In_2Se_3 thin film deposited onto pyrex at 220°C for different annealing temperatures. a) as-deposited, b) T_a = 240°C, c) 260°C, d) 280°C and e) 300°C.

     In the case of InSe and In_2Se_3 films, it has been shown that the grain size, morphology and chemical composition are largely influenced by process conditions (Julien et al., 1986 ; Guesdon et al., 1987b).

The effect of annealing In_2Se_3 films on their electrical conductivity has been studied (Julien et al., 1986) and fig.24 shows a plot of the electrical conductivity versus the inverse absolute temperature for five samples which were annealed for 6 h at temperatures ranging from 220 to 300°C. The conductivity and carrier concentration measurements show an exponential behaviour as a function of the temp- erature and the activation energy decreases with increasing annealing temperature.

The Hall mobility, μ_H, shows two distinct domains which occur at different temperatures. Above 180 K, $\ln(\mu)$ increases linearly with the temperature. However, at low temperatures μ_H becomes constant when the temperature decreases and remains between 0.5 and 30 cm^2 $V^{-1}s^{-1}$. The increasing dependence of μ_H on the temperature could suggest a diffusion mechanism caused by ionized impurities but the Conwell and Weisskopf (1950) relation gives a result which is two orders of magnitude larger than the experimental value. Similar data have been measured for InSe films obtained by flash evaporation (fig.25).

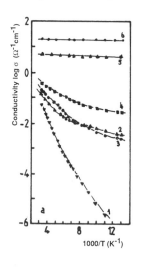

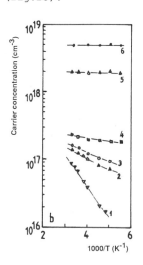

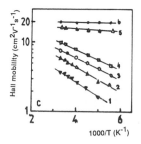

Fig.25 a) electrical conductivity, b) carrier concentration, and c) Hall mobility vs.10^3/T of In_xSe_{1-x} thin films evaporated at T_s = 463 K (1 µm thick) for different annealing temper- atures ranging from 300 K to 573 K : (o) RT, (□) T_a = 493, (o) 513, (Δ) 533, (x) 553, and (Δ) 573 K.

In order to apply the grain boundary model, we assume that polycrystalline films are composed of identical crystallites having a grain size L(cm). We also assume that there is only one type of impurity atom present and that the impurity atoms are totally ionized and uniformly distributed with a concentration N(cm^{-3}). The traps are assumed to be initially neutral and to become charged by trapping a carrier. Using the above assumptions, an abrupt depletion approximation is used to calculate the energy band diagram in the crystallite. In this approximation all the mobile carriers in a region of width W from the grain boundary are trapped by the trapping states, resulting in a depletion zone.

A logarithmic plot of $\mu_B T^{1/2}$ against the reciprocal temperature shows that the Hall mobility μ_B follows approximately an exponential law represented by :

$$\mu_B T^{1/2} = [eL/(2\pi km^*)^{1/2}] \exp(- \Phi_B/kT) \qquad (15)$$

where m^* is the effective mass of carriers. This law is indicative that charge carriers in film are scattered by the potential barrier associated with the intergrain boundaries as proposed by Petritz. Fig.26 shows the variation in $\ln(\mu T^{1/2})$ versus the inverse absolute temperature for In_2Se_3 films. At very low temperature it is obvious that the theoretical values do not agree well with the experimental values. It seems that another fact must be taken into account : a tunneling contribution appears for lower potential barrier heights (Kubovy and Janda, 1977). The grain boundary contains high interface state densities which trap the free carriers in the grains. In this case, the interface charges modify the band structure diagram of the grain and this is the origin of the interface barrier.

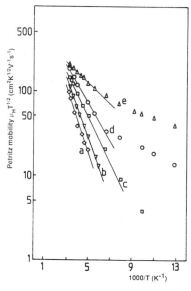

Fig.26 Temperature dependence of $\mu_B T^{1/2}$ at various annealing temperatures for In_2Se_3 films. The solid lines indicate the values given by the Petritz model. The annealing temperatures are identical to those of fig.24.

From the average carrier concentration equation (Seto, 1975), the depletion region width W is given as a function of the potential barrier height ϕ_B :

$$W = (2\phi_B/e^2N)^{1/2} \qquad (16)$$

where N is the doping concentration in cm^{-3}. For a given crystallite size there exist two possible limiting conditions : (i) $W \ll L/2$, the depletion zone extends partially over the grain, and from the Hall measurement the carrier concentration in the grain is obtained from the Petritz model ; (ii) $W = L/2$, the depletion region extends over all the grain and the Hall measurement is lower than the carrier concentration in the grain.

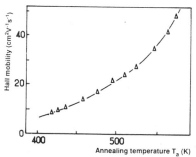

Fig.27:Hall mobility μ_H versus annealing temperature T_a of In_xSe_{1-x} films evaporated at $T_S = 403$ K (1.6 μm thick).

In In_2Se_3 films, for instance, at high annealing temperatures this width is ten times less than the crystallite size measured using an electron microscope.

The set of examined In_xSe_{1-x} polycrystalline films prepared by

vacuum flash evaporation show a relatively low Hall mobility of the order of 10 cm^2 $V^{-1}s^{-1}$. However, the increase of the Hall mobility up to 50 cm^2 V^{-1} s^{-1} is again an effect of thermal annealing, but 600 K seems an upper limit because of the film decomposition appears above (fig.27). Photo-transport parameters have been determined for this
set of thin films. The photocarrier lifetime is of the order of 0.1 ms. The diffusion length is of the order of 60 μm, and the photovoltaic responses of two different photocells show a strong peak at 1.2 eV.

e. WO_3

WO_3 has a nearly cubic structure in which the tungsten atoms are octahedrally coordinated by oxygen (Brown and Banks, 1954). The interstitial space around the cube centre can accomodate ionized metal atoms (M) thereby giving rise to range of non-stoichiometric solids of formula M_xWO_3 which are called tungsten bronzes.

Amorphous tungsten oxide (a-WO_3) films have been known to exhibit electrochromic (EC) properties (Deb, 1973). The physical properties of this material have been studied a great deal in recent years because of its ferroelectric properties and its unusual ability to form tungsten bronzes when metal ions are incorporated in the WO_3 lattice. Films of WO_3 made by different techniques, e.g. thermal evaporation, electron beam deposition (Agnihotry et al., 1986), sputtering (Akram et al., 1987), etc., and with insertion of different metal ions such as Li^+, Na^+ or K^+ instead of H^+, have essentially the same colour.

Films with optimum properties for EC coloration are expected to be fabricated by the sputtering method but interrelations between film properties and sputtering conditions have not yet been clarified in detail.

Tungsten oxide films have been prepared onto substrate at 100°C in the sputtering atmosphere of the Ar-O_2 mixture. The as-deposited films are transparent but for a total pressure lower than 10^{-2} Torr and for an oxygen content of about 5% the films become blue-coloured.

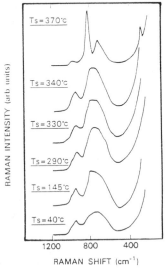

Fig.28 Raman spectra of WO_3 thin films prepared on glass substrate at different temperature T_s (from Shigesato et al., 1987).

Electrical measurements show that the film resistivity is of order of 10^4 Ω cm and optical band gaps determined from absorption measurements are found to be E_g = 3.26 eV and 2.9 eV for amorphous and crystalline films respectively. The electro-optical response time is very sensitive to the preparation conditions, and recently, an aging process by color-bleach cycling is necessary to obtain stable EC characteristics (Akram et al., 1987). The bleaching (or electrochemical colouration) is caused by the simultaneous injection (or extraction) or protons and electrons and in order to obtain the reversible color-bleach cycling, protons and electrons are introduced into the film during the aging process, and they combine with the unsaturated bonds of oxygen. Raman scattering (RS) spectroscopy is a good proble to investigate the WO_3 film crystallinity (Shigesato et al., 1987). Fig.28 shows the RS spectra of WO_3 films prepared on glass substrate at different temperature ranging from 40 to 370°C by electron beam evaporation of powder treated at 700°C for 8 h. prior to evaporation. The broad peak which is observed in the range 600-900 cm^{-1} is attributed to the W-O stretching vibrations which in the crystalline phase are situated at 719 and 807 cm^{-1}.

In an other hand the FTIR spectra of O-H stretching modes before and after the post-annealing can inform on the water content and its bonding state to the framework structure, this with stoichiometry being of crucial importance for EC applications.

f. Other compounds

GaSe and Ga_2Se_3 thin films have been obtained by thermal evaporation of synthesized material with substrate temperature in the range 100°C < T_s < 500°C (Yudasaka and Nakanishi, 1987). The mechanisms of film formation resulting from the evaporation of Ga_2Se_3 and GaSe are considered to be similar to those from the evaporation of In_2Se_3 or InSe. Electrical measurements show that GaSe films have a p-type semiconducting character and a conductivity of 10^{-5} $\Omega^{-1}cm^{-1}$ and an activation energy of 0.21 eV for T_s = 200°C.

Thin film of InTe which have been formed by flash evaporation on different substrate materials at room temperature are amorphous. They transform into the crystalline phase when they are heat treated (Sastry and Reddy, 1980). The mobilities measured on the annealed films varied from 10 to 40 cm^2 $V^{-1}s^{-1}$ for different thickness and substrate temperatures, but the thermoelectric power of annealed InTe films was weakly dependent on temperature.

Sb_2Te_3 as Bi_2Se_3 is a narrow band gap layered semiconductor. Thin films of Sb_2Te_3 have been prepared by a flash evaporation technique (Patel and Patel, 1984). The properties of Sb_2Te_3 thin films were found to be critically dependent on the source and substrate temperature. Films deposited at substrate temperature from RT to 423 K and at source temperature above 1000 K are found to be stoichiometric. The films grown at 423 K have a minimum electrical resistivity (fig.29) while films grown above this temperature are observed to have higher resistivity from a Te-deficiency and formation of Sb_2O_3 in the films.

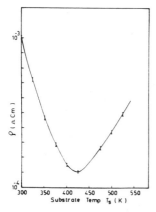

Fig.29 Variation of electrical resistivity ρ of Sb_2Te_3 films with the substrate temperature. (Patel and Patel, 1984)

Bi_2Te_3 films in the thickness range from 300 Å to 3 μm have been prepared by sputtering from a cathode of same composition in a pure argon atmosphere (Francombe, 1964). At low substrate temperatures the stoichiometric Bi_2Te_3 composition is retained in the sputtered deposits, but with increasing temperature, above 250°C, progessive loss of Te occurs, until at 420°C the film composition corresponds approximately to

BiTe. The Te-defficient film structures are identified as metastable β-type phases which are the same structure as Bi_2Te_3.

On amorphous substrates fibre textures are developed which change from a (015) to a (001) type as the temperature is raised. The corresponding single crystal orientations are obtained by epitaxy on the (100) and (111) planes respectively of NaCl crystals. The texture changes are accompanied by striking changes in the crystal habit. Hexagonal platelets appear at low and very high temperatures, while well-developed octahedra are observed at intermediate temperatures. These effects are interpreted by referring the structure to a reduced pseudo-cubic unit cell which assumes a random distribution of Bi and Te atoms on a cubic close-packed lattice. In term of this unit, the low- and high-temperature textures are identified respectively as (100) and (111) nucleation orientations.

Thin films of $SnSe_2$ were prepared by vacuum evaporation at a rate of 50 nm/mn onto substrate of pyrex at RT (Quan and Le Fevre,1982). A two-source evaporation technique was used to prepare doped thin film. Films of $SnSe_2$ having a semiconductor character show a resistivity of 140 Ω.cm at 300 K and are stable below 130°C. The variation in the resistivity with the preparation conditions and with temperature (100-500 K) was studied and an activation energy of 0.2 eV is found. By doping with tin impurities the resistivity is reduced to a few Ω.cm. The high resistivity after annealing or heat treatments and in the case of high substrate temperatures (T_s>170°C) was interpreted as due to ionization and redistribution of impurities and to a partial transformation of $SnSe_2$ to SnSe.

2. PHYSICS OF THIN FILM INTERCALATED CATHODES

Various guest atoms M can be reversibly inserted into layered host thin fims using electrochemical technique. For a univalent ion, the electrode reaction is written :

$$xM^+ + xe^- + < H > \longrightarrow M_x < H > \tag{17}$$

where x is the molar intercalant, < H > the host thin film. This process is accompanied by varying but small volume changes and small, but possibly important changes in the typical characteristics of the film matrix. The inserted atoms behave as donor centres, which are deep or shallow depending upon the temperature, the nature of the atom and the value of x. Not surprisingly, profound changes in optical, electrical and thermodynamic properties occur as the range 0 < x < 1 is traversed. It is the ability to vary x precisely, reversibly and moderately rapidly together with the changes in physical and physicochemical properties, that makes this class of compounds highly interesting from an application viewpoint.

2.1. Basic considerations on interaction process

Thermodynamics

For an electrochemical cell (fig.30), the open circuit voltage V is related to the chemical potential of the intercalation cathode by the expression :

$$V = - (1/ze) (\mu_c - \mu_A) \tag{18}$$

where ze is the charge of the ions in the solution and μ_c, μ_A are the chemical potential of the cathode and the anode respectively.

If N is the number of the accessible sites per unit volume in the cathode, the composition of the cathode is measured as x = n/N where n is the number of the intercalated atoms per unit volume. If we assume that μ_A is contant then the variation of the cell voltage upon intercalation can be understood under the variation of the chemical potential of the cathode.

A simple model to describe the variation of the chemical potential in intercalation compounds has been proposed first by Armand (1980) and recently the best of theories at a quantitative level has been investigated by McKinnon and Haering (1983). This model, the lattice gas model, is adopted by a number of authors and reasonably applies to the most of the MX_2 intercalates.

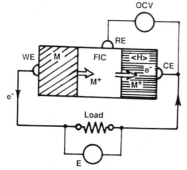

Fig.30 Mode of operation of cell incorporating insertion electrode <H> during the discharge process.
WE = working electrode,
RE = reference electrode
CE = cathodic electrode.

The model is based on the following assumptions :
- the host lattice does not change appreciably during intercalation,
- the number of the host atoms, and hence N, remains constant,
- the host lattice expands freely as the intercalant is added.

Under these assumptions, small changes in the host will lead to effective interactions between the intercalated atoms, which will appear as parameters in the lattice gas model.

We shall suppose, at first, that there is no interaction between the intercalated atoms. Let n_a to be the occupation number of the site a with energy E_a. Since no more than one can be put in a given site, $n_a = 0$ or 1. The energy $E \{n_a\}$ of some distribution $\{n_a\}$ of the atoms over the sites N, is then :

$$E \{n_a\} = n_a . E_a \tag{19}$$

If all the sites have the same energy $E_a = E_o$, than the total energy $E \{n_a\}$ is independent of the distribution $\{n_a\}$, and hence is equal to the average energy E,

$$E \{n_a\} = E = nE_o \tag{20}$$

where n is the number of the intercalated atoms. The entropy S, is just the Boltzman constant k times the logarithm of the number of different ways to place n in distinguishable particles over N sites :

$$S = k \ln [N! / n! (N-n)!] \tag{21}$$

the energy of the system relative to x = 0, is F = E - TS, and the chemical potential $\mu = (\partial F / \partial n)_T$ is :

$$\mu = E_o + kT \ln [x / (1 - x)] \tag{22}$$

If two-body interactions $U_{aa'}$ are also incountered between the intercalant in sites a and a', the energy of the intercalation system of a distribution $\{n_a\}$ has the form :

$$E \{n_a\} = \sum n_a E_a + \sum U_{aa'} n_a n_{a'} \tag{23}$$

If we again assume that all the two-body interactions are identical ($U_{aa'} = U_o$), then the chemical potential of the cathode $\mu_c (x)$ becomes :

$$\mu(x) = E_o + Ux + kT \ln [x / (1-x)] \tag{24}$$

where U is the averaged intercalant-intercalant interaction. It is interesting to note that for repulsive interaction (U > 0), the cell voltage (eq.18) drops more rapidly with x if we encounter the two-body interactions $U_{aa'}$ (eq.23) than for U = 0 (eq.22).

Diffusion

A simplest treatment of the diffusion of M^+ species in thin films can be expressed as following. In the absence of an electric field and at relatively high concentrations (so that space charge effects may be ignored) the particle current j_M of species M in the y direction is given by the relation :

$$j_M = -D (dc/dy) \tag{25}$$

where dc/dy is the particle concentration gradient and D is the chemical diffusion. D is defined as :

$$D = D* (d \ln a_M / d \ln c_M) \tag{26}$$

where $D*$ is the self diffusion coefficient and a_M is the absolute molecular activity of species M. Since the cell (eq.18) potential is given by :

$$V = V_o - (kT/ze) \ln a_{M'}, \qquad (27)$$

where $a_{M'}$ is the relative activity of M in the host material, i.e. the standard state is taken at $x = 1$, and z is the valency of the M^+ in the electrolyte, we have :

$$- dE/dx = (kT/ze) (d \ln a_M/dx) \qquad (28)$$

or

$$-dE/dx = (kT/ze) (d \ln a_M/dx) x^{-1} \qquad (29)$$

when (eq.26) can be rewritten as :

$$D(x) = - D* (x z e /kT) (dE/dx) \qquad (30)$$

Typically $D(x)$ is ten times greater than $D*$. Up to this point the species M have been taken to be electrically neutral. However, in point of fact they are positive ions in a smeared-out sea of electrons, it is most likely that the electron density peaks is the vicinity of A^+ ions of the host matrix. $D*$, and consequently D is expected to fall in value with increasing dilution since there will be less screening of the A^+ nearest neighbour ions of the M^+ and hence a larger Coulomb barrier for M^+ ions on passing through a cube face. Such a simple notion is also consistent with the increasing enthalpy of solution with decreasing x which may be deduced from a discharge curve of the cell (eq.24).

2.2. Performance of thin film cells

Let us, in this paragraph discuss the engineering point of view of thin film cells.

The reactions that occur in charging a battery, as for almost any electrochemical reaction, involve a direct competition between several possible electrode processes. Successful charging of a battery, therefore, involves the suppression of the undesired reactions in favour of the desired reaction. A measure of the ability of a battery to function in a reversible manner can be judged in a practical sense by the current and voltage characteristics of the cell during repetitive charge and discharge. The generalized behaviour of batteries on charge and discharge is shown in fig.31. The departure from the open circuit voltage is the result of irreversible behaviour of the electrode process. The overall cycle energy efficiency of a cell includes the effects of electrode polarization and internal resistance as well as the current efficiency. In fig.31, the increased polarization or energy loss with increased current drain is noted. Also, it may be noted that, on charge, the voltage and total coulombic passage may be signigicantly larger than that on discharge.

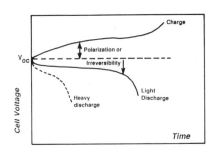

Fig.31 Typical charge-discharge curves of cells.

The theoretical capacity of a cell may be calculated as :

$$Q_{th} = x n F \qquad (31)$$

where x is the theoretical number of moles of reaction associated with the complete discharge of the cell, n is the number of electrons transferred is one mole and F is the Faraday constant F = 26.8 Ah. Some numerical applications are very useful in order to compare three different systems including thin film cells :

1. MoS_2 / Li cell : this "Molicell" which is well developed by Moli

Energy Ltd, delivers a charge capacity of 4 Ah with a voltage range from 2.3 V to 1.3 V. This charge capacity $Q = 4$ Ah x 3600 s = 15.10^3 Cb. corresponds to a theoretical mass m of active cathode of :

$$m = Q \ M/n \ F = 1.5.10^3 \times 140 / F = 22.5 \ g$$

in fact the total mass of this system is given to be 65 g.

2. TiS$_2$ / Li laboratory cell : as it is well-known TiS$_2$ is one the most popular cathode material and its very low atomic weight is favourable in order to obtain a high density storage energy. For a cell including m = 100 mg of active material, the theoretical capacity will be :

$$Q = x \ nF \ m/M = 1 \times 0.1 \times 26.8/110 = 30 \ mAh$$

The measurements on the laboratory cell give a practical capacity of 18 mAh for a discharge deep of 75%.

3. InSe thin film cell : with a cell of 1 cm^2 having a mass of 2 mg the theoretical capacity will be $Q = 0.056$ mAh corresponding to a volumic capacity of $Q_v = 60$ Ah/l.

2.3. Physics of intercalated films

During part two decades, the intercalation of layered compounds have been extensively investigated but until now few studies have been reported onto intercalation of thin films of these layered materials.

a. TiS$_2$

TiS$_2$ thin films cells were fabricated by sequentially depositing a solid electrolyte Li$_{3.6}$Si$_{0.6}$P$_{0.4}$O$_4$ amorphous film (Kanehori et al., 1983). We saw in precedent paragraph that two kinds of films were obtained : type I which has a random orientation of crystallite and type II which is a strong c-axis parallel substrate orientation. The lithium diffusivity in the TiS$_2$ (I) and (II) were 4 x 10^{-13} cm^2 s^{-1} and 1.1 x 10^{-11} cm^2 s^{-1} respectively. These values were too small when compared to that in a TiS$_2$ single crystal, which is about 10^{-8} cm^2 s^{-1} (Whittingham, 1978). The D_{Li} in TiS$_2$ film is influenced by the intrinsic properties of the film, whether orientation, non-stoichiometry or crystallographic imperfection. It is also influenced by a porosity and a contact area between the cathode film and the electrolyte film. Using a liquid electrolyte cell, Li/1M-LiClO$_4$ in PC/TiS$_2$ with PCVD film, the measurements of the apparent chemical diffusion coefficient of Li were evaluated at various Li concentration : x in Li$_x$TiS$_2$ (Kanehori et al., 1986). Furthermore, the activation energies of diffusion at various x were calculated from D value measured in a temperature range of 10-60°C. The results are shown in fig.32. It is found that, in the lower x region, D increased steadily as x is increased. The maximum value, $D = 2 \times 10^{-9}$ cm^2 s^{-1}, is obtained when x is 0.5 ~ 0.6. In higher x region, e.g., x > 0.6, D decreased slightly as x is increased. The chemical diffusion coefficient of Li in TiS$_2$ was previously studied using liquid electrolyte cells. Whittingham (1978) and Winn et al.(1976), who used single crystals as samples, reported that D was about 1 x 10^{-8} cm^2 s^{-1} at a composition of Li$_{1.0}$TiS$_2$ and D was 2.0 x 10^{-8} cm^2 s^{-1} at a composition of Li$_{0.55}$TiS$_2$ respectively. Basu and Worell (1979) who used powder material reported that D was independent of Li concentration and was nearly 5 x 10^{-9} cm^2 s^{-1}. Comparing the data of Kanehori et al. and those measured by using powder material does not seem particularly significant since evaluated D values were affected by electrolyte penetration into the cathode. The D value in the PCVD-film is nearly one order of magnitude lower than that of single crystals. Since the plasma-enhanced CVD is a non-equilibrium process, the PCVD-film might have higher defect concentration in the crystal. The lower D in PCVD-film compared with that of the single crystal may be due to the higher defect concentration in the specimen. In addition, it appears that excess titanium in the PCVD-film may lower D values.

The activation energy of diffusion decreases steadily as x is increasing in the lower x region, e.g. 55 kJ/mol at x = 0.1 and 30 kJ/mol at x = 0.57, and increases slightly in the higher x region. Previous studies which all used the NMR method, reported that the activation energy is less than 30 kJ/mol in a x range of 0 ~ 1 (Silbernagel, 1975 ; Berthier, 1979 ; Matsumoto, 1984).

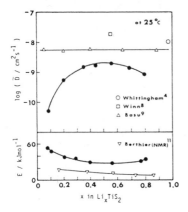

Fig.32 Chemical diffusion coefficient D and activation energy of
diffusion E_a.
(from Kanehori et al., 1986).

The a-TiS$_2$ films produced by the ARE technique were able to both intercalate and deintercalate lithium (Zehnder et al., 1986). The OCV of an a-TiS$_2$/Li cell is 2.7 V while full battery discharge results in a potential of 1.7 V vs lithium. In-situ four-wire resistivity measurements revealed very little change in the electrical resistivity of thin film a-TiS$_2$ as a function of lithium content. Using the constant current pulse method the chemical diffusion coefficient of lithium within the a-TiS$_2$ was estimated (fig.33). The highest diffusion coefficients, 1×10^{-12} to 1×10^{-13} cm^2 s^{-1} occured at open circuit voltages of 1.7 to 2.3 V vs lithium.

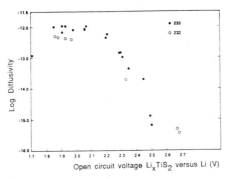

Fig.33 Lithium diffusivity as a
function of open circuit
voltage in thin film TiS$_2$.
(from Zehnder et al., 1986).

Much lower diffusion coefficient were calculated in the voltage range of 2.4 to 2.7 V. These chemical diffusion coefficients are quite identical to those of a CVD film. The fact that amorphous or microcrystalline films yield comparable diffusion coefficients suggests that for thin films of TiS$_2$, a factor other than crystallinity must be significant for providing rapid lithium ion diffusion. At the present time it is not clear whether the contributing factors are related to film chemistry, (e.g., stoichiometry), to aspects of film microstructure (e.g. porosity, grain boundaries, grain morphology) or to a combination of features.

Fig.34 shows the discharge properties of thin film cells (Kanehori et al., 1986) employing a 7 μm thick PCVD-film (full lines) and a 3.7 μm thick CVD film (dasted lines). The discharge capacities (2.5 ~ 1.5 V) or the PCVD-film were nearly 90% of the theoretical capacity at 160-300 mA/m^2, which are large compared to the capacity of CVD-film cell, i.e. 55% - 160 mA/m^2. The apparent chemical diffusion coefficients of PCVD and CVD films into the Li/Li$_{3.6}$Si$_{0.6}$P$_{0.4}$/TiS$_2$ cell were found to be ~ 10^{-10} cm^2 s^{-1} at x = 0.3 from current pulse measurement using these cells. This D value in PCVD-film is about one order of magnitude lower than that measured in a liquid electrolyte cell. This might be due to the interfacial problem in the small contact area between TiS$_2$ and thin film electrolyte.

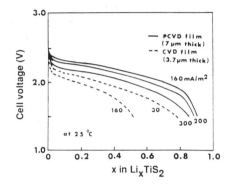

Fig.34 Discharge curves of thin
film solid-state cells
(from Kanehori et al.,1986).

b. InSe

The InSe films evaporated on pyrex are electrochimically active
giving higher initial voltage than that of the bulk InSe (Samaras et al.,
1987). Galvanic cells with the following geometry : Li/LiClO$_4$ - PC/ InSe
were constructed. Each cell, containing three different cathodes with the
same Li reference electrode, gave initial EMF values : E = 3.3 V for an
a-InSe film prepared at T$_s$ = RT and E = 2.9 V for an InSe bulk crystal.
The insertion of Li ions into films was studied first by a
galvanostatic electrochemical method using a three-electrode cell :
lithium metal was used for both counter and reference electrodes. Fig.35
shows the discharge curve of a cell which includes an a-InSe film prepared
at T$_s$ = 433 K and annealed for 50 h. at 475 K. Each step was consisted of
a current pulse of I$_p$ = 1μA during 435 nm, which corresponds to x = 0.05
mole of Li for the estimated cathode mass. When the cell's voltage become
0.5 V lower than the last EMF value, the current pulse was switched for 5
min. Finally the predifined charge (x =0.05) was transferred to the
cathode and the cell was left for relaxation while the voltage of this
cell was recorded up to the point where the voltage of this cell was
recorded up to the point where the slope in OCV became less than 1 mV/h.
The open circuit voltage versus lithium curve of fig.35 shows that a
plateau appears at 1.8 V when the discharge depth reaches x = 0.4. The
cathodes appeared unchanged after electrochemical cycling and no cracking
or exfoliation was observed. Although long term cycling studies were
performed onto similar cell and the ohmic drop was measured as a function
of the number of discharge-charge cycle (fig.36).

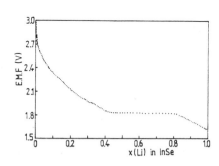

Fig.35 EMF versus x(Li) of InSe
film cathode prepared at T$_s$=433 K
and annealed during 50 h. at 475 K.

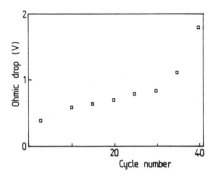

Fig.36 Ohmic drop in a thin film
InSe/Li cell as a function
cycle number.

In order to investigate the cause of the cell's polarization, the
diffusivity of lithium ions within InSe determined as a function of Li
composition. Using the constant current pulse method a plot of overvoltage
versus 1|√t allows a calculation of the chemical diffusion coefficient of

Li. Fig.37 represents the Li diffusivity in InSe film annealed at 433 K for 50 h. as a function of OCV. The highest D value is four orders lower than the corresponding value of bulk InSe.

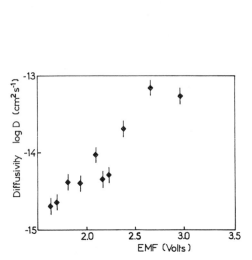

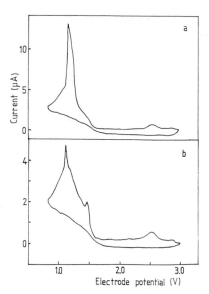

Fig.37 Chemical diffusion coefficient versus cell's EMF using
an InSe thin film as cathode.

Fig.38 Linear sweep voltametry
spectra of InSe thin films:
(a) as-deposited film,
(b) annealed film at T_a = 475 K
for 64 h.

For comparison of annealing treatment between films prepared at T_s = RT we used the linear sweep voltametry (LVS) technique (Dahn and Haering, 1981).

Fig.38 shows the LVS of as-deposited film (curve a) and an annealed film at T_a = 475 K during 64 h.(curve b). We observe that in as-deposited film the peak at 1.2 V is very strong compared to that the annealed film. The apparent non-reversibility of the current peak is partly due to slow kinetics in this lower voltage range.

The data can be analysed by using a simple model. This model treats the grain boundary regions of the film as fast diffusion pathways and the grains of the films as spheres of diameter equal to the mean grain size. The corresponding relation at $Dt/l^2 \geq 0.25$ (Atlung et al., 1979) is given by the following equation :

$$C(l, t) - C_o = (J_o t /l) + (J_o r^2 / 15 l D_r) \qquad (32)$$

where $C(l, t)$ is the surface concentration of M particles at time t, C_o is the initial concentration of M, r is the sphere radius, i.e. half the mean grain size. Thus the intercept of the overvoltage curve versus time is $J_o l/D_r$ and the slope is $J_o l$. The corresponding diffusion coefficient are 2 x 10^{-13} cm^2 s^{-1} at beginning of the discharge of the simple model discussed here, as yet incomplete experiments on films different thickness and grain sizes favour that the distinction between a polycrystalline and an amorphous description of the film is very important. Once the important role of grain boundary diffusion has been established recently on composite electrodes (Dalard et al., 1975) and can be formulated onto films.

Far infrared reflectivity spectra of InSe thin films were performed (Samaras et al., 1987). Fig.39 represents the spectra of an as-deposited film prepared at T_s =RT (curve a), an annealed film at T_a = 475 K during 15 h. (curve b) and an intercalated Li$_x$InSe film with x = 1 (curve c). We observe that the spectrum (a) is that of an amorphous material while in the spectrum (b) a broad band appears in the phonon modes region

$(150-250 \text{ cm}^{-1})$ which is characteristic of a polycrystalline InSe. In both spectra, the large band at higher frequency is mainly due to the electronic disorder in the film. The spectrum of the Li intercalated film (curve c) shows a new broad band appearing at about 400 cm^{-1}. General features of the alkali cation vibrationnal bands are expected in this region involving cooperative cation motion in the matrix.

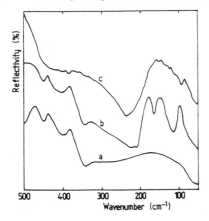

Fig.39 Far-infrared spectra of InSe thin film prepared at T_s = RT (a) as-deposited, (b) annealed at 475 K for 15 h., (c) annealed and intercalated at x = 1.

As it has been observed in TiS_2 films, the morphology and stoichiometry of InSe films are of crucial importance (Julien and Samaras, 1988). Fig.40 shows three different discharge curves of InSe thin films cells having a cathode prepared by flash evaporation. These results allow easy comparison. The $a-In_2Se_3$ film discharge presents a cell voltage lower than those of InSe, and the non-stoichiometric InSe film (curve c) shows that its capacity is about 30% lower and the cell voltage decreases rapidly with x increasing.

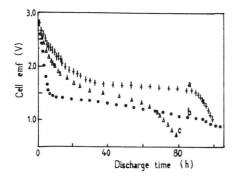

Fig.40 Discharge curves of In-Se thin film cells using 0.1M $LiClO_4$-PC as electrolyte and Li foil as anode : (+) InSe film, (Δ) $InSe_y$ film and (o) In_2Se_3 film.

c. <u>WO_3</u>

Electrochemical behaviour of WO_3 thin films in H^+ or Li^+, Na^+ electrolytes has been studied by many authors (Grandall et al., 1976 ; Dautremont-Smith et al., 1977; Bohnke and Robert, 1982). In their early work, Grandall et al. demonstrate that chemical potential of hydrogen in amorphous H_xWO_3 is a function of the stoichiometric parameter x plus an entropy term which is small for x > 0.4. The generaly accepted mechanism of electrochromic phenomena into WO_3 thin films involves the simultaneous injection of cations and electrons to form a tungsten bronze M_xWO_3, e.g. H_xWO_3, Li_xWO_3. Electrochromic cells of the type $ITO/WO_3/LiClO_4$ (M) in PC/Pt, where WO_3 is an amorphous thin film have been used (Bohnke and Robert, 1982). Different amounts of charges have been injected through these cells by electrochemical means at room temperature. The variations of the WO_3 electrode potential with the injected charge are in good agreement with the assumption of the formation of Li_xWO_3 compounds, according the following reversible reaction :

263

$$WO_3 + xM^{n+} + x\ ne^- \longrightarrow M_xWO_3$$

transparent   blue

Influences of the thickness of the WO_3 film, of the water content in the electrolyte and of the applied potential to the electrode on the optical density of the thin film and on coloration time of the EC display have been discuted by Bohnke et al., (1982a; 1982b).

WO_3 is a pale yellow insulating solid (energy gap E_g = 3 eV), while the tungsten bronzes are strongly coloured and highly conducting since each guest alkali metal atom contributes one conduction electron (Dicken and Whittingham, 1968). The tungsten bronzes of low x value are all blue in transmission since the optical absorption band has a maximum in the near infrared and the absorption falls as we move to higher photon energies associated with the Fermi electron gas : the position of the plasma edge is fairly well predicted using Drude theory. As x increases beyond ~ 0.5 other colours effect become apparent. Fig.41 shows the dependence of colouration i.e. optical density change, on a voltage when a constant charge of Q = 10 mC/cm^2 is injected in the $Au/WO_3/ZrO_2/Au$ cell. In this cell there is no sign of colouration below a threshold voltage of about 1 V.

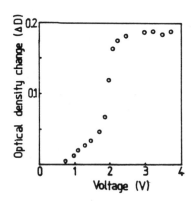

Fig.41 Optical density change versus voltage of the a-WO_3 thin film cell.

d. Others

Thin film fabrication techniques have been employed to produce enlarged prototype cells (area = 0.12 cm^2, thickness < 5 μm) using the V_2O_{5-x} thin film cathode in conjunction with a solid polymeric electrolyte (Upton et al, 1986). V_2O_5 has been described as a pseudo two-dimensional Van der Waals host lattice for lithium (Whittingham, 1976) which contains perovskite-like cavities.

Initial results of the Li/ V_2O_{5-x} cell indicate power densities in excess of 0.05 $mA.cm^{-2}$ and energy densities of approximately 0.4 $J.mm^{-3}$ can be obtained.

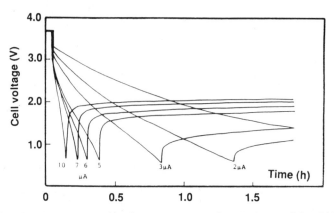

Fig.42 Constant current discharge curves for V_2O_{5-x} thin film cells. (from Upton et al., 1987).

Fig.42 shows the constant current discharge curves, for Li/(PEO)$_8$ LiClO$_4$/ V$_2$O$_{5-x}$ cell, obtained at various current (Upton et al., 1987). Each curve has the form of an immediate IR drop followed by a steady decrease in voltage according to the charge passed. At very low current a maximum capacity corresponding to insertion of about 4 Li per V$_2$O$_5$ is obtained as the voltage decrease to 1.5 V. At greater discharge currents, the curve deviate sharply downwards, such deviations at critical charge values are typical of diffusion limitation.

REFERENCES

Akram, H., Tatsuoka, H., Kitao, M., and Yamada, S., 1987, J.Appl.Phys., 62:2039.

Ando, K., and Katsui, A., 1981, Thin Solid Films, 76:141.

Agnihotry, S.A., Saini, K.K., Saxena, T.K., and Chandra,S., 1986, Thin Solid Films, 141:183.

Armand, M., 1980, in Materials for Advanced Batteries, D.W.Murphy, J.Broadhead and B.C.H.Steele, eds, (Plenum,N.Y) p.145.

Atlung S., West K. and Jacobsen, 1979a, J.Electrochem.Soc.,126:1311.

Atlung S, West K. and Jacobsen, 1979b, in Materials for Advanced Batteries, D.W.Murphy, J.Broadhead and B.C.H.Steele,eds,(Plenum,N.Y) p.275.

Bachrach, R.Z., 1980, in Crystal Growth, B.R.Pamplin, ed., (Pergamon, Oxford), ch.6, p.221.

Bagley, B.G., Aspnes, D.C., and Mogab, C.J., 1979, Bull. Amer. Phys. Soc., 24:363.

Balkanski, M., Kambas, K., Julien, C., Hammerberg, J., and Schleich, D., 1981, Solid State Ionics, 5:384.

Basu, S., and Worrell, W.L., 1979, in Fast Ion Transport in Solid, P.Vashista, J.N.Murdy and G.Shenoy, eds., (North-Holland, Amsterdam), p. 149.

Berthier, C., 1979, in Fast Ion Transport in Solids, P.Vashista,J.N.Mundy and G.Shenoy, eds., (North-Holland, Amsterdam) p. 171.

Bichsel, R., and Levy, F., 1985, Thin Solid Films, 124:75.

Bichsel, R., Buffat, P., and Levy, F., 1986, J. Phys. D : Appl. Phys. 19:1575.

Bichsel, R. and Levy, F., 1986, J. Phys. D : Appl. Phys., 19:1809.

Bichsel, F., Levy, F., and Mathieu, H.J., 1985, Thin Solid Films, 131:87.

Brown, B.W., and Banks, E., 1954, J. Am. Chem. Soc., 76:963.

Bohnke, O., and Robert, G., 1982, Solid State Ionics, 6:115.

Bohnke, O., Bohnke, C., Robert, G., and Carquille, B., 1982, Solid State Ionics, 6:121.

Bohnke, O., Bohnke, C., Robert, G., and Carquille, B., 1982, Solid State Ionics 6:267.

Chaudhari, P., 1972, J.Vac.Sci.Technol., 9:520.

Conwell, E. and Weisskopf, V.F., 1950, Phys. Rev., 77:388.

Crandall, R.S., Wojtowicz, P.J., and Faughhan, B.W., 1976, Solid State Commun., 18:1409.

Dahn, J.R., and Haering, R.R., 1981, Solid State Ionics, 2:19.

Dalard, F., Derso, D., Foscallo, D., and Merienne, J.L., 1985, J. of Power Sources, 14:209.

Dautremont-Smith, W.C., Green, M., and Kang, K.S., 1977, Electrochem. Acta, 22:751.

Davis, G.J., and Williams, D., 1985, in The Technology and Physics of Molecular Beam Epitaxy, E.C.H.Parker, ed.,(Plenum, N.Y), ch.2, p.15.

Deb, S.K., 1973, Phil. Mag., 27:801.

Dickens, P.G., and Whittingham, M.S., 1968, Quart. Rev. Chem. Soc., 22:30.

Dimigen, H., Hubsch, H., Willick, P., and Reichelt, K., 1985, Thin Solid Films, 129:79.

Eddrief, M., Julien, C., Balkanski, M., and Kambas, K., 1984, Materials Letters, 2:432.

Emery, J.Y., Julien, C., Jouanne, M., and Balkanski, M., 1988, Applied

Surface Science, in press.

Flahaut, J., 1978, in Handbook on the Physics and Chemistry of Rare Earth, K.A. Gscheidner, Jr and L. Eyning, eds. (North-Holland, Amsterdam).

Francombe, M.H., 1964, Phil.Mag., 10:989.

Garcia-Cuenca, M.V., and Morenza, J.L., 1985, J.Phys.D:Appl.Phys.,18:2081.

Guesdon, J.P., Kobbi, B., Julien, C., and Balkanski, M., 1987, Phys. Stat.Sol.(a), 102:327.

Guesdon, J.P., Julien, C., Balkanski, M., and Chevy, A., 1987, Phys. Stat.Sol.(a), 101:495.

Goldberg, A.M., Beal, A.R., Levy, F.A., and Davis, E.A., 1975, Philos. Mag., 32:367.

Haering, R.R., Stiles, J.A.R., and Brandt, K., 1980, U.S.Patent, No.4,224, 390.

Honing, R.E., and Kramer, D.A., 1969, RCA Rev., 30:285.

Julien, C., Hatzikraniotis, E., and Balkanski, M., 1987, Solid State Ionics, 22:199.

Julien, C, and Samaras, I., 1988, unpublished.

Julien, C., Eddrief, M., Balkanski, M., Hatzikraniotis, E., and
Kambas, K., 1985, Phys.Stat.Sol.(a), 88:641.

Kambas, K., Julien, C., Jouanne, M., Likforman, A., and Guittard, M., 1984, Phys.Stat.Sol.(b), 124:K 105.

Kanehori, K., Matsumoto, K., Miyauchi, K., and Kudo, T., 1983, Solid State Ionics, 9-10:1445.

Kanehori, K., Ito, Y., Kirino, F., Miyauchi, K., and Kudo, T., 1986, Solid State Ionics, 18-19:818.

Kazmerski, L.L., 1980, in Polycrystalline and Amorphous Thin Films and Devices, Ed. L.L. Kazmerski (Academic, N.Y.) chap. 3.

Kubovy, A. and Janda, M., 1977, Phys.Stat.Sol.(a), 40:225.

Kukkonen, C.A., Kaiser, W.J., Logothetis, E.M., Blumenstock, B.J.,
Schroeder, P.A., Faile, S.P., Colella, R., and Gambold, J., 1981, Phys.Rev., B24:1691.

Lince, J.R., and Fleischauer, P.D., 1987, J. Mat. Res., 2:827.

Liang, C.C., Epstein, J., and Boyle, G.H., 1969, J.Electrocem.Soc., 116:1452.

Matsumoto, K., Nagai, R., Miyauchi, K., Asai, T., and Kawai, S., 1984, Abst.Jpn. Solid State Ionic Meeting, B 103.

McKinnon, W.R. and Haering, R.R., 1983, in Modern Aspects of Electrochemistry, vol.15, R.E. White, J.O.M. Bockris and B.E. Conway, eds,(Plenum, N.Y.) p. 235.

Mott, N.F., and Davis, E.A., 1971, in Electronic Processes in Non-Crystalline Materials, (Clarendon Press, Oxford).

Parker, E.C.H., 1985, in Technology and Physics of Molecular Beam Epitaxy, E.C.H.Parker ed. (Plenum, N.Y).

Patel, T.C., and Patel, P.G., 1984, Materials Letters, 3:46.

Petritz, R.L., 1956, Phys. Rev., 104:1508.

Ploog, K., 1980, in Crystals,Growth, Properties and Applications, H.C.Freyhardt ed.,vol.3 (Springer, Heidelberg) p.73.

Py, M.A., and Haering, R.R., 1983, Can. J. Phys., 61:76.

Quan, D.T., and Le Fevre, J.J., 1982, Thin Solid Films, 98:165

Samaras, I., Guesdon, J.P., Tsakiri, M., Julien, C. and Balkanski, M., 1987, Proc. of 6th Int. Conf. on Solid State Ionics, Garmish, Ext. Abst. P4-7 p. 526.

Sastry, D.V.K. and Reddy, P.J., 1980, Phys.Stat.Sol.,(a), 58:145.

Seager, C.H. and Pike, G.E., 1982, Appl.Phys.Lett., 40:471.

Seto, J.Y.W., 1975, J.Appl.Phys., 46:5247.

Shigesato, Y., Marayama, A., Kamimori, T., and Matsuhiro, K., 1987, Proc.of the 4th Int. Conf. on Solid Films and Surfaces, Hamamatsu (Japon).

Shirley, D.A., 1972, Phys. Rev., B5:4709.

Silbernagel, R.G., 1975, Solid State Commun., 17:361.

Souder, A.D. and Brodie, D.E., 1972, Can. J. Phys., 50:2724.

266

Spalvins, T., 1982, Thin solid Films, 96:17.

Steele, B.C.H., 1984, Solid State Ionics, 12:391.

Thompson, A.H., 1979, in Fast Ion Transport in Solids, P.Vashishta, J.N.Mundy and G.Shenoy (North-Holland, Amsterdam), p.47.

Upton, J.R., Owen, J.R., Rudkin, R.A., Steele, B.C.H., and Benjamin, J.D., 1986, J. Electrochem. Soc., 133:293 C.

Upton, J.R., Owen, J.R., Tufton, P.J., Benjamin, J.D., Steele, B.C.H., and Rudkin, R.A., 1987, Proc. 6th Int. Conf. on Solid State Ionics, Ext. Abstract B7-1, p. 520.

Winn, D.A., Shemilt, J.M., and Steele, B.C.H., 1976, Mat. Res. Bull., 11:559.

Whittingham, M.S., 1978, Prog.Solid St.Chem., 12:41.

Whittingham, M.S., 1976, J. Electrochem. Soc., 123:315.

Yudasaka, M. and Nakanishi, K., 1987, Thin Solid Films, 155:145.

Zehnder, D., Deshpandey, C., Dunn, B., and Bunshah, R.F., 1986, Solid State Ionics, 18-19:813

INSERTION CATHODES FOR SOLID STATE MICROBATTERIES

M.BALKANSKI and   C.JULIEN

Laboratoire de Physique des Solides, associé au CNRS
Université Pierre et Marie Curie
4, place Jussieu 75252 Paris Cedex 05, France

1. INTRODUCTION

The insertion cathode is the place where the electron exchange occurs. In order for this process to be efficient it is necessary to assure an easy access of the ions and a good electron conductivity. The materials which constitute the insertion cathodes have to be, therefore, electron and ion conductors. Good electron conductors are metals and semiconductors and in order to assure good ion conductivity such materials have a specific structures, or channels in which the ion mobility is high[1].

Crystalline host materials can be classified by the dimensionality of the passage way into which the guest ion is inserted (fig.1). The simplest configuration is drain like tunnels which present essentially one-dimensional migration path for guest ions.

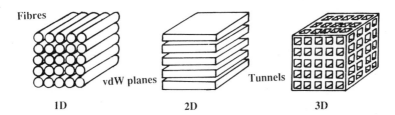

Fig.1. Types of insertion compounds with different structural dimensionality.

Examples of one-dimensional structure are $NbSe_3$, MoS_3, TiS_3, $(CH)_x$.

The most extensively investigated insertion materials are two-dimensional layered compounds.

Example of such materials are MX, MX_2, M_2X_3 compounds with M = In, Ga or V, Ti, Mo, Hf... and X = S, Se and also MPX_3 with M= Fe, Ni, Co and X = S, Se, Te.

These compounds form alternate structures of covalently bounded layers and Van der Waals gaps.

These layer materials are however three-dimensional solids with varying degree of anisotropy in their physical properties. One can establish a general hierarchy of the degree of anisotropy, on the basis of

Solid State Microbatteries
Edited by J. R. Akridge and M. Balkanski
Plenum Press, New York, 1990

force constants determined from phonon studies, for some of these materials, and such a series would be graphite, MoS_2, $MoSe_2$, HfS_2, $HfSe_2$, $SnSe_2$, CdI_2 in decreasing order of anisotropy.

The weak interlayer forces offer the possibility of introducing foreign atoms or molecules between the layers and this is the process of intercalation. In this process the intercalant species occupy interstitial sites, and is generally accompanied by charge transfer between the intercalant species and the host layer. Indeed the tendency for this charge transfer process is the driving force for the intercalation reaction.

Some insertion compounds have intersecting paths that form a three dimensional array and are referred to as framework compounds.

Examples for such insertion compounds are MnO_2, WO_3, V_2O_5, V_6O_{13}, MoO_2, Mo_6S_8.

2. PROPERTIES OF THE HOST MATERIALS

2.1. Structures of insertion materials

2.1.1. Graphite

Pristine graphite is a natural material which consists of honeycomb layers of carbon atoms strongly linked by covalent bonds with only weak interlayer forces as evidenced by the difference between the in-plane and out of plane carbon-carbon distances shown in fig.2. These layers form stacks of hexagonal symmetry with a sequence ABAB. The honeycomb carbon layers have a space group $P6_{3/m} m_c$ and $a_G = 2.46$ Å, $c_G = 6.70$ Å.

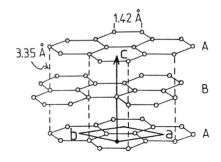

1.42 Å

3.35 Å

Fig.2.Structure of hexagonal graphite showing the ABAB stacking honeycomb carbon layers (from ref.2).

2.1.2. Layered III-VI semiconductors

The III-VI family includes essentially only combinations of type MX and M_2X_3 with M = In, Ga and X = S, Se.

a. *Indium selenide*

Indium monoselenide is a layered semiconductor in which the layers are formed of two InSe formula having the Se atoms on the outside and the In atom on the inside of the layer as shown in fig.3a.,3b which represent the hexagonal structure of the top view of the upper half of the layer.

According to the method of crystal growth different polytypes can be obtained. In fig.4 are shown the polytypes β, ϵ-InSe, with the corresponding first Brillouin zone[3].

The unit cell of the β-structure is of hexagonal type and extends over 2 layers which are rotated by 60° with respect to each other. The crystal space group is C_{6v}^4. The unit cell of the ϵ structure is also of hexagonal type and also extends over two layers but this time these layers are translated in parallel to the Oxy plane with an amplitude of a/2. The ϵ space group is D_{3h}^1 group. These two polytypes comprise eight atoms in the unit cell.

The γ unit cell is rhombohedric and contains four atoms distributed in four adjacent layers. The γ space group is C_{3v}^5.

The rhombohedric representation of γ-InSe is shown in fig.5. The unit cell is rhombohedral with a hexagonal symmetry in the plane of the layers. The layers are stacked in the sequence CBA, ABC separated by the Van der Waals gap.

When InSe crystals are grown by the Bridgman method the polytype γ-3R is obtained with the parameters : a = 4 Å and c = 24.946 Å. The space group of this structure is R3 m.

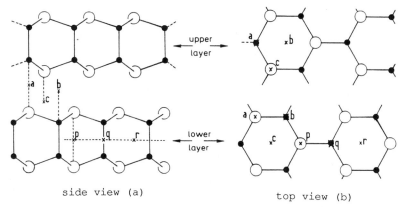

side view (a)   top view (b)

Fig.3.Hexagonal structure of InSe in side view (a) and top view (b) with corresponding possible insertion sites.

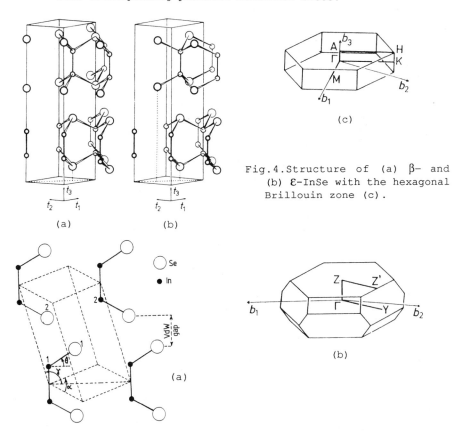

Fig.4.Structure of (a) β- and (b) ε-InSe with the hexagonal Brillouin zone (c).

Fig.5.Elementary cell of γ-InSe (a), and the corresponding first Brillouin zone (b). In the rhombohedral representation a=b=4.002 Å, c=8.75 Å, α=30°, γ=105°, and Θ=29°. The main atom distances are In-In=2.82 Å, In(1)-Se(1)=2.64 Å, and Se(1)-Se(2)=3.86 Å. The van der Waals gap width is 3.05 Å.

b. $III_2^A\ VI_3^B$ _compounds_

A typical example of this group of compounds is In_2Se_3 which exists in three different crystalline varieties α, β and γ with the corresponding phase transition temperatures of 473 K and 923 K for the α-β and β-γ transitions respectively. The α-phase of In_2Se_3 stable below 473 K is a hexagonal layered compound with space group $P6_3$ mc (C_{6v}^4). The alternating selenium and indium layers along the c-axis have a repeat unit of the form Se-In-Se-In-Se and two such units form the unit cell with a hexagonal packing of the form BbCcBCcBbC... ; B, C denote selenium and b, c indium positions. The bonding inside the layers is strongly covalent, while the interlayer interaction (Se-Se) is of the Van der Waals type. The unit cell parameters at room temperature are a = 4.025 Å and c = 19.235 Å.

The γ-phase, stable above 923 K is also a hexagonal structure with space group $P6_1$ and with unit cell parameters a = 7.11 Å, c = 19.34 Å. It can be considered as a distorted wurtzite-like structure in which the In atoms are either tetrahedrally or pentagonally coordinated[4]. In this modification Se and In layers alternate and there are no neighboring Se-Se layers.

c. $V_2^B\ VI_3^B$ _compounds_ Bi_2Se_3

Bismuth selenide, Bi_2Se_3, is a layered narrow-gap, semiconductors[5], E_g = 0.29 eV, which crystallizes in the rhombohedral lattice with D_{3d}^5 (R3m) space group[6] and is isostructural with Bi_2Te_3, and Sb_2Te_3.

The crystal structure of Bi_2Se_3 consists of layers perpendicular to the trigonal c-axis formed by alternating planar arrangements of identical atoms. A layer contains five planes of atoms in the sequence:Se-Bi-Se-Bi-Se. The unit cell is formed by three such layers. The primitive unit cell , containing one formula unit Bi_2Se_3 has the dimensions[6] a_o = 4,14 Å, b_o = 9,84 Å and c_o = 28,64 Å.

The stacking sequence in a close packed structure is given as

... : CaBcA : BcAbC : AbCaB : CaBcA : ...

<---- one unit cell ---->

where capital letters stand for the chalcogen atoms (ABC for Se) and small letters for the positions of Bi atoms (abc for Bi).

Fig.6 presents the crystal structure[7] of Bi_2Se_3. The small circles represent selenium atoms and the large circles bismuth atoms.

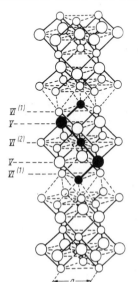

$VI^{(1)}$
V
$VI^{(2)}$
V
$VI^{(1)}$

Fig.6.Structure of Bi_2Se_3.

In the fivefold layers covalent bonding is assumed to dominate the ionic contribution, with the valence states forming highly excited hybrids of the type sp^3d^2 or p^3d^3 around the metal positions. The easy cleavage perpendicular to the trigonal axis is attributed to the weak Se-Se interlayer bonding of Van der Waals character. The energetically accessible sites in the Van der Waals gap, are the usual found between pairs of closely packed atomic layers[8], i.e., one octahedral site and two tetrahedral sites per primitive unit cell.

2.1.3. *Transition metal dichalcogenides*

Almost all transition metals from groups IV, V, VI, VII and VIII of the periodic table form layered structure dichalcogenides. However, only the disulfides and diselenides of groups IV, V and VI have been studied in any detail. Layer metal dechalcogenides, MX_2 are formed with X = S, Se and M = transition metal of the following groups.

Group	IV	V	VI
	Ti	V	Cr
	Zr	Nb	Mo
	Hf	Ta	W

In addition of the hierarchy of the degree of anisotropy, there is also varying degree of ionicity in the bonding between metal atoms and chalcogen atoms. In general the sulfides are the most ionic with the sequence being S > Se > Te. Similarly there is a sequence for the transition metal ions where the order of increasing ionic character, for the group IV for example, is Ti < Zr < Hf.

The layer of transition metal dichalcogenide crystals MX_2 is made of a sheet of metal atoms M, sandwiched between two sheets of chalcogen atoms, X. Each sheet consists of atoms in a hexagonally close-packed network as shown in fig.7.

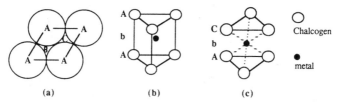

(a)   (b)   (c)

Fig.7.The octahedral and trigonal primatic structures of transition metal dichalcogenides, MX_2.

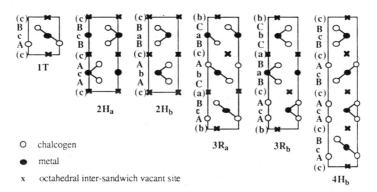

O   chalcogen

● metal

x   octahedral inter-sandwich vacant site

Fig.8.Unit cells of simple polytypes in the 1120 projection
(from ref. 9).

In fig.7a are also shows the three inequivalent hexagonally close-packed positions A, B and C. Two types of coordinations structure are possible, either one or both can form the basic unit of the crystal. In the trigonal prismatic coordination the sanwdich of X-M-X follows the AbA sequence as shown in fig.7b and in the octahedral coordination it follows the AbC sequence shown in fig.7c. Capital letters denote the chalcogen atoms and lower care letters the metal atoms.

In addition to the two intersandwich packings, and a large number of polytypes are known[9]. The six most known polytypes are shown in fig.8.

In designating the polytypes one first indicates the number of sandwishes required to obtain a unit cell perpendicular to the plane of the layers then the overall symmetry of the structure : trigonal T, hexagonal H or rhombohedral R. Lower case subscripts are used to distinguish polytypes otherwise similarly labelled e.g. 2Ha and 2Hb. Thus the simplest polytypes with octahedral coordination, labelled 1T, has a repeat perpendicular to the layer, of the sandwish, and trigonal symmetry and the simplest trigonal prismatic coordination polytypes have two sandwish repeats, and are designated 2Ha and 2Hb. The 2Ha structure is adopted by metallic group V materials such as NbS_2 and stocks the metal atoms directly above each other along the c-axes, whilst in the 2Hb structure, adopted by semiconducting group VI materials such as MoS_2, the metal atoms are staggered. Octahedral interstitial sites between sandwishes are shown in fig.8 in parentheses. There are vacant sites into which intercalation atoms may enter.

2.1.4. *Transition metal phosphorus trisulfides MPS₃*

The MPS_3 compounds can be considered either as metal phosphorus trichalcogenides or as metal triphosphates (fig.9).

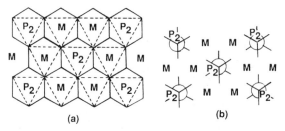

(a)   (b)

Fig.9.(a) Octahedral cation ordering within a MPS_3 layers showing the honeycomb metal network. (b) MPS_3 structure considered as built from M^{2+} and $(P_2S_6)^{4-}$ ions (from ref. 10).

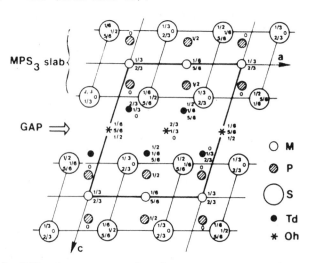

Fig.10. MPS_3 structure projection along the b axis showing slabs and Van der Waals gap sites. T_d and O_h sites for intercalants are shown (from ref. 11).

In the first case, the phases are taken as double chalcogenides of metal and phosphorus and using the formula $M_{2/3}(P_2)_{1/3}S_2$ they can be considered as layered disulfides[10] in which a third of the cation sites are substituted by phosphorus atom pairs P_2 (fig.9a). In the second case the phase is viewed more as a salt constituted from M^{2+} cations $(P_2S_6)^{4-}$ anions (fig.9b).

The MPS_3 crystallizes in the monoclinic symmetry, space group C2/m. The structure is constructed from an essentially sulfur cubic close packed array. Every other layer of octahedral sites are completely filled by the transition metal ions and the phosphorus atoms as P_2 pairs are ordered in a 2:1 ratio[11] (fig.10).

2.1.5. Chevrel-phase compounds

The class of molybdenum chalcogenides known variously as ternary molybdenum chalcogenides or as Chevrel phases are the compounds $A_xMo_6X_8$. Specific feature of these phases[12-13] is a Mo_6X_8 cluster with X (Se or S) atoms approximating a cube and Mo atoms being slightly above the centers of the cube faces. Each cube is rotated approximately 25° about its body diagonal (3 axis). This arrangement of the Mo_6X_8 units leaves large cavities between them. The manner in which the cavities are filled by the A (ternary) atoms depends mainly on the 3 axis, but small ternary atoms occupy tetrahedral sites displaced away from the 3 axis (fig. 11). Lithium chevrel compounds $Li_xMo_6X_8$ can be prepared in electrochemical cells[14] and have four phases for $0 < x < 4$. A first-order transition from x = 1 to 3 between rhombohedral structures has been found[15], and a transition from a rhombohedral structure near x = 3.8 to a triclinic structure near x = 4. This latter transition occurs in two steps ; the intermediate phase has an incommensurate lattice distorsion with a wave vector that varies with x.

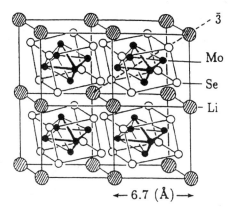

Fig.11. Structure of the Chevrel compound $LiMo_6X_8$.

2.2. Electronic properties

2.2.1. Electronic structure of graphite

Graphite has four valence electrons per carbon atom : three of them ($2s$, $2p_x$, $2p_y$) form covalent in plane s bonds, the fourth electron in the $2p_z$ state gives rise to the conduction and valence π bands. The two dimensional Brillouin zone (BZ)[16] is represented in fig.12.

At the points U, U' related by time reversal operation, the valence and conduction π bands are degenerated by symmetry. In the tight binding approximation, the electron wavefunctions are chosen as linear combinations of two functions built from atomic $2p_z$ orbitals centered at two neighbouring sites. The degenerate π bands are of particular interest since in 2D graphite the Fermi level lies at the degeneracy point U, U'. The dispersion relations for principal directions[16] in the Brillouin zone are shown in fig.13.

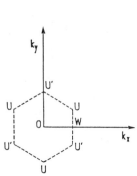

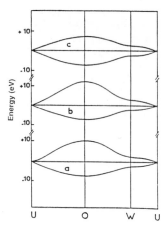

Fig.12.2D-Brillouin zone graphite (from ref. 16).

Fig. 13 Energy band structure of 2D-graphite along the principal directions of the BZ (from ref.16)

A simple representation of the Fermi surface located near the Brillouin zone edges is shown in fig.14.

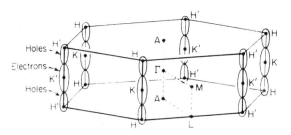

Fig. 14 Graphite BZ showing several high symmetry points and hole Fermi surfaces located along the HK axes.

2.2.2. Electronic structure of III-VI layered compounds

In this group of compounds GaS, GaSe and InSe have a common crystallographic structure : they are all hexagonal layer semiconductors. The theoretical study of the electronic properties of GaS, GaSe and InSe has been developed since longtime on the basis of some drastic simplifications. Early tight binding calculations[17-19] were based on two dimensional approximation. The layer-layer interaction was introduced later using the empirical pseudopotential[20] method with the hope to improve the agreement with the optical stuties. It is generally easy to develop comparative studies on the band structure of compounds within the same family then to study a particular crystal. The absolute precision of a band structure calculation often does not exceed 0.5 •to 1 eV while the relative precision, for example the relative position of two given bands, in two different structures, can be estimated within 10 meV. A discussion on the electronic properties of the layer III-VI compound is carried by Depeursinge[21].

The relative ionicity of the cation-anion bond is in the order InSe, GaS and GaSe. GaSe is the least ionic whereas the In-Se bond is the most ionic.

The band structures of GaSe, InSe, and GaS have been obtained by the "simultaneous band structure calculation" method of Depeursinge[21] and electronic band structure of β-InSe (a), and ε-InSe (b) in the spinless case are shown in fig.15.
For these three compounds the unit cell spreads over two layers.

276

Consequently each electronic state in one layer has a replica in the
adjacent layer . These states are said to be "equivalent electronic
states". The spliting of the bands which correspond to such equivalent
states is due to the interaction between the layers. Some electronic
states are very sensitive to this interaction ; they are said
"three-dimensional states". In particular the two uppermost valence bands
which are strongly split at Γ corresponds to a combination of s and p_z
states over the atoms in the unit cell. As the p_z orbitals of Se spread
far in between the layers, it is not surprising that the corresponding
bands are sensitive to the interlayer interaction. On the contrary, the
"two-dimension bands" are those which are weakly dispersed by the
interaction. The eight bands, for example, situated right under the two
uppermost valence bands at Γ, correspond to a symmetrized configuration of
p_x and p_y orbitals over all the atoms in the unit cell. As the p_x and p_z
orbitals lie parallel to the layers, the corresponding crystal states will
not be very sensitive to the interlayer interaction.

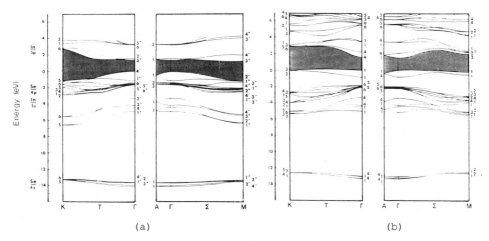

Fig.15. Electronic band structure of β-InSe (a), and $\mathcal{E}$-InSe (b)
in the spinless case (from ref. 21).

The value for the direct gap in this three compounds follows the
rule[22] that the direct fundamental gap is a decreasing function of the
dimension of the unit cell.

Table 1. Direct gap energy and unit cell dimension of III-VI
compounds.

Compound	InSe	GaSe	GaS
V (Å^3)	240.49	194.64	172.52
E_g (eV)	1.4	2.2	3.1

2.2.3. <u>Electronic structure of transition metal dichalcogenides</u>

The electronic structure deduced from the optical properties
and band structure calculations for the group IV, V and VI transition
metal dichalcogenide can be schematized in a density of state
representation[9] shown in fig.16.

The group IV metal, such as Zr, adopts the 1T structure with
octahedral coordination of the metal by the chalcogens. Zr d electrons
are used in bonding to S so that a d^o configuration corresponds to Zr^{4+}
and S^{2-} for a fully ionic model. The crystal field splitting of the d
orbitals gives the actual form of the conduction bands. The valence band
formed by sulfur p orbitals. The p-d semiconduction gap is large for the

more ionic compounds ZrS_2, $ZrSe_2$, HfS_2 and $HfSe_2$ small for TiS_2 and there is a small band overlap for $TiSe_2$ to give semimetallic properties.

The group V metals form polytypes with octahedrally and trigonal-prismatically coordinated sandwishes. The resultant d^1 configuration should give metallic properties with a partially-field d band. However such materials as $1T-TaS_2$ and $1T-TaSe_2$ are particularly susceptible to Fermi-surface driven structural distortions, and even at room temperature only weak metallic properties are observed. In the trigonal prism coordination, the 2H structures show rather better metallic properties and some material, are formed to be superconductors.

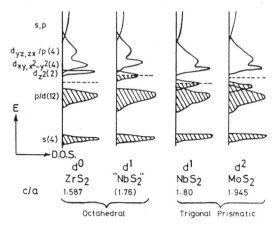

Fig.16.Density of states diagrams for IV, V and VI transition metal dichalcogenides (from ref. 9).

The Fermi surface for the undistorted 1H and 2H polytypes[23] are shown in fig.17. Whilst the Fermi surface for the 1T polytypes is electron-like, with cylindrical electron pockets at the center of the zone faces, the 2H polytype has a hole-like character, with hole surfaces at the zone center and at the vertices of the zone boundary.

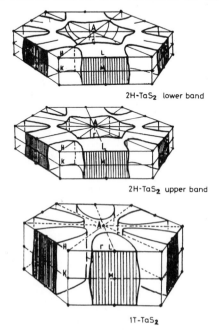

Fig.17. Fermi surfaces for undistorted 1T and 2H-TaS$_2$ (from ref. 23).

The group VI metals usually adopt trigonal prismatic coordination structures, eg. MoS_2, WS_2. These materials are semiconductors.

2.2.4. Electronic structure of Chevrel-phase compounds

Using muffin-tin orbitals and the atomic-sphere approximation, the band structures of Chevrel-phase molybdenum chalcogenides, $A_xMo_6X_8$ have been reported[24] and level schemes are computed for a range of Mo and X potentials, for three Mo_6X_{14} clusters appropriate for the crystal structures of Mo_6S_8 and $PbMo_6S_{7.5}$ respectively. The clusters levels give the positions of the Mo 4d-like bands, while the widths and dispersion are estimated analytically in the tight-binding approximation taking the covalent mixing with the X p states into account. The Fermi level falls in a doubly degenerate E_g band with Mo wave functions of x^2-y^2 character and the E_g bandwidths vary between 65 and 35 mRy in the compounds considered. The E_g band is probably crossed by a five times wider, singly degenerate A_{1g} band of predominantly $3z^2-r^2$ character. The E_g and A_{1g} bands are the only ones crossing the Fermi level in the ternary compounds.

2.3. Transport properties

2.3.1. Electronic properties of graphite

Up to some limiting concentrations the graphite intercalation compounds can be considered as a simple modification of the electronic properties of graphite, it is therefore interesting to consider first pure graphite.

Measurements of the room temperature basal plan conductivity σ_a as a function of the concentration x of different intercalated species suggest two distinc regimes : a dilute regime where σ_a (x) is a simple monotomic function and a concentrated regime where changes in band structure c-axis interactions become important.

The electrical conductivity in graphite shows a high degree of anisotropy. For the parent graphite the anisotropy of metallic conductivity at ambiant temperature is of the order 10^4. This appears to be among the highest known for any crystals.

On introducing various intercalate molecules between the carbon hexagonal network, the electrical conductivity parallel to the layers is always increased. In direction perpendicular to the layers of the host crystal, introduction of donors such as alkali metal atoms usually increases the electrical conductivity likewise often sufficiently to reduce the anisotropy ratio of conductivity σ_a/σ_c below that of the parent graphite. By contrast, introduction of acceptor molecules in graphite nearly always decrease the electrical conductivity σ_c, in the direction of c-axis, as though the layers, when separated by this insertion, becomes more nearly insulator the parent graphite.

2.3.2. Electrical properties of n-type InSe

In layered semiconductors like InSe the crystal anisotropy significantly affects the carrier scattering and consequently the electrical conductivity. The weakness of the binding forces gives rise to a kind of planar defects : stacking faults or interlayer precipitates of impurities parallel to the layers, which do not affect, or affect only very weakly the carrier mobility $\mu_\perp$ in the layer or perpendicular to the c-axis, but strongly reduce the mobility $\mu_{//}$ parallel to the c-axis.

In the temperature range in which lattice scattering predominates, the carrier mobility μ changes with absolute temperature T according to a law : $\mu \propto T^{-\gamma}$, with an exponent γ higher than the value 3/2 corresponding to acoustic phonon scattering. A model[25] has been proposed in which, carriers in layered semiconductors are mainly scattered by homopolar optical phonons polarized along the layer normal. Strong coupling between carriers and homopolar optical phonons arises from the low site symmetry typical of layered structures. The temperature dependence of carrier mobility can be calculated with the use of relaxation-time approximation. The exponent of $\mu \propto T^{-\gamma}$ is than related to the optical phonon energy. For n-type InSe the relation for the In layer mobility is $\mu_\perp \propto T^{-2}$ between 300 and 500 K and the energy of the optical phonon which scatters the electron is 28 meV. Below 300 K one finds $\mu \propto T^{-3/2}$ which is typical for impurity scattering.

Electron mobility in the layers $\mu_\perp$ in InSe reaches values of the order of 10^3 cm^2/V.s at room temperature and higher than 10^4 cm^2/V.s at 60 K which makes possible the investigation of the magnetic field dependence of the Hall coefficient.

The electron mobility in doped n-type InSe above 120 K is explained[26] by the coupling of electrons with a 14.3 meV phonon which may be identified as the low energy A'_1 phonon.

At temperature below 20 K in n-type InSe, the electron behaves as a typical 2D system.

2.3.3. Transport properties of transition metal dichalcogenides

a. semiconductors

The small gap semiconductors, ZrSe$_2$ and TiS$_2$, can only be grown with a small excess of metal present and this results in partial filling of the d conduction band and "extrinsic" semiconductor properties. Crystals, of ZrSe$_2$ [27] show electron concentrations of ~ 5×10^{19} cm^{-3} and TiS$_2$ shows a range from 1×10^{20} up to 3×10^{21} cm^{-3} for crystals with an excess titanium of ~5%.

The temperature dependence of the electron mobility is usually formed to be much stronger than expected for acoustic phonon scattering. The resistivity for materials with a temperature independent extrinsic carrier concentration is usually fitted by the empirical relation $\rho = \rho_o + AT^n$ where ρ_o is the temperature-independent contribution form impurity scattering. The exponent n is found to be as large as 2.3 for samples of TiS$_2$ with low carrier concentration ($1-2 \times 10^{20}$ cm^{-3}) falling to 1.6 for samples with higher concentration (3×10^{21} cm^{-3}) [28]. The model which explains these behaviours considers scattering both within and between pockets enlosed within the Fermi surface. The combination of normal temperature dependence from intrapocket scattering from this interpocket scattering is formed to model the observed behaviour. As the size of electron pockets is increased normal metallic behaviour is restored.

b. semi-metals

TiSe$_2$, TiTe$_2$ and HfTe$_2$ show semimetallic properties with overlap between the top of the chalcogen p valence band and the bottom of the transition metal d conduction band[29] as shown in fig.18.

The transport properties of these materials are determined by both the electrons and holes present, and a more complicated behaviour in the temperature dependence of the Hall coefficient, than in the extrinsic semiconductors, is observed.

At temperatures below 202 K, TiSe$_2$ undergoes a structural distortion to give a low temperature superstructure with doubled in-plane and out-of-plane cell parameters[30-31]. The transition is associated with on anomaly in the resistivity[27] as it can be seen in fig.19.

The superstructure that develops below the transition temperature couples the p band hole pocket at Γ with the d band electron pockets at the zone boundary, and many of the carriers present above the transition are removed. The only remaining carriers which are n-type are those present as a result of excess, intersandwish titanium, as in the case of extrinsic semiconductors. The removal of electrons and holes as result of p-d hybridization following the appearance of the superlattice is shown schematically in fig.18. Different other models have been also proposed to explain the results of structural distortions which are a general feature for many of the transition metal dichalcogenides.

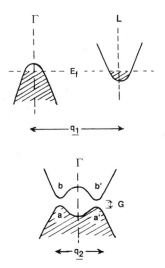

Fig.18.Schematic representation of the effect of electron-hole coupling following superlattice formation in TiSe$_2$ (a) shows semimetallic behaviour, (b) shows semiconducting in the presence of the superlattice (from ref. 29).

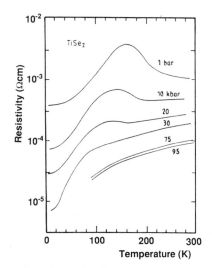

Fig.19.Resistivity versus temperature for TiSe$_2$ at ambient pressure and at elevated pressures (from ref. 27).

3. INTERCALATION

3.1. Graphite

The most widely investigated and the best known intercalation material is graphite[32].

Graphite intercalation compounds are formed by the insertion of atomic or molecular layers of a different chemical species called the intercalant between layers in a graphite host materials. The graphite compounds are of particular physical interest because of their relatively high degree of structural ordering. The most characteristic ordering property is the staging phenomenon characterized by intercalate layers that are periodically arranged in a matrix of graphite layers. Graphite intercalation compounds are thus classified by a stage index n denoting the number of graphite layers between adjacent intercalate layers as shown in fig.20.

Intercalation provides to the host material a means for controlled variation of many physical properties over wide ranges. Because the free carrier concentration of the graphite host is very low (~ 10^{-4} free carriers/atom at room temperature) intercalation with different chemical species and concentrations permits wide variation of the free carrier concentration and thus of the chemical, thermal and magnetic properties of the host material. The electrical conductivity seems to be the most interesting of the fabrication of intercalation compounds with room temperature conductivity exceeding that of copper. The large increase of conductivity in graphite results from a charge transfer from the intercalate layers where the carriers have a low mobility to the graphite layers where the mobility is high.

The graphite intercalants can form donor and acceptor compounds. The most common of the donor compounds are the alkali metal compounds with K, Rb, Cs and Li. The acceptor compounds are those formed with halogen Br$_2$ or halogen mixtures. The intercalation generally causes

281

crystal dilatation along the c-axis. Many of these compounds are unstable in air. They usually require encapsulation.

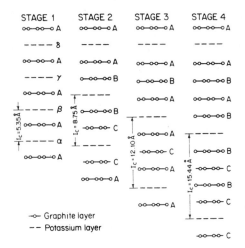

Fig.20.Schematic diagram illustrating the staging phenomenon in graphite-potassium compounds for stages $1 \leq n \leq 4$.

The charge transfer in graphite intercalation compound (GICs)[33], can be discussed on bases of the diagram shown in fig.21. If the intercalant species is an alkali metal and if the electron of the metal is completly transferred to the host the intercalate layer becomes positively charged. Of the various types of layers within the unit cell, shown in fig.22, the conductivity of graphite bounding layers, adjacent to the intercalant, is dominant because of the high carrier density in these layers relative to the graphite interior layers and because of the much higher carrier mobility in the graphite bounding layers relative to the intercalate layers.

The electrical conductivity anisotropy ratio in GICs can be as high as 10^6. The wide range of behaviour in the electrical conductivity occurs because intercalation increases the carrier density while decreasing the carrier mobility.

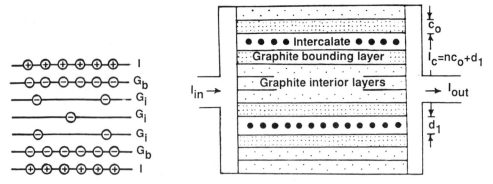

Fig.21.Schematic diagram of the electronic charge distribution in the graphite layers of a stage 5 donor compound.

Fig.22.Additive conductance model for graphite intercalation compounds.

The in-plane conductivity, for a simple stage sample of stage n, can be described qualitatively in terms of a simple phenomenological model[34]. The total conduction per unit cell of length I_c is equal to the sum of the conductance of the constituant layers contained within the unit cell, with the notations in diagram shown in fig.22, we have for $n \geq 2$:

$$I_c(\sigma_a/\sigma_a^\circ) = d_i(\sigma_i/\sigma_a^\circ) + 2c_o'(\sigma_{gb}/\sigma_a^\circ) + (n-2)c_o(\sigma_{gi}/\sigma_a^\circ)$$

and for stage 1

$$I_c(\sigma_a/\sigma_a^\circ) = d_i(\sigma_i/\sigma_a^\circ) + c_o''(\sigma_{gb}/\sigma_a^\circ)$$

where c_o is the graphite interlayer separation and I_c is the repeat distance.

The graphite interieur layer thickness is c_o and d_i is the intercalate layer thickness, c'_o is the thickness of a graphite bounding layer except for stage 1 where is denoted by c''_o. σ_a ' σ_a° ' σ_i ' σ_{gb} and σ_{gi} are respectively the in-plane conductivities of the intercalators compound, pristine graphite, the intercalated layer, the graphite bounding layer and the graphite interior layer whereas the addition of both donor and acceptor intercalants increases the in-plane conductivity σ_a the effect on the c-axis conductivity is different : donor intercalation tends to increases σ_c while acceptor intercalation tends to decrease σ_c. There is a considerable desagreement with regard to the interpretation of the c-axis measurements. It is believed that stacking faults rather than electron-phonon interaction dominate the room temperature scattering. A simple phenomenological model based on the same diagram in fig.22 suggest the addition of the resistance of the various layers of the sample.

$$\rho_c = \Sigma \rho_{ci} t_i / I_c$$

where ρ_{ci} and t_i are respectively the resistivity and the thickness of layer i, I_c, is the repeat distance.

3.1.1. *Temperature dependence of the conductivity*

The temperature dependence of the conductivity is qualitatively different in the intercalation compounds as compared to the graphite host material. In pristine graphite the carrier concentration decreases by a factor of 5 on cooling from room temperature to liquid nitrogen. A large increase in the in-plane mobility is, however, achieved on cooling graphite. As a result the in-plane conductivity σ_a° shows an increase as the temperature is lowered with a dominance of electron-phonon scattering. The temperature dependence of σ_c° is very weakly dominated by defect scattering.

For intercalated material the temperature dependence of the conductivity is directly associated with the temperature dependence of the mobility since the carrier density is dominated by the carriers arising from the charge transfer from the intercalate layers and it is essentially temperature independent. Electron-phonon and electron-electron scattering mechanisms are considered.

3.1.2. *Hall effect*

For a single carrier metal the carrier density is readily determined by measurements of the Hall coefficient R_H where R_H is related to the carrier density N_H by : $N_H = 1/ e c R_H$.

A second method for determination of the carrier density is by measurement of the magnetoresistance $[\rho(H) - \rho(0)] / \rho(0)$ yielding an average Hall mobility $\langle \mu_H \rangle$ given by :

$$\langle \mu_H \rangle = (\Delta\rho c^2 / \rho_o H^2)^{1/2}$$

which when combined with the zero field conductivity $\sigma_o = 1/\rho_o$ yields the carrier density $N_{\mu R} = \sigma_o / e \langle \mu_H \rangle$.

For simple metals that obey a one-carrier model, the carrier densities as determined by the two models are equal $N_{\mu R} = N_H$ other methods such as de Haas-Van Alphen and Shubmikov-de Haas techniques are also used to obtain carrier densities as well as thermopower measurements, optical reflectivity measurements near the plasma edge, electron energy loss spectroscopy, and angle resolved photoemission.

In the case of pristine graphite the Hall densities and mobilities are approximately equal. Acceptor compounds show a positive Hall coefficient and R_H for donors is negative. The high in-plane conductivity for acceptor compound relative to that of donor compounds is associated with much larger value found for the in-plane mobility of the acceptor compounds.

3.1.3. *Free carrier effects on the optical properties in donor and acceptor compounds*

For almost all the graphite intercalation compounds the optical reflectivity exhibits the characteristic Drude edge at the plasma edge ω_p and the optical transmission is characterized by a transmission window near ω_p typical for metallic free carrier absorption. As shown in fig.23 the reflectivity[35] is high for $\omega < \omega_p$ and a sharp edge and low minimum reflectivity is found for $\omega = \omega_p$. In some cases $\varepsilon_{interband}$ makes an important contribution to the dielectric constant and shifts the plasma edge towards the interband frequency. The behaviour of the reflectivity and transmission measurements can be modelled approximately by a single carrier Drude model where the plasma frequency for the polarization $\mathbf{E} \parallel \mathbf{a}$ is given by :

$$\omega_{p,a}^{2} = (4\pi \, N \, e^{2}/m_{opt,a} \, \varepsilon_{core,a}) = (4\pi \, \sigma_{opt,\,a}/\, \tau \, \varepsilon_{core,a})$$

where $m_{opt,a}$ and $\sigma_{opt,a}$ are respectively the in-plane optical effective mass and the real part of the complex optical conductivity, τ is the relaxation time, N is the carrier density and $\varepsilon_{core,a}$ is in plane dielectric constant. If interband effects can be neglected, the reflectivity minimum and the transmission maximum occur near ω_p. The plasma frequency is upshifted with increasing intercalate concentration or decreasing stage[35-36] as illustrated in fig.23 and fig.24.

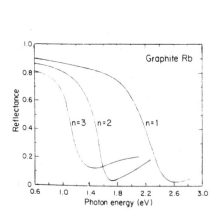

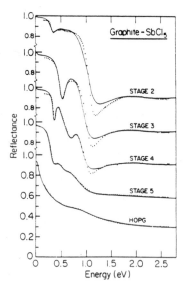

Fig.23. Reflectivity measurements for stage 1, 2 and 3 graphite-Rb donor compounds (from ref.35)

Fig.24. Reflectivity measurements for graphite-SbCl$_5$ acceptor compounds of various stages (from ref.36).

The absolute magnitude for the reflectivity, both for donor and acceptor compounds, increases with decreasing stage, while the magnitude of the reflectivity at the reflectivity minimum decreases with decreasing stage. ω_p for donor compounds occurs at higher frequencies than for acceptor compounds of the same stage this shows a significantly higher carrier generation per intercalant for donors then for acceptors. A quantitative model for optical reflectivity should include free carrier contributions from multiple carrier pockets as well as contributions from interband contributions[37].

3.1.4. Use of graphite for battery cathode material

The use of graphite intercalation compounds for battery materials and electrodes have been discussed[38] and specific application have been attempted. Graphite intercalation compounds are attractive for cathode materials because of their potential for low weight relative to energy density.

Graphite fluorides have been studied[39] for use with lithium anode and non-aqueous electrolyte. The most serious technical problems associated with batteries based on intercalated graphite electrodes concern the metastability of the electrolyte near the anode or cathode surface (where graphite could act as catalyst). The high conductivity of intercalated graphite is an attractive material property for this application.

3.2. Intercalation in InSe and In$_2$Se$_3$

Layered III-VI compounds such as InSe and In$_2$Se$_3$ also provide interesting host materials for intercalation. For example, lithium or silver can be intercalated into these semiconductors by a spontaneous reaction in n-butyl lithium or via an electrochemical process[40-41]. These systems are of interest as a Li solid-state battery. The Li ions in InSe are intercalated on tetrahedral sites, and optical measurements have shown new features in the electronic band structure[42].

3.2.1. Li intercalated indium selenide

Lithium intercalation in the van der Waals gap of InSe is still under investigations. Absorption and luminescence spectra of the two excitons bound to the direct transition between the two uppermost valence bands and the conduction band are not drastically modified after intercalation. The absorption spectra of intercalated single crystals taken at low temperatures are shown in fig.25 and in the photoluminescence spectra of the second transition at 2.55 eV in fig.26. These results are in good agreement with the conductivity measurements on Li intercalated InSe where a variation of three or four orders of magnitude is observed[43]. In these conditions, the Li-intercalated crystal is still a semiconductor, and the observation of the exciton peaks in the Li$_{0.3}$InSe shows in this case only a weak electronic charge transfer in the compound.

The different shifts of the excitonic peaks observed in absorption as well in Raman scattering indicate a modification of the electronic structure. We observe a significant evolution with increasing lithium concentration in the host as it is shown in fig.27. The first exciton E_0 at 1.352 eV shifts towards higher energies while the second exciton E'_1 at 2.52 eV shifts towards lower energies with increasing Li dose. In the photoluminescence spectrum, in the lower energy region of the first set of emission bands, we observe the appearance of new bands upon lithium intercalation. The analysis of the dependences of these peaks on temperature, incident excitation power, and lithium dose, leads to the conclusion that these bands can be attributed to transition associated with strain deformation and optical transition related to Li donor centres.

The reflectivity spectra recorded in far-infrared show that the contribution of the free carriers appears very strongly below 40 cm^{-1}. Taking into account, in first approximation, $\omega_{min} = \omega_p$, a free carrier concentration of about 10^{19} cm^{-3} is found for Li$_2$InSe.

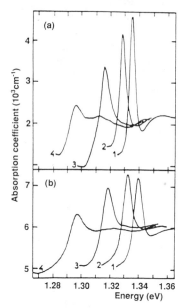

Fig.25. Absorption spectra of pure InSe (a) and Li chemically inter-calated InSe (b), at different temperatures : (1) 10 K, (2) 60 K, (3) 100 K and (4) 150 K.

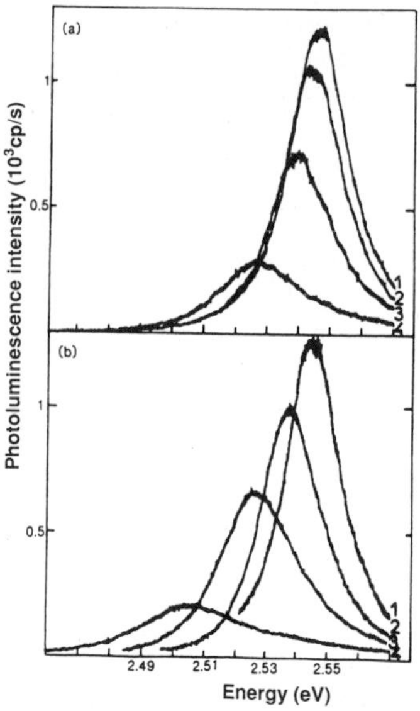

Fig.26. Photoluminescence spectra excited by the 457.9 nm line of an Ar$^+$ laser of InSe (a) and Li$_{0.3}$InSe (b) at different temperatures : (1) 20 K, (2) 40 K, (3) 60 K and (4) 90 K.

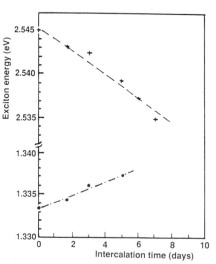

Fig.27. Energy position of excitonic peaks versus Li concentration in Li$_x$InSe.

Preliminary theoretical calculations of the band structure of Li intercalated InSe show significant results compared with those of the pristine material. A simple tight bonding calculation gives a lithium bonding energy of 2.1 eV and shows that the effect of intercalation is to screen out the electrostatic interaction between layers. Obtained results are compared with experimental data in table 2.

By approximating the temperature dependence of the excitonic position with the simple analytical expression :

$$E_T = E_o - B \left[\exp \left(\hbar\omega_{ph} / kT \right) - 1 \right]^{-1}$$

where T is the sample temperature and $h\omega_{ph}$ is the interacting phonon energy, and taking the value of the phonon energy related to the lowest Γ_1^2 optical mode as 117 cm^{-1}, into account, one obtains increasing values for the E_o and B parameters.

Table 2. Experimental and theoretical data of
energy variation in $Li_x InSe$.

	experimental	theory
$\partial E_o / E_o$	3.4×10^{-3}	4.4×10^{-3}
$\partial E_1 / E_1$	-3.1×10^{-3}	-2.3×10^{-3}

In addition, the Raman spectrum of $Li_{0.3}InSe$ shows modifications because the stong resonance of the vibrational modes along the x-axis when the incident energy is in coincidence with the second exciton shifted by the presence of Li ions.

3.2.2. Li intercalated In_2Se_3

The effets of Li insertion in In_2Se_3 have been studied using different methods. Electrochemical properties of the non-stoichiometric $Li_x In_2Se_3$ phase are given in the range $0 \leq x \leq 1$ for either quenched or annealed compounds[44].

The variation of open-circuit-voltage and inverse derivative voltage $-\partial x/\partial V$ as a function of x(Li) content in the host are represented in fig.28. The data are normalized to x = 0 at 2.8 V. The discharge curve is showing horizontal segment at 1.6 V for $0.1 \leq x \leq 0.25$; thus the voltage is continuouly decreasing down to 1.2 V from $0.25 \leq x \leq 1.45$. The minimum in the vicinity of x = 0.25 observed in the inverse derivative voltage suggests an ordering of Li atoms which has been recently confirmed by electron microscopy.

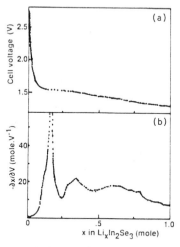

Fig.28. (a) Typical open-circuit-voltage, and (b) Inverse derivative voltage as a function of $Li_x In_2Se_3$.

3.3. Transition metal dichalcogenides

3.3.1. *Intercalation procedure and structures obtained*

The alkali metals Li, Na, K, Rb, Cs and also metals such as Ag and Cu readily enter between the layers to form intercalate complexes. For high concentration of alkali metals and low temperature ordered structures can be obtained.

For the process of intercalation, the lithium versus TiS_2 electrochemical cell has been particularly carefully studied, since it has possible applications for high-energy-density secondary batteries. The variation of the open-cell voltage as a function of Li molar fraction x is shown[45] in fig.29.

The open cell voltage is above 2.4 V at x = 0.1 and falls only to 1.8 V as x approaches 1. The fall is in part due to the increase in the Fermi energy in the Ti d band ; that it falls by only 0.4 V is a consequence of the high value of the density of states in this d band.

As a result of the intercalation there are changes in the lattice parameters, and an expansion in the c direction. To a first approximation this expansion can be attributed to an increase Δc in the van der Waals gap. This assumes that the sandwich height does not alter intercalation, an assumption that is not always exact.

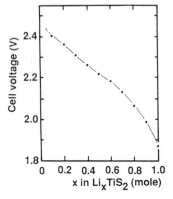

Fig.29. Variation of open-circuit-voltage with x for the system Li_xTiS_2 versus Li (from ref.45).

The occupation of well defined inter-sandwich sites, together with the relatively high diffusion coefficients usually found for alkali metal intercalation complexes allows the possibility of forming ordered or disordered layers of intercalants, and of achieving order-disorder transitions as a function of composition, temperature and pressure. Observation of ordered intercalant superlattices in lithium complexes has been difficult to achieve. High-resolution electrochemical titration of Li_xTiS_2 has revealed[45] peaks in the inverse derivative of the cell EMF at concentration x = 1/9, 1/4, 5/7, 4/5 and 6/7 and associated these with a degree of superlattice ordering at these concentration. Diffuse X-ray scattering experiments[46] have shown superlattices at x=1/4 and x=1/3.

3.3.2. *Electronic properties*

Alkali metals donate most or all of the outer s valence electron to the host material d band. Semiconductor like MoS_2 turns into a good metal[47] after intercalation with Na. Optical spectra for Na_xMoS_2, fig.30 show the disappearance of the exciton levels labelled A and B with increasing x due to Thomas-Fermi screening by the carriers, and the onset of free carrier reflectivity at lower phonon energies.

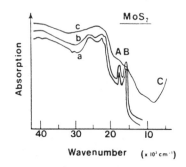

Fig.30. Optical absorption spectra for MoS_2 (a) and Na_xMoS_2 (b,c) (from ref.47)

Electrical and other transport measurements show that the intercalate complex becomes metallic and a superconductor[48].The Hall coefficient for Li_xTiS_2 is almost independent of temperature and the electron concentration, n, estimated from the Hall coefficient R_H (n = $1/R_H$ e) follows closely the value of x confirming that charge transfer is close to complete. Temperature dependent resistivity measurements are shown in fig.31.

In all cases good metallic behaviour is observed with resistivity ratio $\rho(300K)/\rho(4.2\ K)$ of between 6 and 12. Also ploted in the figure is $\log\ (\rho-\rho_o)$ versus $\log$ T where ρ_o is the resistivity obtained by extrapolatation to T = 0.

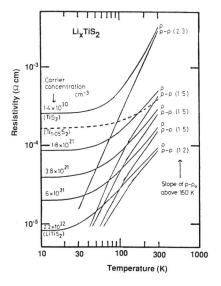

Fig.31. Resistivity versus temperature for Li_xTiS_2 (from ref.27)

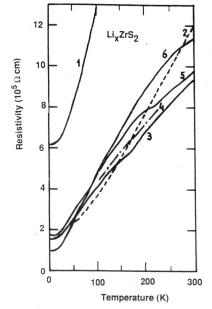

Fig.32. Resistivity versus temperature for Li_xZrS_2. Carrier concentration determined from room temperature Hall coefficient are (1) 2.2×10^{21}, (2) 6.8×10^{21}, (3) 1.16×10^{22}, (4) 1.23×10^{22}, (5) 1.23×10^{22}, and (6) 1.15×10^{22} cm^{-3} (from ref.49).

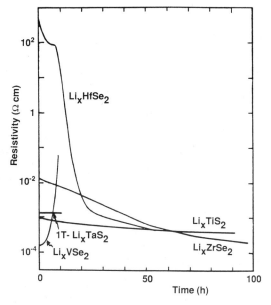

Fig.33. Time dependence of electrical resistivities of TiS_2 $ZrSe_2$, $HfSe_2$, TaS_2 and VSe_2 single crystals immersed in a 1.6 M n-butyl lithium solution. (from ref.50).

For TiS$_2$ the slope of 2.3 has been modelled by a combination of longitudinal acoustic phonon scattering both within and between very small electron pockets at the L points of the Brillouin zone boundary. As the size of the electron pockets is increased with increasing x in Li$_x$TiS$_2$ the temperature dependence of the resistivity above ~ 100 K falls towards the linear dependence expected for a simple metal, since the distinction in size of the phonon wave-vector between intra- and inter-pocket scattering is rapidly lost as the size of the electron pockets becomes comparable to the dimensions of the Brillouin zone.

The temperature dependence of the resistivity[49] for Li$_x$ZrS$_2$ is shown in fig.32. The Hall measurements are weakly temeprature dependent and observations of small transport anomalies may be indicators of lattice distortions.

Similar behaviour is observed in Li$_x$ZrSe$_2$[50]. After a few day's reaction in n-butyl lithium, the electrical resistivity of semiconducting ZrSe$_2$ ($\rho = 10^{-2}\,\Omega$.cm) changes into metallic intercalates ($\rho = 10^{-4}\,\Omega$.cm). But the most spectacular modification in resistivity occurs in HfSe$_2$ which falls by more than six decades due to Li intercalation[50] (fig.33).

3.4. Chevrel-phase compounds

The intercalation of Li in the Chevrel-phase compound Li$_x$Mo$_6$Se$_8$ near room temperature is accurately described by a lattice-gas model for $0<x<1$[51] (fig.34).

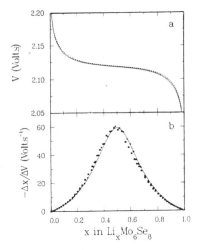

Fig.34. Voltage V and derivative $-\partial x/\partial V$ versus x in a Li/Li$_x$Mo$_6$Se$_8$ electrochemical cell at 38°C (from ref.51).

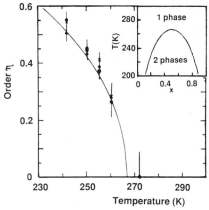

Fig.35. The order parameter η versus temperature. The data points were determined from the (411), (532) and (600) peaks. The solid curve is the prediction. The inset shows the theoretical phase diagram for T_c=267 K (from ref.52).

This model predicts that the Li atoms are uniformly distributed throughout the Mo$_6$Se$_8$ host in the temperature range studied, but they should separate into regions of high and low Li concentration at lower temperatures. The onset temperature varies with x and reaches a maximum value T_c at x = 1/2.

This phase separation, which is driven by the attractive energy of interaction between any two Li atoms in the host, has been observed in metal-hydrogen systems, but there lattice-gas models describe the experiments only qualitatively. The phase separation predicted by mean-field theory in Li$_x$Mo$_6$Se$_8$ is found for the critical temeprature T_c=267 K[52]. The order parameter η, e.g., the difference between the Li concentrations of the coexisting phases has been determined by in-situ X-ray measurements and its variations[52] are shown in fig.35.

Moreover, the size of the interaction energy is consistent with the expansion of the lattice by the ions. Thus all the variations in chemical potential can be explained accurately with a model that completely ignores the electron contribution μ_e. This implis that μ_e is constant over this range of x which is consistent with the interpretation of screening, where the Fermi energy remains fixed as the electrons are added.

REFERENCES

1. see for example:
   a) Intercalation in Layered Materials, M.S.Dresselhaus, ed., NATO-ASI Series, B148 (Plenum, N.Y.,1986).
   b) Chemical Physics of Intercalation, A.P.Legrand and S.Flandrois, eds., NATO-ASI Series, B172 (Plenum, N.Y.,1987).
2. R.Moret, in ref.1a, p.185.
3. S.Nagel, A.Baldereschi, and K.Maschke , J.Phys.C.:Solid State Phys., 12:1625 (1979).
4. A.Likforman, D.Carre, and R.Hillel, Acta Crystal., B34:1 (1978).
5. H.Gobrecht, S.Seeck, and T.Klose, Z.Physik, 190:427 (1966).
6. J.Black, E.M.Conwell, L.Seigle and C.W.Spencer, J.Phys.Chem.Solids, 2:240 (1957).
7. G.R.Hyde, H.A.Beale and I.L.Spain, J.Phys.Chem.Solids, 35:1719 (1974)
8. A.F.Wells, Structural Inorganic Chemistry, (Oxford Univ. Press, London, 1975).
9. W.Y.Liang, in ref.1a, p.31
10. R. Brec, in ref. 1a, p.93.
11. R.Brec, Solid State Ionics, 22:3 (1986).
12. R.Chevrel, M.Sèrgent, and J.Prigent, J.Solid State Chem., 3:515 (1971).
13. K.Yvon, in Current Topics in Materials Science, E.Kaldis, ed., (North-Holland, Amsterdam, 1979), 3:53.
14. P.J.Mulhern, and R.R.Haering, Can.J.Phys., 62:527 (1984).
15. W.R.McKinnon, and J.R.Dahn, Phys.Rev., B31:3084 (1985).
16. C.Rigaux, in ref.1a, p.235.
17. G.Fisher, Helv. Phys. Acta, 36:314 (1963).
18. F.Bassani, and G.Pastori, Nuovo Cimento, 50:95 (1967).
19. H.Kamimura, and K.Nakao, Phys.Rev., 22:1313 (1969).
20. M.Schlüter, Nuovo Cimento, B13:313 (1973).
21. Y.Depeursinge, E.Doni, R.Girlanda, A.Baldereschi, and K.Maschke, Solid State Commun., 27:1449 (1978).
22. F.Bassani, D.Greenaway and G.Fisher, Proc.of the Int. Conf. on Physics of Semiconductors,(Dunod, Paris, 1964) p.51.
23. J.A.Wilson, F.J.Di Salvo, and S.Mahajan, Adv.Phys., 24:117 (1975).
24. O.K.Andersen, W.Klose, and H.Nohl, Phys.Rev., B17:1209 (1978).
25. R.Fivaz, and E.Mooser, Phys. Rev., 163:743 (1967).
26. A.Segura, F.Power, A.Cantarero, W.Krause and A. Chevy, Phys., B29:5708 (1984).
27. P.C.Klipstein, and H.R.Friend, J.Phys.C:Solid State Phys., 17:2713 (1984).
28. P.C.Klipstein, A.G.Bagnall, W.Y.Liang, E.A.Marseglia and R.H. Friend, J.Phys C:Solid State Phys., 14:4067 (1981).
29. R.H.Friend, and A.D.Yoffe, Adv.Phys., 36:1 (1987).
30. J.A. Wilson, and S. Mahajan, Commun. Phys., 2:23 (1976).
31. F.J.DiSalvo, D.E.Mougton and J.W.Waszczak, Phys. Rev., B14:4321 (1976).
32. M.S.Dresselhaus, and G.Dresselhaus, Adv.Phys., 30:139 (1981).
33. M.S.Dresselhaus, in ref.1a, p.1.
34. M.S.Dresselhaus and S.Y.Leung, Ext. Abstr. of the 14th Bien.Conf. on Carbon, (Pensilvania St.Univ.) p. 496.
35. D.Guérard, G.M.T.Foley, M.Zanini and J.E.Fisher, Nuovo Cimento, B38:410 (1977).
36. P.C.Eklund, D.S.Smith and V.R.K.Murphy, Synth. Metals, 3:111 (1981).

37. J.Blinowski, H.H.Nguyen, C.Rigaux, J.P.Vieren, R.Le Toullec, G. Furdin, A.Hérold and J.Mélin, J. Phys.(Paris), 41:47 (1980).
38. M.Armand, and P.Touzain, Mater.Sci.Eng., 31:319 (1977).
39. M.S.Whittingham, J. Electrochem. Soc., 121:1308 (1975).
40. C.Julien, E.Hatzikraniotis, K.Paraskevopoulos, A.Chevy, and M.Balkanski, Solid State Ionics, 18/19:859 (1986).
41. C.Julien, E.Hatzikraniotis, and M.Balkanski, Solid State Ionics, 22:199 (1987).
42. C.Julien, P.A.Burret, M.Jouanne, and M.Balkanski, Solid State Ionics, 27 (1988) in press.
43. C.Julien,and E.Hatzikraniotis, Mater.Letters, 5:134 (1987).
44. C.Julien, I.Samaras, and M.Balkanski, Mat.Sci.Eng., B2 (1988).
45. A.H.Thompson, Phys. Rev. Lett., 40, 1511 (1978).
46. T.Hibma, Solid State Commun., 33:445 (1980).
47. J.V.Acrivos, W.Y.Liang, J.A.Wilson and A.D.Yoffe, J.Phys.C:Solid State Phys., 4:418 (1971).
48. R.B. Somoano and A. Rembaum, Phys. Rev. Lett., 27:402 (1971).
49. N.Ahmad, P.C.Klipstein, D.S.Obertelli, E.A.Marseglia, and R.H.Friend, J.Phys.C:Solid State Phys., 20:4105 (1987).
50. Y.Onuki, R.Inada, S.Tanuma, S.Yamanaka, and H.Kamimura, J.Phys.Soc.Jpn, 51:880 (1982).
51. J.R.Dahn, W.R.McKinnon, and S.T.Coleman, Phys.Rev., B31:484 (1985).
52. J.R.Dahn, and W.R.McKinnon, Phys.Rev., B32:3003 (1985).

KINETICS STUDIES OF LITHIUM INSERTION IN In-Se THIN FILMS

I.Samaras, C.Julien and M.Balkanski

Laboratoire de Physique des Solides, associé au CNRS
Université Pierre et Marie Curie
4, place Jussieu 75252 Paris Cedex 05, France

Insertion process kinetics have been studied by potential sweep
voltammetry in InSe and In_2Se_3 thin films. The influence of the growth
conditions of the host material has been determined between the limits of
the diffusion with the cell In-Se film/$LiClO_4$-PC/Li. The diffusion
coefficient of lithium in thin film is estimated and compared with
galvanostatic measurements.

INTRODUCTION

In the last years, the search for stable, efficient, low cost
electrode materials for Li-based batteries has led to the chalcogenides.
Within such a large family, In-Se layered semiconductor as InSe and In_2Se_3
are found to be suitable intercalation host structures[1,2] and thin films of
these materials have been used as positive electrode in lithium
microbatteries[3].

Thin film insertion materials could offer significant
technological advantages due to amorphous or polycrystalline structure and
thinning of layers gives a lower resistance in transversal direction.
Although the disordered or spongy nature of the films which could be
favourable for fast movement of inserted atoms in the host, compared with
the single crystalline phase, generally shows a low diffusion rate of
ions[4,5] which is the most important limiting factor for thin films as
cathodes in microgenerators.

For example, the rate at which Li atoms can be inserted into a WO_3
film to form a tungsten bronze[4] has been estimated to be 4×10^{-13} $cm^2 s^{-1}$ at
300 K and recently[5], the silver diffusion measurements in thin film of
V_2O_5-P_2O_5 lead to a value of 10^{-15} $cm^2 s^{-1}$ with an activation energy of the
diffusion of $E_D = 0.88$ eV.

We report on electrochemical insertion of Li into electronically
conductive InSe and In_2Se_3 thin films grown by flash evaporation.
Voltammetric investigations are given as a function of growth conditions
and behaviour of the lithium diffusion is compared with results of
coefficient diffusion obtained by galvanostatic measurements.

PREPARATION OF THIN FILMS

Thin films (TF) of the In-Se system, e.g. InSe and In_2Se_3 have been
prepared by conventional flash evaporation technique which has been

reported elsewhere[6,7]. Films having a thickness of 5 μm are deposited on pyrex substrate and appeared amorphous for a substrate temperature below T_s = 400 K. The polycrystalline films are formed by subsequent thermal annealings with T_a ranging from 300 to 600 K, in a purified flowing argon atmosphere. All flash evaporated films have a n-type semiconducting character. For both InSe and In_2Se_3 films prepared at T_s = 433 K the optimum value of the Hall mobility remains stable and constant at about 50 cm^2/V.s, while the carrier concentration does not change notably above a thermal annealing at T_a = 553 K for 15 h. The Hall mobility increases consequently with the annealing temperature, but the temperature of about 600 K seems to be an upper limit because higher annealing temperature induces chemical reaction. The apparent decomposition can be attributed to the dissipation of the volatile selenium.

The electrical conductivity shows an exponential behaviour as a function of temperature and the activation energy decreases as annealing temperature increases[6]. The room temperature conductivity increases by three orders of magnitude when the annealing temperature rises from 300 up to 575 K; for T_a > 575 K the electrical conductivity becomes temperature independent.

The Li-TF galvanic cells are constructed using the $LiClO_4$-PC as non-aqueous electrolyte which is prepared, dried and kept in an anhydrous argon atmosphere. Its ionic conductivity is 4.76×10^{-3} $\Omega^{-1}cm^{-1}$ at 25°C for a 1M solution of $LiClO_4$ in PC[8]. Films of InSe and In_2Se_3 evaporated under different conditions are electrochemically active giving higher initial EMF that of the corresponding bulk material.

RESULTS AND DISCUSSION

Electrochemical properties of InSe films

The insertion of Li ions into films was studied first by a galvanostatic electrochemical method using a three-electrode cell which utilizes a lithium metal foil for both counter and reference electrodes. The cell voltage decreases smoothly from E = 3.3 V for a film prepared at T_s = 433 K and E = 3.4 V for an amorphous film (T_s = RT) in the compositional range $0 \leq x \leq 0.4$; then a pseudo-plateau appears at 1.8 V and thereafter the voltage falls down from x = 0.8. The cathodes appeared unchanged after electrochemical cycling and no cracking or exfoliation are observed on films for depth of discharge above E = 1.3 V.

In order to investigating the cause of the cell's polarization, the diffusivity of lithium into InSe film was determined as a function of the Li composition. During the introduction of Li into the film cathode, the variation of E as a function of the time t and insertion ratio x is used to calculate D, assuming that diffusion is the controlling process, according to the equation[9]:

$$D = 4/\pi \ (iV_m/zFA)^2 [(dE/dx)/(dE/d\sqrt{t})]^2$$

where A is the contact surface area, i the current through this surface, V_m the molar volume of the cathode material. This equation assumes that the electrode volume variation is neglected for any insertion ratio x, and that the diffusion progresses in a semi-infinite medium taking into account t much smaller than l^2/D where l is the cathode thickness.

Fig.1 shows the diffusion coefficient of Li in InSe film annealed at 433 K for 50 h as a function of OCV. Note that the higher D is four orders lower than the corresponding value of bulk InSe.

The data can be analysed by using a simple model[10] which treats the grain boundary regions of the film as fast diffusion pathways and the grains of the film as spheres of diameter 2r equal to the mean grain size. The corresponding relation at $Dt/r^2 \geq 0.25$ is given by:

$$C(l,t) - C_o = (J_o t/l) + (J_o r^2/15 \ lD)$$

where C(l,t) is the surface concentration of Li at time t, C_o the initial concentration. The corresponding diffusion coefficient for Li at 300 K in InSe film is 2×10^{-13} $cm^2 s^{-1}$.

Thermodynamic factor of Li insertion in InSe films

It is well-known that in the approximation of no nearest neighbor interaction, the EMF cell is given by the lattice gas model equation :

$$E(x) = E^{\circ} - RT/F \ln a(x)$$

where $a(x)$ is the activity of the inserted Li^+-ions in the cathode, assuming the anode activity as 1. The concentration dependent thermodynamic factor K_t associated with small concentration changes $|\Delta x| << x$ can be found[11] when the EMF-x relation is known :

$$K_t(x) = (\partial \ln a / \partial \ln c) = (xF/RT)(\Delta E / \Delta x)$$

The thermodynamic factor can also be obtained from kinetic data resulting from electrochemical perturbation methods, such as current steps.

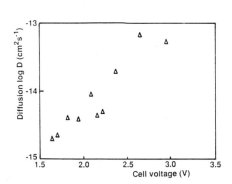

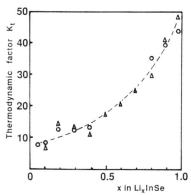

Fig.1.Diffusion coefficient of Li in a InSe film annealed at 433 K for 50 h as a function of OCV.

Fig.2.Thermodynamic factor K_t of InSe film. (o) values from EMF-x measurements, (Δ) values from diffusivity measurements.

Fig.2 shows the Li compositional dependence of the thermodynamic factor of InSe film. The values for K_t are obtained from the slope $\Delta E / \Delta x$ of the OCV discharge curve. In the pseudo-plateau region, values for K_t are only determined from the diffusivity measurements. Kinetic and stationary data of the measurements agree well and are consistent with the limiting values of the model.

Voltammetry studies of InSe films

In linear sweep voltammetry, the cell voltage is swept linearly with time while the cell current is recorded; and measurements can be of greatest values in clarifying the nature of the insertion system[12]. In the case of a two-phase system, the insertion is characterized by a phase boundary which separates two regions of different concentration of intercalant. Dahn and Haering[12] showed that in such system the current peak I_p appearing in the voltage domain V_1-V_2 is given by :

$$I_p = (2\alpha Q_0/R)^{1/2}$$

where α is the sweep rate, R a series impedance which modelled the internal losses and Q_0 the transferred charge.

Fig.3 represents the linear sweep voltammogram of an as-deposited InSe film obtained at T_s = RT (curve a) and an InSe film deposited at T_s = 433 K and annealed at 475 K for 64 h (curve b). Two peaks are observed, the first one at 2.6 V weaker than the second at 1.2 V. The thermal annealing process has mainly two effects : an increase of the first peak and a broadening of the strong peak at 1.2 V. Also, the kinetics play an important role in the voltammetry process, for an amorphous film (curve a) a strong polarization is superimposed to the cathodic peak. It is evident that the apparent non-reversibility of the current peak is also partly due to slow diffusivity in the lower voltage range.

Fig.4 shows the variation of the cell current as a function of composition of an In_2Se_3 film deposited at T_s = 423 K and annealed at 450 K for 7 h. The corresponding voltammogram was carried out using a sweep rate of 4.6 μV/s. We observe a broad band with two current maxima at x = 0.75 and 0.98. This behaviour corresponds to the formation of a two-phase system and plateau in the discharge voltage curve is then expected in the range 0.45 ≤ x ≤ 1.2.

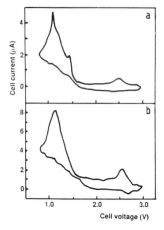

Fig.3.Linear sweep voltammogram of an as-deposited InSe film obtained at T_s = RT (a) and an InSe film deposited at T_s = 433 K and annealed at 475 K for 64 h (b).

Fig.4.Variation of the cell current as a function of composition of an In_2Se_3 film deposited at T_s = 423 K and annealed at 450 K for 7 h.

CONCLUSION

The present electrochemical investigations suggest that the InSe and In_2Se_3 films have very slow kinetics for lithium insertion, as it is found generally in amorphous structures. The morphology plays an important role but the grain sizes obtained here are enough small and grain boundaries can be the cause of the low diffusion coefficient. A future study is planned to relate the kinetics and grain size in the films.

REFERENCES

1.C.Julien, and E.Hatzikraniotis, Mater.Letters, 5:134 (1987).
2.C.Julien, E.Hatzikraniotis, and M.Balkanski, Solid State Ionics, 22:134 (1987).
3.M.Balkanski, C.Julien, and J.Y.Emery, 4th Int.Meeting on Lithium Batteries, Vancouver (1988).
4.M.Green, Thin Solid Films, 50:145 (1978).
5.L.Jourdaine, M.Bonnat, and J.L.Souquet, Solid State Ionics, 18/19:461 (1986).
6.C.Julien, M.Eddrief, K.Kambas, and M.Balkanski, Thin Solid Films, 137:27 (1986).
7.J.P.Guesdon, C.Julien, M.Balkanski, and A.Chevy, Phys.Stat.Sol.(a), 102:327 (1987).
8.S.Atlung, K.West, and T.Jacobsen, J.Electrochem.Soc., 126:1311 (1979).
9.W.Weppner, and R.A.Huggins, J.Electrochem.Soc., 124:1569 (1977).
10.M.Green, W.C.Smith, and J.A.Weiner, Thin Solid Films, 38:89 (1976).
11.A.Honders, J.M.der Kinderen, A.H.van Heeren, J.H.W.de Witt, and G.H.J.Broers, Solid State Ionics, 15:265 (1985).
12.J.R.Dahn, and R.R.Haering, Solid State Ionics, 2:19 (1981).

THE ROLE OF NiPS$_3$ AS A CATHODE IN SECONDARY

LITHIUM SOLID STATE BATTERIES

I.Saikh, C.Julien and M.Balkanski

Laboratoire de Physique des Solides, associé au CNRS
Université Pierre et Marie Curie
4, place Jussieu 75252 Paris Cedex 05, France

INTRODUCTION

The transition metal phosphorus trisulfides have received considerable attention in recent years as a cathode element for the solid state secondary batteries. Detailed investigations into the structure[1-3], physical properties[4,5] of intercalates and electrochemical intercalation studies[6-9] have contributed to the fundamental understanding of the chemical physics of Li-NiPS$_3$ system. It has been shown[7] that NiPS$_3$, a material of structure similar to that TiS$_2$, has on intercalation with Li a higher energy density. Three Li ions per molecular unit can be intercalated, two of them reversibly while keeping the electrode potential of the secondary cell at considerable value and the process of Li intercalation is completely reversible without causing any lattice expansion of the unit cell. Evidence of some intercalated phases was seen as a function of lithium composition also[10].

In this paper, we describe the performance of solid state batteries using NiPS$_3$ as a cathode and also the effect of lithium intercalation in its structure. Our studies have been carried out at ambient temperature using the asymmetric cell configuration Li/glass/NiPS$_3$ and our data show that NiPS$_3$ can be intercalate with high degree of reversibility.

EXPERIMENTAL TECHNIQUE

Both the electrolyte and the positive electrode were taken in powder form in the solid state cell construction. The positive electrode NiPS$_3$ was powdered with grain size of 80 μm[11] under inert atmosphere. The glass with its composition B$_2$O$_3$-0.57Li$_2$O-0.18LiCl has been chosen as electrolyte for its higher ionic conductivity of the order of $10^{-4}\Omega^{-1}$cm^{-1} for pressed pellets[12]. All processing except isostatic compression was done in an argon filled dry box. The cathode was prepared by mixing NiPS$_3$, graphite and teflon[13] powder in 70:15:15 of the total weight. The cathode and electrolyte were pressed under pressure of 10^4 kg/cm^2 in successive layers to form a pellet in a steel die. The pellet was extracted and transferred to the argon box where a pure lithium disc was placed on the electrolyte-cathode composite being careful not to short the cell. The lithium anode was lightly pressed onto that composite by a spring load and mounting the whole unit in a special air-light container taken out from the dry box for testing.

Testing experiments were conducted by galvanostatic method in a computer controlled system using a programmable constant current source (Keithley 220).

RESULTS AND DISCUSSION

Discharge characteristics

Small area (0.2 cm^2) cells have been fabricated and discharged under constant current at room temperature. The discharge was termined when the cell voltage fell to 1.0 Volt. Fig.1 shows the effect on discharge efficiency of successively heavier drains. It is clear from these curves that battery voltage remains unaltered during the initial period of discharging and thereafter the gradual discharge reaction is further enhanced when the current drain is increased. The rapid decrease in cell potential at high current density ($\sim$ 150 $\mu A/cm^2$) may be due to the lithium/solid electrolyte interfacial disruption. In such cells, the practical capacity which is about 1.8 mAh reaches the theoretical value. Capacity decline on cycling is observed but can be reduced by the selection of appropriate voltage limits[10].

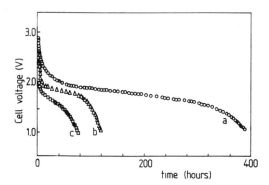

Fig.1 Curves of discharge of a $NiPS_3/B_2O_3$-0.57Li_2O-0.18$LiCl$/Li cell under three current densities : (a) 5, (b) 100, (c) 150 $\mu A/cm^2$.

Internal resistance of the cell

The internal resistance of the $NiPS_3$/Li cell was investigated in two systems : (a) a Li/$LiClO_4$-PC in paper/$NiPS_3$ cell, and (b) a Li/ B_2O_3-0.57Li_2O-0.18$LiCl$/ $NiPS_3$ cell.

The ohmic drop has been calculated over a range of discharge for $0 < x < 1$ and the kind of variation is represented in the fig.2 for the a-type cell. When the battery discharge occurs continuously the discharge product at the cathode electrolyte interface increases and hence the rise in the internal resistance occurs.

The same test on cell of b-type has been done in the normal cycling test where the cycling period contains 1 hour discharge, 1 hour charge and 2 hours relaxation. It was found that the ohmic drop increases almost linearly at the beginning for the first few cycles and thereafter it maintains a stable value without showing any large increment of the IR drop for the rest of the cycles as shown in fig.3. After 100 cycles, the internal resistance was about twice the initial value.

The kinetics of Li intercalatoin into $NiPS_3$ have been reported elsewhere[10]. The diffusion coefficient D has been determined by galvanostatic measurements and values of 2.4×10^{-9} and 4×10^{-10} cm^2/s have been found for $x = 0.3$ and $x = 1.5$ respectively, while the thermodynamic enhancement factor is between 120 and 470 for the same compositional range.

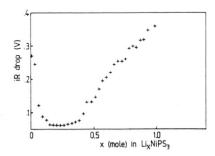

Fig.2 Ohmic drop as a function of Li content in $NiPS_3$ powdered cathode during a discharge under a current density of 50 $\mu A/cm^2$.

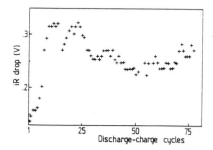

Fig.3 Ohmic drop as a function of number of discharge-charge cycles in a cell with the $NiPS_3/B_2O_3$-0.57Li_2O-0.18LiCl/Li configuration.

Reversibility

In addition, from the cyclic voltametry test the strong reversibility of $NiPS_3$ is shown in fig.4 where the cell was repeatedly cycled with the increment function "x" of Li_xNiPS_3.

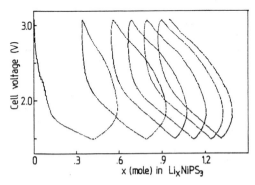

Fig.4 Curve of the cell voltage over Li composition in the cathode of a $NiPS_3/B_2O_3$-0.57Li_2O-0.18LiCl/Li cell.

CONCLUSION

The aim of the composite positive electrode is to enhance the migration of lithium atoms between the grains of the lamellar compound. The used technique which includes a mixture with teflon allows to utilize a bigger amount of $NiPS_3$ with a higher discharge current. Weak polarization still appears during discharge and increases slowly during the following charge process but the reversibility of our system seems under a medium rate of discharge.

REFERENCES

1. R.Brec, Solid State Ionics, 22:3 (1987).
2. G.Ouvrard, R.Frecour, R.Brec, and J.Rouxel, Mat.Res.Bull., 20:1053 (1985).
3. G.Ouvrard, R.Brec, and J.Rouxel, Mat.Res.Bull., 20:1181 (1985).
4. B.E.Taylor, J.Steger and A.Wold, J.of Solid St.Chem., 7:461 (1973).
5. M.Barj, C.Sourisseau, G.Ouvrard and R.Brec, Solid State Ionics, 11:179 (1983).
6. A.Le Mehauté, C.R.Acad.Sc.Paris, C287:309 (1978).
7. A.Le Mehauté, G.Ouvrard, R.Brec, and J.Rouxel, Mat.Res.Bull., 12:1191 (1977).
8. P.J.S.Foot and B.A.Nevett, Solid State Ionics, 8:169 (1983).
9. R.Brec, G.Ouvrard, A.Louisy, J.Rouxel and A.Le Mehauté, Solid State Ionics, 6:185 (1982).
10. I.Samaras, I.Saikh, C.Julien and M.Balkanski, J.Mat.Sci.and Engineer., (1988) in press.
11. The powdered $NiPS_3$ material was prepared by G.Ouvrard (Laboratoire de Chimie des Solides, Nantes, France) who is acknowledged.
12. A.Levasseur, J.C.Brethous, J.M.Reau and P.Hagenmuller, Solid State Ionics, 1:177 (1980).
13. Teflon or PTFE (polytetrafluoroethylene) is an Aldrich product in powder form and has a density of 2.

THEORY AND APPLICATIONS OF AMORPHOUS SOLID

FOR ELECTROCHEMICAL CELLS

J.L. Souquet, A. Kone, M. Levy

Laboratoire d'Ionique et Electrochimie des Solides

BP 75 - 38402 Saint Martin d'Hères - France

PART ONE : GLASSES AS SOLID ELECTROLYTES

In the history of solid state electrochemistry, inorganic glasses were the first solids shown to exhibit electrical conductivity of an ionic nature. Oxide-based glasses can be easily formed into impervious thin layers, a property which enabled Warburg in 1884 (1) and then Burt in 1925 (2) to obtain accurate measurements of their cation transport number. Subsequently, the understanding of the electrolytic properties of these materials progressed at a slower rate than for traditional and crystallized solid electrolytes. The first predicting theories pertaining to ionic transport in solids were based on the crystalline defect concept and were conceived between 1940 and 1960, accompanying the rapid development of X-ray crystallography techniques applied to the study of ionic crystal structures. Further interest in the electrolytic properties of glasses re-emerged only in the 1970's as part of the multidirectional research into new materials suitable for electrochemical energy storage systems. In the meantime, powerful investigation techniques were developed in traditional electrochemistry and progress was made in the understanding of non-aqueous liquid electrolytes. Benefitting from these advances, research into vitreous electrolytes moved ahead rapidly and led the way to new materials of equal or better performances than the best known solid electrolytes.

Today, many ionic conductive glasses have reached cationic conductivities of between 10^{-4} and 10^{-2} (Ω.cm)$^{-1}$ at their utilisation temperature as solid electrolytes in electrochemical cells. By hollow fiber pulling (3) isostatic compression (4) or RF sputtering (5) (6) they may be shaped in thin layer from 1 μm to 1 mm.

At the time scale of a voltammetry sweep they present large Red/Ox windows (over 3 to 4V) but, nevertheless, many glassy electrolytes slowly react with the reducing or oxidizing species used as anode or cathode materials (7-9). A good conductivity associated with a good chemical stability strongly reduces the number of available glass compositions. Good performances have been obtained from borate glasses for sodium - sulfur batteries (3) and phosphorous sulfide based glasses for room temperature lithium primary cells (4-10). Recently vitreous thin films of lithium glassy oxides as solid electrolyte associated with vitreous or crystalline host materials have been associated to form microbatteries for potential application in microelectronic (5) (6).

Solid State Microbatteries
Edited by J. R. Akridge and M. Balkanski
Plenum Press, New York, 1990

These applications reflect the important role of vitreous materials in modern electrochemistry. Their main thermodynamic and structural characteristics will be outlined below.

I EXPERIMENTAL AND IDEAL GLASS TRANSITION TEMPERATURES

Figure 1 compares vitreous materials out of thermodynamic equilibrium with the same compounds in crystallized form in thermodynamic equilibrium. The glass transition temperature Tg separates the high temperature domain where the material possesses the thermodynamic and mechanical properties of a liquid from the low temperature domain where the material possesses the properties of a solid. The definition of the glass transition temperature is closely linked to the structural relaxation time concept. Roughly speaking, the structural relaxation time is less than a second above Tg and much greater than a second below Tg. From a macroscopic viewpoint, the structural relaxation time is the time constant characteristic of the strain rate of the material when subjected to a mechanical stress and is proportional to the viscosity. On a microscopic level, it corresponds to the time required by the elements of the macromolecular chains constituting a glass to move a distance comparable to their size. The glass transition temperature is thus closely related to kinetic parameters and to the duration of experiments conducted on the material. Tg therefore has a kinetic significance.

When a liquid is cooled rapidly enough to lead to the formation of a glass, the glass transition temperature is a function of the quenching rate q. Over a limited temperature range, a relationship of the type $q = q_o \exp [-E_a/RTg]$ is observed (11). Tg thus decreases with the quenching rate. A lower limit nevertheless exists for the values of the glass transition temperature and is referred to as the ideal glass transition temperature To. The ideal glass transition temperature may be expressed as a function of thermodynamic quantities alone, involving no kinetic considerations.

Kauzmann defined the ideal glass transition temperature on the basis of a thermodynamic paradox (12) (13) : To is the temperature below which the properties of the liquid cannot be extrapolated since the liquid phase would have an entropy less than that of a crystal of the same composition (figure 1). The difference in entropy between the liquid and the crystal is essentially configurational, i.e. in a liquid state, the atoms can explore a large number of configurations of equal energy within a reasonable time. In a glass, the increase in the structural relaxation time leads to the impossibility of real exploration of these configurations and maintains, for kinetic reasons, an entropy greater than that of the crystal of the same composition. The analysis developed by Gibbs and Di Marzio (16) (17) and then by Adam and Gibbs (18) associates the ideal glass transition temperature with the structural relaxation time of the macromolecular chains, proportional to $\exp [-W/R(T-To)]$, where W has the significance of an activation energy.

From a thermodynamic viewpoint, a glass is a material out of equilibrium ; it does not possess, for its chemical composition, minimum entropy and enthalpy. However, these two quantities do not change significantly over the duration of a measurement, on a human time scale, and are thus definable for a given glass. To assign a single vitreous state to a given composition would be erroneous. Indeed, depending on the synthesis technique used, a glass may present different thermodynamic characteristics (internal energy and enthalpy) from one sample to another for the same composition. They can none the less be defined.

This thermodynamic characterization of glass reveal the importance of the glass transition temperature. A glass can thus be defined as a non-crystalline solid presenting the glass transition phenomenon (11). This definition

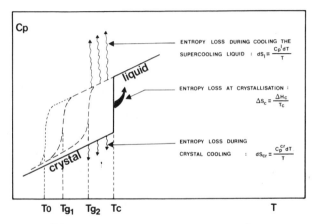

Figure 1   The Kauzmann paradox

imposes no preparation conditions and excludes all non-crystalline solids not presenting sufficient stability with respect to time.

It is impossible to observe a glass transition temperature (T_g) less than the ideal glass transition temperature (T_o). If such a case did exist, we would have a liquid of a less ordered structure than the crystal of lower entropy. A glass possesses a less ordered structure than the crystal, however, for kinetic reasons, it is impossible for the constituent atoms to explore the high geometrically available configurations (the structural relaxation time is too high). T_o corresponds to the temperature below which the Kauzmann paradox would exist. At To, the entropy lost upon crystallisation and crystal cooling down to T_o is equal to the entropy lost during the cooling of the supercooled liquid :

$$\int_{T_c}^{T_o} C_p^{Cr} \ \frac{dT}{T} + \Delta S_c = \int_{T_c}^{T_o} C_p^l \ \frac{dT}{T}$$

If $\Delta C_p = C_p^l - C_p^{Cr}$ # constant, then $T_o = T_c \exp \left(\frac{\Delta S_c}{\Delta C_p} \right)$

II IONIC CONDUCTION IN THE GLASSY AND OVER COOLED STATES

A glass exhibiting ionic conductivity, most often cationic, may be considered as the association of an anionic and macromolecular sublattice with a sublattice formed by all the cations. Each sublattice has its own structural relaxation time. The displacement mechanism of an element of the macromolecular chain is very different from that of a cation. The mechanical properties are fixed by the relaxation time of the anionic sublattice. For the electrical properties, two cases must be considered (19) :

First of all, at high temperatures and above Tg, the cationic relaxation time is less than the relaxation time of the macromolecular chain. Ion transport is co-operative and facilited by the movement of the chain elements. The conductivity varies with temperature according to a relationship analogous to that which represents the structural relaxation time of the macromolecular chain (V.T.F. equation). This behavior is illustrated in figures 2 and 3.

At lower temperatures, i.e. below Tg, cationic displacement is decoupled with respect to the excessively slow movements of the elements of the macromolecular chains and conductivity is an Arrhenius plot phenomenon.

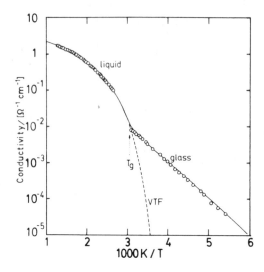

Figure 2   Electrical conductivity versus temperature, represented in the 0.7 AgI 0.3 Ag_2MoO_4 mixture in the overcooled (T > Tg) and the glassy (T < Tg) states. Data from Reference (14), extrapolation using V.T.F. equation.

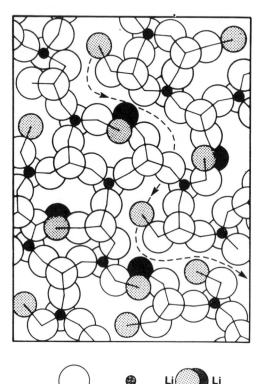

Figure 3.1.   Representation of a charge carrier's path (Li^+) in an anionic macromolecular network of tetraedra chains.

304

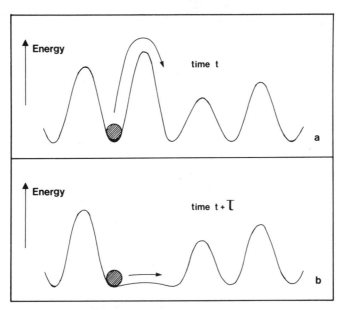

Figure 3.2.   Coulomb potential wells (in the absence of an electrical field) in vitreous or supercooled liquid state (a). In (b), two potential wells are brought together by the relative displacement of segments of the network former chain, thereby lowering the activation energy. There is co-operation between the chain movement and ionic transport.

III CONSTITUENTS OF HIGH IONIC CONDUCTIVITY GLASSES AND THEIR INTERACTIONS

Figures 4 and 5 show conductivity variations as a function of temperature for various silver or lithium cation conducting glasses chosen among the most conductive (20-29).

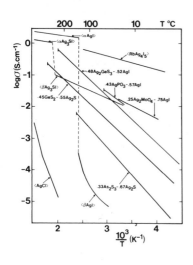

Figure 4   Electrical conductivity versus temperature, represented as an Arrhenius plot, for various silver cation conducting crystallized (< >) and vitreous solid electrolytes.

305

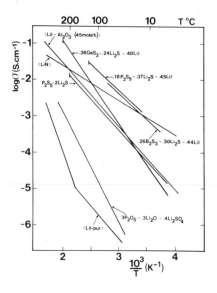

Figure 5 Electrical conductivity versus temperature, represented as an Arrhenius plot, for various lithium cation conducting crystallized (< >) and vitreous solid electrolytes.

For comparison purposes, these figures also show the results of several crystallized solid electrolytes conducting via the same cations. Note that the conductivity of the glasses is comparable if not better (in the case of lithium) than those of the best crystallized solid electrolytes known. Note also that these glasses have complex chemical compositions which can nevertheless always be expressed as the resultant of mixing the three following types of compounds in different proportions :

     a) network formers (eg. SiO_2, P_2O_5, B_2S_3)
     b) network modifiers (eg. Ag_2O, Li_2O, Na_2S)
     c) doping salts (eg. AgI, LiI, LiBr).

III.1. Network formers

     Network formers are compounds of a covalent nature such as SiO_2, P_2O_5, B_2O_3, GeS_2, P_2S_5, B_2S_3, etc... They form macromolecular chains which are strongly cross-linked by the assembly of tetrahedra (SiO_4, PO_4, BO_4...) or triangles (BO_3...). When pure, network formers easily give glasses by cooling of the liquid phase or alcoholate polymerisation (30). A certain dispersion of valence angles and bond lengths characterizes the disorder existing in the vitreous state. The existence of local order which provides the stability of the tetrahedral or triangular entities is the result of the covalent character of the bonds. The possibility of deforming this local order is on the other hand the results of the partially ionic character of these same bonds. STANWORTH (31) observed in this respect that the former oxides are all characterized by a difference in electronegativity between the oxygen and the former cations amounting to 0.4 to 0.5 on the PAULING scale.

III.2. Network modifiers and doping salts

     Network modifiers and doping salts are compounds of a marked ionic character. Their structure, when dissolved in a given former, is entirely different than when in pure and crystallized form.

306

III.2.1. Network modifiers

Network modifiers include oxides or sulfides (eg. Ag_2O, Li_2O, Ag_2S, Li_2S, etc...) which interact strongly with the structure of network formers. A true chemical reaction is involved, leading to the breaking of the oxygen or sulfur bridge linking two former cations. The addition of a modifier introduces two ionic bonds. For instance, the reaction between silica and lithium oxide may be expressed schematically as :

$$
\begin{array}{c}
\mid \\
- Si - O - Si - \\
\mid
\end{array}
+ Li_2O
\qquad
\begin{array}{c}
\mid \quad\quad Li^+ \quad\quad \mid \\
- Si - O^- \quad ^-O - Si - \\
\mid \quad\quad Li^+ \quad\quad \mid
\end{array}
\qquad (1)
$$

The increasing addition of a modifier to a given former leads to the progressive breaking of all oxygen bridges according to a type (1) reaction.

As the number of non-bridging oxygen atoms increases, the average length of the macromolecular chains decreases. The chemical reaction symbolized in (1) is strongly exothermic, and the mixing enthalpies are in the order of several hundreds of kiloJoules (32). The magnitude of these values is difficult to account for on the basis of the energy balance of the bonds described in (1). The origin could be the stabilization of the negative charge carried by the non-bridging oxygen atom by interaction of the oxygen p orbitals and the silicon d orbitals. The result is a reinforcement of the bond, representing the probable origin of the increase in the force constant observed by IR and Raman spectroscopy, and the shortening of this same bond observed by X-ray crystallography on the recrystallized glasses (33).

The case of boron as a former cation is somewhat particular in that the element has no available d orbitals. However, a p orbital is available when the boron has a coordination number of 3, which allows stabilization of an electronic duplet of the oxygen or sulfur introduced by the modifier. This oxygen or sulfur giving up a duplet to another boron atom increases the cross-linking by the formation of two BO_4 tetrahedra. In hybridization terms, the boron is altered from the sp^2 configuration to the sp^3 configuration. The coordination change of boron has been especially well observed by RMN (34).

Although the environment of the former cation is relatively well known today, that of the modifier cation is much less so due to the lack of appropriate spectrocopic techniques. The absence of direct experimental data has given rise to the coexistence in present literature of hypotheses as different as those based on a totally random distribution of ionic bonds as well as, on the contrary, those based on very modifier cation rich zones alternating with less rich zones (35).

III.2.2. Doping salts

These are generally halide salts of general formulation MX or in some cases sulfates. From a structural viewpoint, it is almost certain that the halide anions are not inserted in the macromolecular chains. Indeed, no modification in the vibrations of the macromolecular chain has been revealed by spectroscopic analysis. The only certainty today is the absence of chemical reactions with the macromolecular chains. The arrangement of halide salt molecules with respect to one another is yet unknown. Proposed hypotheses range from the formation of salt clusters (24) to an uniform distribution throughout the mass of the glass. From a thermodynamic viewpoint, the absence of chemical reactions likely leads to low mixing enthalpies in the order of a few kiloJoules per mole, as measured by Reggiani et al. (36).

IV. IONIC CONDUCTIVITY VERSUS GLASS COMPOSITION AND THE WEAK ELECTROLYTES THEORY

Figures 6 and 7 show the extreme sensitivity of ionic conductivity to the overall alkali or silver cation concentrations in inorganic glasses.

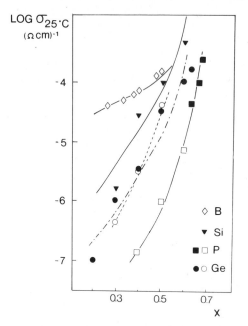

Figure 6   Logarithm of electrical conductivity at 25°C for glasses of the system $xLi_2S-(1-x)YS_2$

versus M_2S content. Calculated curves using a quasi-chemical model (41). Data from reference (15).

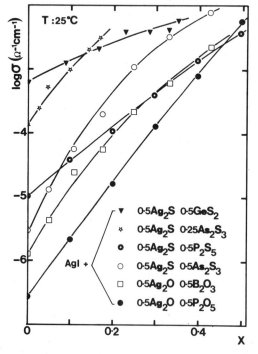

Figure 7   Logarithm of electrical conductivity at 25°C for various silver cation conducting glasses versus the molar fraction of silver iodide. From reference (38).

The conductivity versus cation content relationship is far from linear and presents more of an exponential character. The most quantitative interpreation of this behavior is the so-called weak electrolyte theory, initially proposed by Ravaine and Souquet (37) for the case of binary former/modifier systems and recently extended by Kone and Souquet (38) to ternary former/modifier/doping salt systems. The basic reasoning leads to a correlation of the number of free charge carriers at time t with the thermodynamic activity of the salts rather than with the overall cation concentration since the former shows much greater variations. The reasoning which justifies theses relationships compares glasses to alkali salt solutions within the solvent constituting the network former. When several salts are dissolved together in the glass former the dissociation of the salt is largely predominant. It is for instance the case of lithium iodide as compared to lithium oxide. In that case, assuming constant mobility and activity coefficient for the mobile cation versus composition, the ionic conductivity of a glass σ_{M^+} is simply proportional to the square root of the thermodynamic activity of the more dissociated alkali salt MY (a_{MY}).

$$\sigma_{M^+} = Fu\ K_{(T)}^{1/2}\ a_{MY}^{1/2} = Fu\ K_{(T)}^{1/2}\ \exp\frac{\overline{\Delta G_{MY}}}{2\ RT} \tag{2}$$

In equation (2) u is the mobility and $K_{(T)}$ the salt dissociation constant. The same relationship may be expressed either with the thermodynamic activity or the partial free energy $\overline{\Delta G_{MY}}$. Equation (2) has been experimentally verified associating conductivity and thermodynamic measurements (36) (37). At very low concentration of the ionic salt in a network former ($< 10^{-4}$ mole cm^{-3}) activity becomes proportional to the concentration (Raoult law) and conductivity variation becomes proportional to the square root of the concentration (40). In a given binary system (1-x) glass-x salt experimental conductivity variations are well represented by an Arrhenius law.

$$\sigma_{M^+}(x) = \sigma_o(x)\ \exp - \frac{E(x)}{RT} \tag{3}$$

where $E(x)$ and $\sigma_o(x)$ are respectively the temperature independant activation energy and preexponential term (Figure 8).

At constant temperature, partial differentiation of (2) and (3) with x leads to the following expression :

$$\frac{\delta \ln \sigma}{\delta x} = -\frac{1}{2R}\frac{\delta \overline{\Delta S_{MY}}(x)}{\delta x} + \frac{1}{2RT}\frac{\delta \overline{\Delta H_{MY}}(x)}{\delta x}$$

$$\frac{\delta \ln \sigma}{\delta x} = \frac{\delta \log \sigma_o}{\delta x} - \frac{1}{RT}\frac{\delta E}{\delta x} \tag{4}$$

from which it is easily shown that variations in the preexponential term are to be related with variations of partial mixing entropy of the salt and those of the activation energy to the partial mixing enthalpy variations.

$$2R\frac{\delta \ln \sigma_o}{\delta x} = -\frac{\delta \overline{\Delta S_{MY}}(x)}{\delta x} \tag{5}$$

$$2\frac{\delta E}{\delta x} = -\frac{\delta \overline{\Delta H_{MY}}(x)}{\delta x} \tag{6}$$

Equations (2), (5) and (6) allow the determination of $\overline{\Delta G_{MY}}$, $\overline{\Delta H_{MY}}$, and $\overline{\Delta S_{MY}}$ variations with composition. It is then possible to find the best thermodynamic model fitting with the experimental conductivity data. By such an approach it has been shown that the halogenated salts-glass solutions behave simply like a regular solution (39) as usually found in molten salt mixtures. The mixing enthalpy, a few $kJ.mole^{-1}$ is probably the result of local reorganizations of the dipole-dipole interactions. As it can be suggested from equation (1) a quasi-chemical model is a convenient representation for the network former - network modifier solutions (41).

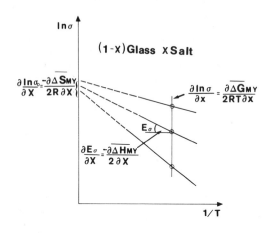

(1-x)**Glass** x**Salt**

Nevertheless, if such models give a simple interpretation of the conductivity-composition relationship, it is actually a lack of available calorimetric data that prevents confirmation of the assumptions' validity.

Figure 8 Relationship between variation in $\ln\sigma$, $\ln\sigma_o$, E with partial thermodynamic function in a (1-x) glass - x salt systems.

V CONDUCTIVITY ENHANCEMENT

For many ionic conductive glasses, conductivity improvement may be explained by optimisation of the dissociation constant K. For this reason, halogenated salts with the largest anions (Br^-, I^-) dissolved in a glass network of high dielectric constant like sulfides are the most conductive (25) (42) (43) (44).

Another possibility could be the thermodynamic activity enhancement of the dissociating salt due to the metastability of the glassy state. Let us consider two glasses with the same network modifier (i.e. Na_2O) and two different network formers (i.e., B_2O_3 and P_2O_5). The mixture of the two glasses give a glass of the general formula $Na_2O.n(xB_2O_3.$ $(1-x)P_2O_5)$. At constant n value let us suppose a negative mixing enthalpy ΔH_m. For a moderate value of ΔH_m (up to 4 kJ) the entropic term, $T\Delta S$, is sufficiently high at high temperature to allow mixing in the liquid state. At lower temperature, the entropic term is too small to counterbalance the enthalpic term and the ΔG_m variations, as a function of x are characterized by two minima at x_1 and x_2. For any values of x between the minima, the glass should phase separate with phases x_1 and x_2 compositions. The phase separation probability is :

$$P = P_o \exp \frac{(\Delta x)^2 \frac{\delta^2\Delta G_m}{x^2}}{2\ RT}$$

where Δx is a local composition fluctuation, P_o a kinetic parameter, and $\frac{\delta^2\Delta G_m}{\delta x^2}$ a thermodynamic one. If $\frac{\delta^2\Delta G_m}{\delta x^2}$ is negative, the phase separation is

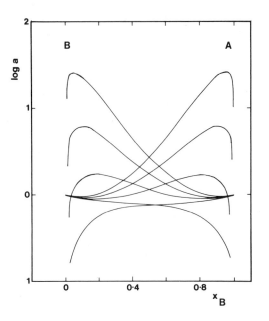

highly probable (spinodal phase separation) but not certain, depending on the P_o value. If the phase separation should occur for a temperature below T_g the structural relaxation time would be too long to allow it. In such a case the thermodynamic activity of the two limiting compositions (Na_2O nB_2O_3 and Na_2O nP_2O_5) should present two maxima as shown in figure 9. We could expect the same behavior for the ionic conductivity.

Figure 9 Partial free mixing energy $\overline{\Delta G_A}$ = RT ln a_A and $\overline{\Delta G_B}$ = RT ln a_B for the two components A and B of an endothermic regular solution ($\Delta H_m = \alpha x_A x_B$). Values of α are : - 1000 ; - 2000 ; - 3000 and - 4000 cal mole^{-1}.

Figures 10 shows isothermal conductivity curves in the Na_2O $n(B_2O_3$ $P_2O_5)$ (45) which could be explained by this thermodynamic approach. Similar results have been obtain in Ag_2O B_2O_3 P_2O_5 and Li_2O B_2O_3 TeO_2 systems (46) (47). In these ternary systems phase separation zones are observed. The microscopic origin of the endothermic mixing enthalpy could be associated with an increase of the boron coordination. Such an "endothermic glass" has been proved to exist in the Na_2O SiO_2 B_2O_3 system by calorimetric measurements (48).

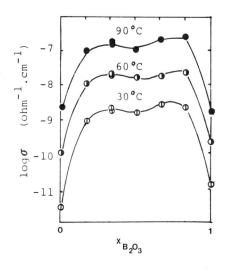

Figure 10 Isothermal variations in ionic conductivity as a function of x for glasses of the 0.4 Na_2O-0.6 (xB_2O_3 $(1-x)P_2O_5$) system. Experimental data from Reference (45).

PART TWO : GLASSES AS ELECTRODE MATERIALS

A large number of investigations are presently focused on the electrochemistry of mixed conductive materials, i.e. materials which simultaneously exhibit ionic and electronic conduction. Used as electrodes in electrochemical devices, these materials serve a dual function.

First of all, they allow charge transfer, i.e. when current is flowing, they allow the change in the nature of the charge carriers between the current leads (electrons) and the electrolyte (ions). This phenomenon is not confined simply to the electrode - electrolyte interface, but extends throughout the material. This delocalization of the electrochemical reaction enhances electrode reversibility and reduces overvoltages.

Secondly, mixed conduction materials can store an electroactive species, often hydrogen or an alkali metal, in the bulk of the material. Within the solid structure, the cations M^+ and electrons are stored separately : the protons or the cations occupy well-defined sites, the associated electrons generally occupy the available d orbitals of a transition metal incorporated in the structure. When dissolved in this manner in the solid structure, the species M is said to be intercalated and its chemical potential is far less than that of the pure species. This difference in chemical potential can be used to form high energy density electrochemical batteries. Manganese dioxide, MnO_2, is without a doubt the most widely used intercalation material. The γ form is used to store hydrogen $(H_3O^+ + e)$ in Leclanché type cells and the β form to store lithium in most commercially available lithium batteries.

Many similar materials exist, all characterized by an "open" structure, lamellar or channel type, offering numerous available sites to the intercalatable cations and including an electron accepting transition metal. Examples include TiS_2, $NiPS_3$, V_6O_{13}, MoO_3, etc. In certain cases, the intercalation of electrons in these structures leads to the occupation of the initially empty electronic levels or bands. This causes the change of colour observed in electrochromic devices used to opacify windows or display information. Tungsten trioxide is the most well known example. Most of the intercalation materials investigated to date are either crystallized sulphides or oxides, the properties of which are now well known [49-52].

Recent investigations have shown that the crystalline materials mentioned above are not the only intercalation materials possible and that several amorphous or vitreous compounds present analogous properties. The initial interest in these materials stemmed from the fact that they could not be prepared in crystalline form. This is the case for MoS_2 [54] and V_2S_5 [55] obtained by precipitation in aqueous solution, or for WO_3 obtained as a thin film by cathodic pulverization [56]. Amorphization may also result from a large number of electrochemical cycles on an initially crystalline material. Such is the case for V_2O_5 [57] or MoS_2 [58].

These amorphous structures were considered merely as "accidents" until only recently. Today, however, they are studied in their own right due to certain interesting characteristics related to the disordered state.

One of these characteristics is the open nature of the amorphous structure, presenting "a priori" a large number of available intercalation sites. This already disordered structure should show practically no change during charge-discharge cycles and very stable performance for high intercalation rates is expected.

Another interesting property of amorphous materials is that the energetic stabilization of the transition metal is without a doubt much lower than in the crystalline state due to the deformation of the environment of the latter. The electronegative character of this metal should therefore be strengthened and the chemical potential of the intercalated species reduced. For use as cathodes in electrochemical batteries, this means a higher stored energy density.

Finally, the mobility of cations can be very high in an amorphous structure. The best lithium cation conducting solid electrolytes known to date are in fact glasses [25]. However, the disordered structure reduces the electronic mobility. The consequences of these two variations on the diffusion coefficient have not yet been clearly established, but there is presently no reason to expect a lower coefficient than for an ordered structure usually found between 10^{-8} and 10^{-12} cm^2 sec^{-1}.

The great potential interest and many unsolved questions involved has encouraged the systematic investigation of intercalation in amorphous solids. The amorphous state can be easily obtained by adding a former oxide (B_2O_3, P_2O_5, etc...) to a transition metal oxide. Such mixtures lead to glasses clearly marked by the existence of a glass transition temperature. This is not the case for other amorphous materials, pure transition metal oxides or sulphides, which decompose or recrystallize before a possible glass transition temperature can be observed.

In any case, although the investigation of the intercalation properties of amorphous transition metal oxides or sulphides is a recent undertaking, these materials have nevertheless already been widely studied in other respects. The difficulty encountered in presenting a comprehensive overview of their properties lies in the fact that the different aspects have been studied by different scientific communities using different scientific languages. Physicists have interpreted electronic conduction by the transfer of electrons between localized states. These localized states correspond to various degrees of oxidation of the transition metal and their relative concentrations can be derived from the notion of redox equilibrium developed by chemists. Finally, electrochemistry and its techniques have more recently thrown light on the thermodynamic and kinetic aspects of alkali cation intercalation. These various contributions are discussed below.

I ELECTRONIC CONDUCTIVITY IN AMORPHOUS MATERIALS : THE PHYSICIST'S CONTRIBUTION

Since amorphous materials present thermally activated variations in electronic conductivity, they have been improperly called semiconductors. In fact, the conduction mechanisms are completely different from those encountered in classical semiconductors for which a description with empty conduction and full valence bands separated by an energy gap is appropriate and for which conduction is obtained by transferring electrons from the valence to the conduction band. For "semiconducting" glasses the electronic displacement occurs by the hopping of one electron between two localized states situated in the band gap. In transition metal containing glasses acceptor and donor levels may be free or empty d - orbitals and for instance a donor could be Fe^{2+} and an acceptor Fe^{3+}.

For temperatures above the Debye temperature, Θ_D, the hopping of one electron from a donor level to an acceptor level is a phonon-assisted tunneling process. The Debye temperature is defined by the energetic identity $h\nu_o = k\Theta_D$ where ν_o is the predominant phonon frequency. The electronic drift mobility μ is a function of the hopping rate P and the site-to-site separation, R :

$$\mu = \frac{eR^2}{kT} P$$

P may be written as the product of three probability terms, $P = P_1 x P_2 x P_3$.

P_1 : Probability of adjacent donor/acceptor levels. In case of transition metal-containing glasses $p \simeq c(1-c)$ where c is the ratio of ion concentration in the low valency state to the total concentration of transition metal ion.

P_2 : Probability of a thermal distortion which brings the electronic level of the occupied site into momentary coincidence with an empty one on a neighboring site :

$$P_2 \simeq \nu_o \exp - \frac{W}{kT}$$

The required energy W is only half of the electron trapping energy since relaxation energy does not have to be performed.

P_3 : Probability of tunneling transfer when coincidence occurs. Assuming a simple exponential wave function with an α rate decay, then $P_3 \simeq \exp(-2\alpha R)$ which is the electronic overlap integral between sites at distance R. A schematic picture of the hopping process is represented in figure 11. Then, in the particular case of a transition metal-containing glass, the following expression for conductivity has been proposed (59) (60).

$$\sigma = \frac{e}{kTR} c (1-c) \nu_o \exp - \frac{W}{kT} \exp(-2\alpha R) \qquad (7)$$

which, in a limited range of temperature, may be compared to an Arrhenius law
$\sigma = \sigma_o \exp - \frac{E_a}{kT}$. Many terms in the relationship (7) are given as a first

approximation and prevent a strict verification of the preexponential term and activation energy with composition. Typical values of σ and E_a are shown on Table 1.

Table 1   Electronic conductivity at 300 K and activation energy for different transition metal containing glasses.

composition	$\log \sigma_{300}$ $(\Omega .cm)^{-1}$	$E_a(eV)$	References
V_2O_5 (thin film)	– 4,3	0.38	61
$Li_{0.1} V_2O_5$ (thin film)	– 4	0.34	61
0.5 V_2O_5 – 0.5 TeO_2	– 6	0.25	62
0.5 V_2O_5 – 0.5 P_2O_5	– 5	0.38	63
0.65 WO_3 – 0.35 P_2O_5	– 6	0.35	64
0.6 TiO_2 – 0.4 P_2O_5	– 9	0.6	65
0.6 MoO_3 – 0.4 P_2O_5	– 9	0.76	66
0.5 FeO – 0.5 P_2O_5	– 10.25	0.61	67

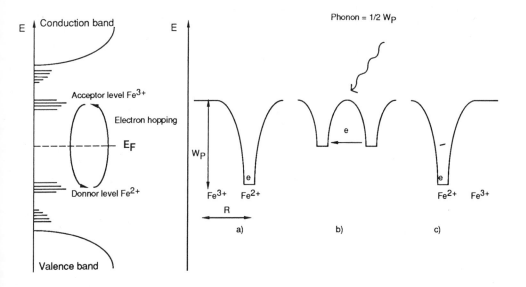

Fe containing glass

A)   B)

Figure 11. Schematic representation of an electron hopping between a donnor level (Fe^{2+}) and an acceptor level (Fe^{3+}).
A) Relative position of the localized donnor and acceptor level between the bandgap. Due to the disordered state, localized state are situated near the conduction and valence bands (band tails).
B) Electron hopping by a phonon assited tunelling process.
a) Before hopping, b) actived transition state, c) after hopping.

The c(1-c) variation is poorly verified (68) and the estimation of R assuming a random distribution of transition metal ion may be criticized (69). A more important unsufficiency is related to the large influence of the nature of the transition metal on the experimental values of the preexponential term and the activation energy. These variations are not explicitly predicted from equation (7). For temperature below the Debye temperature, equation (7) remains valid, but the thermal energy W, needed for a jump to a neighbouring site is rarely provided and the hopping mobility does not come from nearest neighbouring hops, but is likely to result from longer range transitions (R') for which W is smaller. Minimization of the exponential term

$\exp(-2\alpha R' - \frac{W_{R'}}{kT})$ in equation (7) near the Fermi level ($n_{(E)} \simeq$ constant) leads to a most probable hopping distance $R' \simeq \frac{1}{T^{1/4}}$. The resulting conductivity law is then $\sigma = A \exp(-\frac{B}{T^{1/4}})$ as experimentally found.

Such considerations are not essential for electrochemical applications near room temperature, well above the Debye temperature, but are a partial confirmation of the validity of equation (7).

315

II REDOX COUPLES IN ELECTRONIC CONDUCTIVE GLASSES :
THE CHEMIST'S CONTRIBUTION

Electronic acceptor and donor levels allow electronic conductivity. They also fix the Fermi level. From a chemical point of view they are a redox couple with a defined electrochemical potential. In a given glass solvent, generally made of at least one network former and one network modifier, all the possible Red/Ox couples may be accurately and comparatively situated using Red/Ox equilibria. Equilibria may involve oxide ions in the solvent glass such as :

$$V^{5+} + 1/2\ O^{2-} \rightleftharpoons V^{4+} + 1/4\ O_2 \qquad (8a)$$

or two different transition metals :

$$Mn^{4+} + 2\ Fe^{2+} \rightleftharpoons Mn^{2+} + Fe \qquad (8b)$$

For kinetic reasons these equilibria are obtained at high temperature, usually over 1000°C, and it is reasonably expected that the equilibrium concentrations will be maintained during quenching (70-73). Concentrations of various species are then determined by chemical analysis or spectroscopy.

For obvious practical reasons, the reference redox couple is the O_2/O^{2-} couple. but it will be different in each glass since the thermodynamic activity of O^{2-} ions will be different. Figure 12 shows the comparison between different glasses and aqueous solutions (74). Inversion may occur from one glass to another on occasion, but a general behaviour is conserved. For transition metals redox couples, a value of 1.5 V to 2 V separates the most reducing species (i.e., Cr^{2+} / Cr^{3+}) from the most oxidizing species (i.e., Co^{2+}/Co^{3+}).

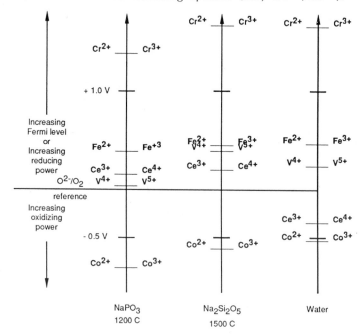

Figure 12   Red/Ox scales referred to the O_2/O^{2-} couple in different glasses compared to those in acid aqueous solution and pure oxides. Data from (74).

It is worth noting that equilibria like 8a fix the molar ratio c and thereby the electronic conductivity of the glass. Equilibria like 8b have been proposed (75) to interpret the influence of a second transition metal dissolved in the glass on its electronic conductivity (76-80).

III MIXED GLASSY AND AMORPHOUS CONDUCTORS AS INSERTION ELECTRODES THE ELECTROCHEMIST'S CONTRIBUTION

The aim of electrochemical studies on intercalation electrodes is to determine theoretical and practical energy densities, to evaluate the diffusion coefficient of the inserted species and, when complete batteries are constituted, to test the reversibility and the cyclability of such systems. All these studies are generally made by classical electrochemical methods such as voltammetry, coulometric titration, impedance spectroscopy and galvanostatic discharges.

The theoretical energy density is a thermodynamic characteristic determined by coulometric titration. This method is based on the measurement of the cell voltage as a function of the intercalation ratio. Experimentally, the intercalation ratio is increased by passing small quantities of current through the cell which results in the modification of the stoichiometry. The voltage is measured after the cell has return to equilibrium (voltage variations less than 1 mV per hour). Under such conditions, it is assumed there is no remaining concentration gradient in the material. The integration of the titration curve allows then the calculation of the theoretical energy density.

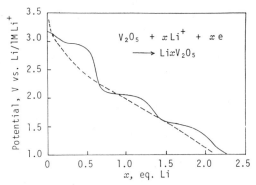

Figure 13   Comparison between discharge curves at 1 mA/cm^2 for vitreous (dashed line) and crystalline (full line) intercalation cathodes lithium cells. From (57).

The energy density is of course, directly proportional to the oxidizing power of the different redox couples involved in the electrode reactions. The density also depends on the slope of the titration curve. For crystalline compounds, titration curves often show several plateaus (Figure 13) corresponding to the coexistence of different phases in equilibrium. For non-crystalline compounds, the decrease in voltage with intercalation ratio is generally monotonous, indicating that the material remains in the same phase even for large intercalation ratios. For amorphous materials, most of the results show higher titration curves than those observed on crystalline analogues leading to higher energy density.

For example, Sakurai and Yamaki (81) have found a value of 900 Wh.kg^{-1} at 1.5 cut-off voltage for a Li/V$_2$O$_5$-P$_2$O$_5$ glass cell with liquid electrolyte. This corresponds to an intercalation ratio of 3 lithium atoms per vanadium atom. The value of the energy density is high compared to the value given for Li/TiS$_2$ cells (500 Wh.kg^{-1}) or for crystalline oxide based cells such as Li/V$_6$O$_{13}$ (666 Wh.kg^{-1}). Very high values, situated between 1200 and 2000 Wh.kg^{-1} have been observed for Li/amorphous Cr$_3$O$_8$ cells [82].

The knowledge of the previously described titration curves allows the determination of the diffusion coefficient by impulse techniques or impedance spectroscopy. The first method was developed by Weppner and al. [83] [84]. Using Fick's diffusion equations in the case of galvanostatic studies, the interfacial concentration in lithium can be approximated by the following expression for small values of time (i.e. $t \ll L^2/D$, where L is the thickness of the electrode and D the diffusion coefficient) :

$$C(0,t) = C_o + \frac{2I_o \sqrt{t}}{zFS \sqrt{\pi D}}$$

(9)

In this expression I_o is the current intensity, S the electrode area and F the Faraday constant. Replacing the unmeasurable concentration by the molar fraction of the intercalated ion, y, $C_o = y/V_m$, with V_m the molar volume of the electrode and differentiating Eq. 9 with respect to the square root of time, the diffusion coefficient can then be expressed by :

$$D = \frac{4}{\pi} \left(\frac{I_o V_m}{zFS}\right)^2 \left(\frac{dE}{dy} \Big/ \frac{dE}{d\sqrt{t}}\right)^2$$

(10)

All terms in this expression are known or can be measured experimentally: dE/dy is the slope of coulometric titration curve and $dE/d\sqrt{t}$ the slope of voltage drop as a function of the square root of time, when a weak direct current is passed through the cell.

Diffusion coefficients can also be determined by impedance spectrocopy [85]. The diffusion of electro-active species at the interface is reflected by a straight line (Warburg impedance) in complexe impedance diagrams, corresponding to the following equation :

$$Z^* = A(j\omega)^{-1/2}$$

(11)

The resolution of the Fick's equation with the approximation $\omega \ll D/L^2$ where ω is the angular frequency of the applied sinusoidal signal, gives :

$$Z^* = \frac{V_m}{2zFS} \frac{dE}{dy} \left(\frac{1}{\omega D}\right)^{1/2}$$

(12)

which leads to :

$$D = \left(\frac{V_m}{zFS}\right)^2 \left(\frac{dE}{dy}\right)^2 \frac{1}{2A^2}$$

(13)

Values of the diffusion coefficient for different systems are summarized in Table 2.

Voltammetry studies are complementary methods used to determine kinetic and thermodynamic parameters. In the case of composite working electrodes, the results are extremely difficult to interpret from a kinetic point of view. Under such conditions, only qualitative informations can be obtained.

Practical energies are obtained by following the voltage drops during discharges under constant current. As would be expected, these discharge curves show that practical energy densities are lower than the theoretical values and decrease with increasing current density [86].

Table 2   Values of different diffusion coefficient for Li^+ and Ag^+ in semi-conductive glasses.

Cathode material	Diffusing species	Diffusion coefficient $cm^2.s^{-1}$	Technique	References
V_2O_5-P_2O_5 films	Ag^+	10^{-15} - 10^{-16} 20°C	Current pulse impedance	85
Cr_3O_8	Li^+	10^{-8}	Current pulse	82
V_2O_5-P_2O_5 films	Li^+	10^{-15}-10^{-16} 20°C	Current pulse impedance	87
V_2O_5-TeO_2	Li^+	10^{-11}-10^{-13} 80°C	Current pulse	86
V_2O_5-P_2O_5	Li^+	10^{-8} - 10^{-9}	Current pulse	88

Depending on the nature of the material, intercalation reactions may be more or less reversible. This reversibility is the most important factor in the use of such electrodes in rechargeable batteries. The more reversible is the intercalation process, the more recyclable the cell may be. Cyclability is tested by discharging and recharging cells, generally under the same constant current. For example, more than 500 cycles have been obtained for a Li/PC-DME-2MeTHF-LiAsF$_6$/V_2O_5-P_2O_5 glass cell [87] under a charging - recharging current density of 0.5 mA.cm^{-2}. Under these conditions, this cell has an average specific capacity of 100 Ah.kg^{-1}. On the other hand, for a V_2O_5 crystalline based cell, only 300 cycles were possible, accompanied by a continuous drop in the specific capacity from 200 to 50 Ah.kg^{-1} (Fig. 14).

Recent comparative studies [89] on vanadium - molybdenum oxides and their P_2O_5 glasses show, once more, that results with amorphous materials display higher energy densities and superior capacity retention on cycling compared to their crystalline analogues.

From a thermodynamic view-point, sulphide based glasses or amorphous materials have similar or slightly inferior properties compared to vitreous oxides. As a matter of fact, the energy density found for aLi/V_2S_5 cell [55] is 950 $Wh.kg^{-1}$, with an intercalation ratio of 2.5 above 1.4 V. These results are quite similar to those obtained on V_2O_5 based cells. Moreover, these sulphide materials generally present a higher electronic conductivity and their cation diffusion coefficients are expected to be better than in oxide glasses. MoS_3 and $MoSe_3S_1$ for example, have a very good electronic conductivity 3×10^{-3} $(\Omega\ cm)^{-1}$ for $MoSe_3S$, and high current densities could probably be obtained [90-92]. Kinetically, these new materials seem, therefore, to be very promising for lithium secondary batteries.

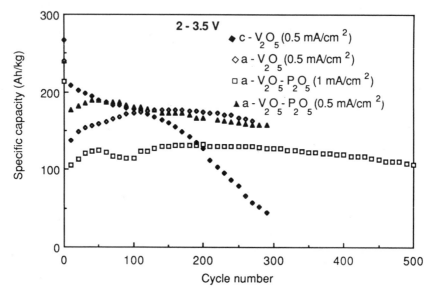

Figure 14   Comparison in capacity changes with cycling for batteries with amorphous and crystalline intercalation cathodes. From (88).

REFERENCES

(1) E. Warburg, Ann. der Physik, 21, 622, (1884)
(2) R.C. Burt, J. Opt. Soc. Am. 11, 87, (1925)
(3) C.A. Levine, R.G. Heitz and W.E. Brown, Proceeding of the 25th Annual Power Source Conference, 50, (1972)
(4) J.R. Akridge and H. Vourlis, Union Carbide Corporation, Solid State Ionics, 18 & 19, 1082-1087, (1986)
(5) L. Jourdaine, V. Delors, J.L. Souquet and M. Ribes, Swiss Patent n° 4764-86-0 (1986)
(6) Y. Ito, K. Miyauchi andT. Kirino, Nipon Kagaku Karishi, 86-3, 445, (1986)
(7) J.H. Kennedy and Z. Zhang, Electrochem. Soc. Meeting, San Diego, (Oct. 1986), n° 746
(8) A. Herczog, J. Electrochem. Soc., 132, 1539, (1985)
(9) C.C. Hunter and M. D. Ingram, Phys. Chem. Glasses 27, 51, (1986)
(10) J.P. Malugani, B. Fahys, R. Mercier, G. Robert, J.P. Duchange, S. Baudry, M. Broussely and J.P. Gabano, Solid State Ionics, 9 & 10, 659-666, (1983)
(11) J. Zarzycki, Les verres et l'état vitreux, Masson Paris, (1982)
(12) W. Kauzmann, Chem. Rev. 43, 219, (1948)

(13) C.A. Angell, J. of Chem. Educat., 47, n° 8, 583-587, (1970)

(14) J. Kawamura and M. Shimoji, J. Non Cryst. Solids 88, 295-310, (1986)

(15) A. Pradel, M. Ribes, To be published in Materials Chemistry and Physics

(16) J.H. Gibbs et E.A. Di Marzio, J. Chem. Phys., 28, 373, (1958)

(17) E.A. Di Marzio et J.H. Gibbs, J. Chem. Phys., 28, 807, (1958)

(18) G. Adam et J.H. Gibbs, J. Chem. Phys. 43, 139, (1965)

(19) C.A. Angell, Solid State Ionics, 9 & 10, 3-16, (1983)

(20) L.W. Strock, Z. Physik. Chem. B, 25, 411, (1934)

(21) T. Takahashi and O. Yamamoto, Electrochim. Acta, 11, 779-789, (1966)

(22) J.N. Bradley and P.D. Green, Trans. Faraday Soc., 63, 424, (1967)

(23) Y. Kawamoto and M. Nishida, J. Non-Cryst. Solids, 20, 393-404, (1976)

(24) T. Minami, H. Nambu and M. Tanaka, J. Amer. Ceram. Soc., 60, n° 5-6, 283-284, (1977)

(25) J.P. Malugani, A. Wasniewski, M. Doreau, G. Robert et A. Al-Rikabi Mat. Res. Bull., 13, 427-433, (1978)

(26) J.P. Malugani, A. Saïda, A. Wasniewski et G. Robert, C.R. Acad. Sc. Paris, C, 101, (9 juillet 1979)

(27) J.P. Malugani and G. Robert, Solid State Ionics, 1, 519-523, (1980)

(28) B. Carette, E. Robinel and M. Ribes, Glass Technology, 24, n° 3, 157-160, (1983)

(29) C.A. Angell, Solid State Ionics, 18 & 19, 72-88, (1986)

(30) S. Sakka and K. Kamiya, J. Non-Cryst. Solids, 42, 403-422, (1980)

(31) J.E. Stanworth, J. Soc. Glass Technol. 30, 54T, (1946) ; 32, 154T, 366T, (1948) ; 36, 217 (1952)

(32) J.L. Souquet, Thèse de Doctorat d'Etat, Université de Grenoble, (1968)

(33) J. Zarzycki et F. Naudin, Verres et Réfractaires, p. 113, (Mai-Juin 1960)

(34) Y.H. Yun and P.J. Bray, J. Non-Cryst. Solids, 27, 363-380, (1978)

(35) K. Kamiya, S. Sakka, K. Matusita and Y. Yoshinaga J. Non-Cryst., Solids, 38 & 39, 147-152, (1980)

(36) J.C. Reggiani, J.P. Malugani and J. Bernard, J. Chim. Phys. 75, n° 9, 849-854, (1978)

(37) D. Ravaine and J.L. Souquet, Phys. and Chem. of Glasses, 18, n° 2, 27-31, (1977)

(38) B. Carette, Thèse de Docteur-Ingénieur, Université des Sciences et Techniques du Languedoc, (1982)

(39) A. Kone and J.L. Souquet, Solid State Ionics, 18 & 19, 454-460, (1986)

(40) M. Tomozawa, J.F. Cordaro and M. Singh, J. Non-Cryst Solids, 40, 189, (1980)

(41) D. Ravaine, J.L. Souquet, Phys. and Chem. of Glasses, 19, 115-120, (1978)

(42) B. Barrau, M. Ribes, M. Maurin, A. Kone and J.L. Souquet, J. Non-Cryst. Solids, 37, n° 1, 1-14, (1980)

(43) A. Pradel and M. Ribes, Solid State Ionics, 18 & 19, 351-355, (1986)

(44) M. Wada, M. Menetrier, A. Levasseur and P. Hagenmuller, Mat. Res. Bull., 18, 189, (1983)

(45) T. Tsuchiya and T. Moriya, J. Non-cryst. Solids, 38 & 39, 323, (1980)

(46) A. Magistris, G. Chiodelli and M. Duclot, Solid State Ionics, 9/10, 611, (1983)

(47) A. Rodriguez, A. Kone and M. Duclot, Solid State Ionics, 9/10, 611, (12983)

(48) R.L. Hervig and A. Navraotsky, J. Am. Ceram. Soc., 68, 314, (1985)

(49) M.S. Whittingham, Progress in Solid State Chemistry, 12, 41 (1978)

(50) D.W. Murphy and P.A. Christian, Science, 205, 651 (1979)

(51) J.I. Yamaki, J. Power Sources, 20, 3 (1987)

(52) M. Armand, in "Electrodes materials for advanced batteries", Ed. D.W. Murphy, J. Broadhead, B.C.H. Steele, Plenum Press, p. 125 (1980)

(53) J. Roussel, R. Brec, Ann. Rev. Mater. Sci., 16, 137-62, (1986)

(54) A.J. Jacobson and S.M. Rich, J. Electrochem. Soc., 127, 779 (1980)

(55) A.J. Jacobson, R.R. Chianelli and M.S. Whittingham, J. Electrochem. Soc, 126, 2277 (1979)

(56) C.M. Lampert, Solar Energy Materials, 11, 1 (1984)

(57) M. Sugawara, Y. Kitada, N. Dato and M. Matsuki, 172th Meeting of the Electrochem. Soc. (18-23 oct. 1987) Honolulu, Hawaï.

(58) D.W. Murphy and F.A. Trumbore, J. Electrochem. Soc., 123, 960 (1980)

(59) N.F. Mott, Adv. Phys. 16, 49 (1967)

(60) I.G. Austin and N.F. Mott, Adv. Phys., 18, 41 (1969)

(61) L. Murawski, G. Gledel, C. Sanchez, J. Livage and J.P. Audières, J. Non Cryst. Solids, 89, 98 (1987)

(62) B.W. Flynn, A.E. Owen and J.M. Robertson, Proc. 10th Int. Conf. on Amorphous and Liquid Semiconductors, Edinburgh, p. 678, (1977)

(63) T. Pagnier, M. Fouletier and J.L. Souquet, Solid State Ionics, 9-10, 649 (1983)

(64) A. Mansingh, V.K. Dhawan, R.P. Tandon and J.K. Vaid, J. Non Cryst. Solids, 27, 309, (1978)

(65) T. Havashi and H. Saito, Phys. Chem. of glasses, 20, 108 (1979)

(66) S. Courant, M. Duclot, T. Pagnier and M. Ribes, Solid State Ionics, 15, 147 (1985)

(67) L. Murawski, J. of Material Science, 17, 2155, (1982)

(68) L. Murawski, C.H. Chung and J.D. Mackenzie, J. Non Cryst. Solids 32, 91-104, (1979)

(69) A.K. Bandyopadhyay, J.O. Isard and S. Parke, J. Phys. D : App. Phys. 11, 2559 (1978)

(70) W.D. Johnston, J. Amer. Ceram. Soc., 48, 185 (1964)

(71) S. Banerjee and A. Paul, J. Amer. Ceram. Soc. 57, 286 (1974)

(72) A. Paul, J. Non Cryst. Solids 71, 260, (1985)

(73) H.D. Schreiber, J. Non Cryst. Solids 84, 129, (1986)

(74) J. Wong and C.A. Angell, in "Glass structure by spectroscopy" (Marcel Dekker Inc. 1976), p. 258

(75) R. Bruckner, J. Non Cryst. Solids 71, 49, (1985)

(76) W. Chomka, D. Samatowitz, O. Gzowski and L. Murawski, J. Non Cryst. Solids 45, 145 (1981)

(77) A.K. Bandyopadhyay, and J.O. Isard, J. Phys. D : Appl. Phys. 10, 199 (1977)

(78) M. Sayer and G.F. Lynch, J. Phys. C : Solid State Phys., 6, 3676 (1973)

(79) A. Mansingh, A. Dhawan, M. Sayer, J. Non Cryst. Solids, 33, 351 (1979)

(80) L.D. Bogomolova, I.V. Filatova, M.P. Glassova and S.N. Spasibkina, J. Non Cryst. Solids, 90, 625 (1987)

(81) Y. Sakurai and J. Yamaki, J. Electrochem. Soc., 132, 512 (1985)

(82) O. Yamamoto, Y. Takeda, R. Kanno, Y. Oyabe and Y. Shinya, J. Power Sources, 20, 151 (1987)

(83) W. Weppner and R.A. Huggins, J. Electrochem. Soc., 124, 1569, (1977)

(84) C. John Wen, B.A. Boukamp, R.A. Huggins and W.Weppner, J. Electro chem. Soc., 126, 2258 (1979)

(85) L. Jourdaine, M. Bonnat and J.L. Souquet, Solid State Ionics, 18-19, 461 (1986)

(86) M. Levy, M.J. Duclot and F. Rousseau, 4th Inter. Meeting on Lithium Batteries, May 1988 Vancouver, to be published in J. Power Sources.

(87) L. Jourdaine, J.L. Souquet, V. Delord, M. Ribes, Solid State Ionic (1988) to be published.

(88) S. Sakurai, S. Okada, J. Yamaki and T. Okada, J. Power Sources, 20, 173 (1987)

(89) M. Uchiyama, S. Slane, E. Plichta and M. Salomon, 172th Meeting of the Electrochem. Soc., (18-23 oct. 1987), Honolulu, Hawaï.

(90) D.M. Pasquariello and K.M. Abraham, Mat. Res. Bull., 22, 37 (1987)

(91) K.M. Abraham, D.M. Pasquariello and G.F. McAndrews, J. Electrochem. Soc., 134, 2661 (1987)

(92) J.J. Auborn, Y.L. Barberio, K.J. Hanson, D.M. Schleich and M.J. Martin, J. Electrochem. Soc., 134, 580, (1987)

SILVER GALLIUM THIOHYPODIPHOSPHATE GLASSES

Claus Wibbelmann and Brian Munro

University of Aberdeen
Department of Chemistry
Aberdeen, Scotland, UK

INTRODUCTION

Silver ion conductors can be used in electrochemical applications [1]. A number of silver sulphide-based glasses have been described in the literature [2,3,4]. Silver iodide-doped thiometaphosphate $(AgPS_3)$ glasses are among those with the highest ionic conductivity at 25°C [3]. The reactions of metals with phosphorous and sulphur or of metal sulphides with phosphorous sulphides $(P_4S_{10}$ or $P_4S_7)$ at high temperatures yield a variety of thiophosphates(V) and (IV) [5,6]. The reaction conditions, stoichiometry, preferred oxidation state and charge of the counter ion(s) control the formation of ortho-(PS_4^{3-}), pyro-$(P_2S_6^{4-})$, meta-di-$(P_2S_6^{2-})$, and hypo-$(P_2S_6^{4-})$ thiophosphates via a number of coupled redox and Lewis acid-base equilibria [7 - 11]:

$$1/4(P_4S_7 + P_4S_9) = 1/2("P_4S_8") \overset{2S^{2-}}{\Longleftrightarrow} P_2S_6^{4-} \overset{1/8S_8}{\Longleftrightarrow} P_2S_6^{2-} + S^{2-}$$

$$\updownarrow$$

$$2\ PS_3^-$$

$$P_2S_6^{2-} \overset{S^{2-}}{\Longleftrightarrow} P_2S_7^{4-} \overset{S^{2-}}{\Longleftrightarrow} 2\ PS_4^{3-}$$

Two and four valent cations form almost exclusively thiohypodiphosphates, whereas trivalent cations will form predominantly orthothiophosphates. In the course of investigations of $M^I M^{III} P_2 S_6$

systems [12] we found that melts of the stoichiometry $Ag^I Ga^{III} P_2 S_6$ will form stable glasses. Sulfide glasses containing octahedral $P_2 S_6^{4-}$ anions have so far not been described.

EXPERIMENTAL

The $AgGaP_2S_6$ glasses were prepared from stoichiometric quantities of the elements in evacuated, sealed quartz glass ampules at $900^{\circ}C$. Air quenching produced homogeneous, clear, amber coloured glasses.

Raman spectra were obtained with a 90° scattering configuration using a Jobin Yvon U1000 double monochromator, a cooled RCA 31034a photomultiplier, a Pacific Instruments model 126 photon-counting system and a Lexel model 77 argon ion laser ($4879\overset{\circ}{A}$, 100mW). Infrared spectra of samples of powdered glass in KBr pellets were recorded on a Perkin Elmer 457 grating spectrometer. Thermal analysis of pieces of glass was performed in sealed aluminum pans using a Perkin Elmer DSC-2C differential scanning calorimeter. Empty aluminum pans were used as a reference.

The electrical contacts for ac impedance measurements were formed on both faces of polished discs by sputter coating with thin films of gold or by applying a silver-metallizing preparation (Johnson Matthey, P-700) followed by baking at $30^{\circ}C$ below T_g for 1 hour. The ac impedance measurements were carried out using the two terminal configuration, a Solartron Frequency Response Analyzer connected to an 1286 Electrochemical Interface controlled by a BBC microcomputer via an IEEE 488 bus (65kHz - 0.01Hz, 0.1V applied across cell and reference R, 10 measurements per decade). The voltage drop over the cell relative to R and the phase angle difference were analyzed to give the real and imaginary impedance.

RESULTS

The glasses prepared in this way were X-ray amorphous, and according to electron microprobe analysis, homogeneous. All of these glasses are quite stable towards hydrolysis. The vibrational spectra of the $AgGaP_2S_6$, $Ag_{1.5}Ga_{0.83}P_2S_6$ and $Ag_2Ga_{0.66}P_2S_6$ glasses are shown in Figure 1. The results of the DSC and ac impedance analysis are summarized in Table 1 and compared with the published data for a silver thiometaphosphate ($AgPS_3$) glass [4].

DISCUSSION

Raman spectroscopy was the main tool for the characterization of the glasses and the monitoring of the response of these systems to variations of the preparation conditions. The stretching frequencies of all species occurring in the equilibria mentioned above are quite characteristic and well separated.

The vibrational spectra of the $Ag_xGa_{(1-1/3x)}P_2S_6$ and $AgPS_3$ glasses shown in Fig. 1 contain a number of relatively sharp lines. The assignment of the features observed in the $AgGaP_2S_6$ spectrum (Fig. 1(d)) is proposed on the basis of a structure model consisting of chains of alternating, edge-sharing P_2S_6 octahedra and GaS_4 tetrahedra. Comparison of the structures of $GaPS_4$ [13], $In_{4/3}P_2S_6$ [14], and $As_2P_2S_7$ [15] supports this model. The vibrations of the P_2S_6 and GaS_4 units are not as strongly coupled as the 'PS_4' and 'GaS_4' [16]. The highest observed fundamental frequency around 660 cm^{-1} is due to a P=S stretch. This compares well with the 653 cm^{-1} found in $As_2P_2S_7$ [17].

The broad band at 550 cm^{-1} is due to the asymmetric stretching vibrations and an intense line at 374 cm^{-1} is due to the symmetric stretching vibration of the $P_2S_6^{4-}$ unit. Both frequencies are close to those observed for $Cs_4P_2S_6$ [9]. The strong band at 316 cm^{-1} is assigned to the symmetric GaS_4 stretching vibration. The deformation vibrations which occur below 300 cm^{-1} will not be discussed in detail. The weak bands at 480 and 420 cm^{-1} in Fig. 1(d) can be assigned to symmetric stretching vibrations of the PS_3^- and $P_2S_6^{2-}$ species [8].

The Raman spectrum of $Ag_{1.5}Ga_{0.83}P_2S_7$ indicates the weakening of the P=S bond, the increased presence of $P_2S_7^{4-}$ (405 cm^{-1}) and most markedly a new band at 340 cm^{-1} which is probably due to a new type of GaS species. In the Raman spectrum of $Ag_2Ga_{0.66}P_2S_{6_{4-}}$ the intensity of the GaS bands is greatly reduced while the $P_2S_6^{4-}$ is still present. The band at 460 cm^{-1} is due either to polysulfide species or PS_3^-.

Table 1. Composition, glass transition temperature, conductivity at $25^{\circ}C$, and activation energy for $Ag_{1+x}Ga_{(1-1/3x)}P_2S_6$ and $AgPS_3$ [4] glasses.

Composition	$T_g(^{\circ}C)$	$T_c(^{\circ}C)$	$\sigma_{25^{\circ}C}(\Omega/cm)^{-1}$	$E_a(eV)$
$AgGaP_2S_6$	291	394	3×10^{-7}	0.47
$Ag_{1.5}Ga_{0.83}P_2S_6$	288	-	5×10^{-6}	0.47
$Ag_2Ga_{0.66}P_2S_6$	268	333	5×10^{-5}	0.48
$AgPS_3$	175	-	2×10^{-5}	0.40

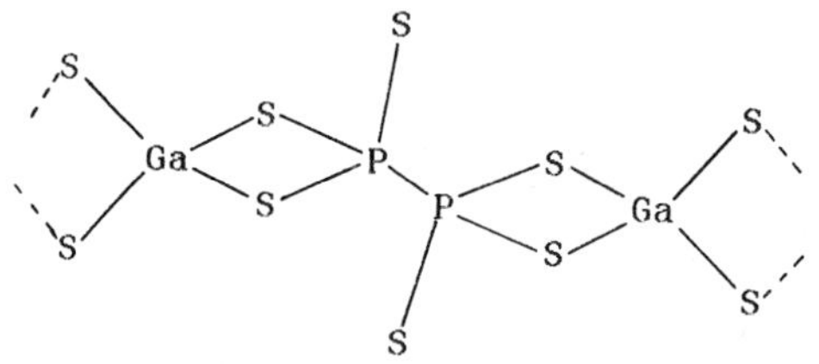

The ionic conductivity of the $AgGaP_2S_6$ glass compares well with other $AgPS_3$ glasses [2]. The substitution of silver for gallium in $AgGaP_2S_6$ increases the ionic conductivity. The silver ion concentration of the limiting composition $Ag_2Ga_{0.66}P_2S_6$ is 23at% compared with 25at% in $AgPS_3$. The former glass has a slightly higher ionic conductivity at $25^{\circ}C$ and a significantly higher glass transition temperature (cf. Table 1). Preliminary results indicate that thiohypophosphate glasses can be doped with AgI.

ACKNOWLEDGEMENTS

Support for this work from the Commission of the European Communities (Contract ST2J-00117) and the British Science and Engineering Research Council (GR/D/56204 and ES 87/51) is gratefully acknowledged. We would also like to thank W.B. Ried and M.D. Ingram, for help with the ac impedance measurements, and the Hoechst AG, Werk Knapsack for a donation of ultrapure red phosphorous.

326

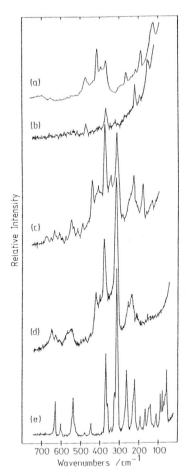

Figure 1.   **Raman spectra of meta- and hypo-thiophosphate glasses.**

(a) $AgPS_3$; (b) $Ag_2Ga_{0.66}P_2S_6$;

(c) $Ag_{1.5}Ga_{0.83}P_2S_6$;

(d) $AgGaP_2S_6$ and a crystallized sample of

(e) $AgGaP_2S_6$

REFERENCES

1. M.D. Ingram, Physics Chem. Glasses, 28(6), 215 (1974)

2. E. Robinel, B. Carette and M. Ribes, J. Non-Cryst. Solids, 57, 49 (1983)

3. J.P. Malugani, G. Robert and R. Mercier, Mat. Res. Bull., 15, 715 (1980)

4. Y. Kawamoto, and M. Nishida, J. Non-Cryst. Solids, 20, 393 (1976)

5. E. Glatzel, Z. Anorg. Chem., 4, 186 ((1893)

6. M.S. Wittingham and A.J. Jocobsen, "Intercalation Chemistry," Academic Press Inc., London (1982)

7. C. Wibbelmann, W. Brockner, B. Eisenmann and H. Schafer, Z. Naturforsch., 38b, 1575 (1983)

8. R. Becker and W. Brockner, Z. Naturforsch., 39a, 1120 (1984)

9. S.J. Cyvin, B.N. Cyvin, C. Wibbelmann, R. Becker, W. Brockner, and M. Parensen, Z. Naturforsch., 40a, 709 (1985)

10. W. Brockner and R. Becker, Z. Naturforsch., 42a, 511 (1987)

11. R. Becker, W. Brockner, and B. Eisenmann, Z. Naturforsch., 42a, 1309 (1987)

12. R. Becker, W. Brockner, and C. Wibbelmann unpublished results Stable glasses are also formed with M^I = Na, K, Cs, Tl.

13. P. Buck and C.-D. Carpentier, Acta Cryst., B29, 1864 (1973)

14. R. Diehl and C.-D. Carpentier, Acta Cryst., B34, 1097 (1978)

15. W. Hoenle, C. Wibbelmann and W. Brockner, Z. Naturforsch., 39b, 1088 (1984)

16. V.S. Dordyai, I.V. Galagovets, E. Yu. Peresh, Yu. V. Voroshilov, V.S. Gerasimenko, and V. Yu. Slivka, Russ. J. Inorg. Chem., 24, 1603 (1979)

17. C. Wibbelmann and W. Brockner, Z. Naturforsch., 36a, 836 (1981)

MODELING, FABRICATION, AND DEVELOPMENT OF

MINIATURE ION SENSORS

Robert F. Savinell, Chi-Jin Chen, Chung-Chiun Liu

Department of Chemical Engineering
Case Western Reserve University
Cleveland, OH 44106

INTRODUCTION

Advances in microelectronic fabrication technology provide a unique approach to the development of chemical and physical sensors. Micro and miniature size sensors can now be produced using these fabrication techniques. Microsensors possess many desirable features which are both fundamentally and practically important. The pioneering work of Fleischmann et al. [1–4] as well as those by Wrightman and his associates [5–10] and by Wrighton et al. [11–14] on microelectrodes provide significant impetus to the development of microsensors.

The advantages of microsensors over conventional size sensors include their obvious compactness, nonlinear diffusion—which could lead to larger current density and quasi–steady state behavior, and the fact that they can be used in media with high specific resistivity [1]. Single microsensors have been used in analytical applications such as the carbon fiber microelectrode used by Wrightman and his associates [5–10] in in vivo biomedical sensing. However, striking progress has been made in the development of microsensors by extending single microelectrodes to an array of many microelectrodes [15–16]. Multiple microelectrodes connected in parallel yield much higher current and thus higher sensitivity for measurement. For example, gold and platinum microelectrode arrays have been fabricated and employed in voltammetry measurements [17]. Other transient techniques such as AC and pulse voltammetry can also be applied to these arrays of microelectrodes [17,18].

Murray et al. [19] combined microelectrode techniques and polymer film modifications and used this polymer–coated array of microelectrodes for measurement of redox electron conduction through polymer film and electrochemical luminescence. A similar, but somewhat different approach, has been developed by Wrighton et al. [11,12,13]. The array of microelectrodes is coated with conducting polymers such as polypyrrole or polyaniline and then used as a molecule–based electronic device based on the potential dependence of the conductance of the polymer. It is also noteworthy that Bard et al. [14] demonstrated that an array of microelectrodes manifests itself similar to rotation ring–disk electrodes (RRDE) and can be used as a RRDE type sensor.

Modeling of various geometric configurations of micro–size electrodes has been investigated by various researchers. The transient current as well as the non–linear (radial) diffusion effects have been demonstrated at micro–disk electrodes [5,22], spherical electrodes [23,24], and cylindrical electrodes [25,26]. Analytical solutions to models of these electrodes are available for these symmetrical geometries [27,28]. In this paper, particular attention will be given to band type electrode geometries. Even

Solid State Microbatteries
Edited by J. R. Akridge and M. Balkanski
Plenum Press, New York, 1990

for single band electrodes, there are no analytical solutions for this non–symmetrical geometry. However, two models have been proposed which describe the transient response of these electrodes under limited conditions. The hemicylindrical model [29,30] works well for very small electrodes while another model is available for large electrodes [31].

From a modeling point of view, arrays of microelectrodes have received even less attention. The current responses of microdisk arrays have been studied by Reller [24,32], assuming the element electrodes are so many that they are all affected by each other. The enhancement of signals from arrays of microband electrodes, relative to single electrodes, in thin–layer flow cells has been modeled by Anderson [33]. These modeling studies emphasize the virtues and limitations of practical applications.

As mentioned, advances in microelectronic fabrication technology have made the production of microsensors a reality. Lithographic reduction techniques, including x–ray and optical ones, permit a clear definition of the geometric configuration of microsensors. Thin and thick film and reactive sputtering metallization processes deposit desired metals, semiconductors or insulating layers. Chemical and plasma etching can properly define the dimensions of microsensors . These technologies result in well–delineated patterning of metallic, insulating and semiconducting surface layers and three dimensional structures. Hence, highly uniform, geometrically well–defined structures can be reproducibly produced–essential elements for microsensor development. This approach also makes it possible to produce an array of identical or different sensors on a single, relatively small substrate. Consequently, on–chip signal processing may be incorporated, thus greatly enhancing the reliability and reproducibility of the sensor. Furthermore, implementation of computational logic for "smart" sensor development using an array of microsensors may become a reality.

This paper describes the general modeling approach and fabrication techniques of microsensors and microsensor arrays. Experimental verification of the efficacy as well as the prospects for sensor development will also be discussed.

THEORIES OF MODELS

Theoretical modeling provides fundamental understanding and insight of characteristics and performance of the microelectrodes of microsensors. Hence, the development of mathematical models to predict the current response from arrays of sensing microelectrodes is undertaken in our research group. In this paper, we will limit the discussion to describing a model of a multiple band electrode array where the kinetics of all charge transfer and associated chemical reactions are relatively fast compared to the mass transfer or diffusion process. We shall first consider the fundamental concepts of describing the electrochemical aspects of the electrode system.

The electrode reaction rates are directly proportional to the ionic flux, N_i [moles/cm^2.sec] of each species i according to Faraday's law. The ionic flux, N_i, in an electrochemical system is established by three driving forces :

1. Migration— Movement of a charged species under the influence of an electric field.

2. Diffusion— Movement of a species because of chemical potential.

3. Convection— Two kinds of convection exist: forced convection, resulting from agitation; and natural convection, resulting from density or temperature gradient.

Therefore, the total ionic flux, N_i, is the summation of diffusional, migration, and convective fluxes, namely for a simple redox reaction,

$$N_i = \frac{i}{nF} = -\frac{z_i F}{RT} D_i C_i \nabla\phi - D_i \nabla C_i + C_i V \tag{1}$$

where i is current density$[Amp/cm^2]$; n is the number of electrons transferred; z_i is the charge of species i; F is the Faraday constant; D_i is diffusivity of species i$[cm^2/sec]$ and V is fluid velocity$[cm/sec]$.

In most practical situations, a supporting electrolyte such as acid, base or salt is added to suppress the ohmic loses. Consequently, the effect of migration becomes negligible. As an initial step, we limit our interests to the diffusion of active species in the absence of convection. Under these conditions, i.e. in a stagnant solution without density gradient, the general flux equation(1) for species i becomes:

$$N_i = -D_i \nabla C_i \tag{2}$$

Combining equation(2) with material balance equation of active species i, the absence of chemical reactions in the solution phase

$$\frac{\partial C_i}{\partial t} = -\nabla \cdot N_i \tag{3}$$

gives Fick's second law for diffusion, namely,

$$\frac{\partial C_i}{\partial t} = D_i \nabla^2 C_i \tag{4}$$

Based on the expression of Eqn.(4), mathematical models for single band electrodes and for band–typed electrode arrays can then be developed.

SINGLE–ELECTRODE MODEL

The analyses of stagnant electrochemical systems presented here assume that the diffusion process is the only mode of mass transport. At the beginning, the reactant species concentration in the fluid is uniform. Then the potential is stepped sufficiently

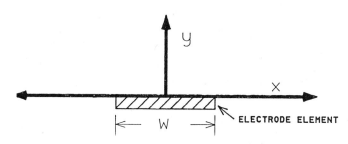

Fig. 1. Coordinates of the Single-Electrode Model.

large so that the rate of the charge transfer reaction is extremely fast and the surface concentration of reactant species drops to zero. The corresponding partial differential equation which describes the unsteady, semi–infinite diffusion to single band electrodes is given by:

$$\frac{\partial C}{\partial t} = D \left(\frac{\partial^2 C}{\partial x^2} + \frac{\partial^2 C}{\partial y^2} \right) \tag{5}$$

where D is the diffusion coefficient and C is the concentration of the electroactive species.(note that the subscripts have been dropped for convenience).

The coordinate configurations are as shown in Fig.1. The initial and boundary conditions can be described by the following:

$t = 0$,	$C = C^{\text{o}}$	(5.1)
$t > 0$, $x \to \infty$,	$C = C^{\text{o}}$	(5.2)
$t > 0$, $x = 0$,	$\frac{\partial C}{\partial x} = 0$	(5.3)
$t > 0$, $y = 0$, $0 \le x \le w/2$,	$C = 0$	(5.4)
$t > 0$, $y = 0$, $w/2 < x$,	$\frac{\partial C}{\partial y} = 0$	(5.5)
$t > 0$, $y \to \infty$,	$C = C^{\text{o}}$	(5.6)

The third boundary condition (5.4) states that, under the conditions of this experiment, the concentration of the active species is reduced to zero immediately upon application of the potential step, i.e., the reaction is purely diffusion controlled. Thus the local diffusion limited current density to the system is written as:

$$i(x) = nFD \left(\frac{\partial C(x)}{\partial y} \right) \Big|_{y=0} \tag{6}$$

While the average diffusion limited current per unit length, i_{avg}, is :

$$i_{\text{avg}} = nFD \int_0^{w/2} \frac{\partial C(x)}{\partial y} \Big|_{y=0} dx \tag{7}$$

This expression is practically useful because only the average current density is measured in experiments.

MULTIPLE–ELECTRODE MODEL

In this case let us consider an array, consisting of N number of band electrodes, which is inlaid on an insulating surface. Each element electrode is parallel, equally spaced and of micrometer dimension in width, w, but relatively macroscopic in length. Fig.2 shows the physical configuration, and the area within the dotted line represents one unit of the electrochemical cell of interest. Because of symmetry, only half of the unit cell is used in this model development. The fraction of the conducting area, r, is defined as $r = w/(w+a)$, where a is the distance between two adjacent electrodes.

Similarly, the corresponding equation which describes the unsteady state, semi–infinite diffusion process at multiple electrodes is the same as that of the Single–Electrode Model(Eqn.(5)). However, the corresponding boundary conditions are slightly different from the Single–Electrode Model. For example, in the multiple electrode model, the x–direction of each electrode is bounded by another element

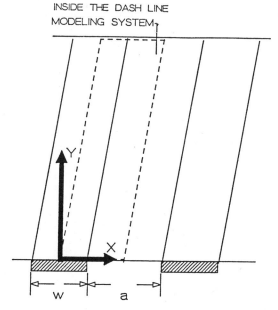

INSIDE THE DASH LINE
MODELING SYSTEM

X AXIS: PARALLEL TO ELECTRODE SURFACE

Y AXIS: PERPENDICULAR TO ELECTRODE SURFACE

Fig. 2.   Coordinate of modeling system
for electrode arrays, W:
electrode width, a: the
distance between two element
electrodes.

electrode; therefore, Eqns.(5.2) and (5.5) are replaced by Eqn.(5.7) and Eqn.(5.8), respectively :

$$t > 0, \quad x = a/2 \qquad\qquad \frac{\partial C}{\partial x} = 0 \qquad\qquad (5.7)$$

$$t > 0, \quad y = 0, \quad w/2 < x < (w+a)/2, \qquad\qquad \frac{\partial C}{\partial y} = 0 \qquad (5.8)$$

The current response of the outer element electrodes will not be the same as the others because the current behavior of those elements which lie in the interior interact with each other.   However, the outer element electrodes are just partly affected by others. Thus, the influence of the outer element electrodes upon total current becomes consequential as N (total numbers of element electrodes) is small, or as r (fraction of conducting area) approaches 1. The combined contributions of the interior element as well as the outer element electrodes is used to predict the current transients at a band–typed electrode array.

For the outer element electrode, it is assumed that the half unit facing inward interacts with others, the same as an interior element electrode, and that the other half unit behaves like a single element electrode.   The total current of the outer element electrode is then defined as the summation of two half units predicted by the Multiple–Electrode Model and two half units predicted by the Single–Electrode Model. The resulting expression can be written as :

$$I = 1 \left[(i_{avg})_{int} + (i_{avg})_{ext} \right] \qquad\qquad (8)$$

where, $(i_{avg})_{int}$ and $(i_{avg})_{ext}$ represent the current per unit length of the interior half

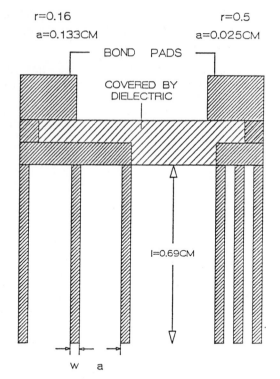

r=0.16   r=0.5
a=0.133CM   a=0.025CM

BOND   PADS

COVERED BY
DIELECTRIC

l=0.69CM

w   a

Fig. 3.   Physical configuration of
          test electrode arrays.
          Electrode width, w=).0254cm
          for both arrays; spacing
          parameter, r=w/ (w∸a).

unit and of the outer half unit, respectively.   Therefore, the total current for an N
element electrode array is given by :

$$I_{total} = l\,[\,2\,(N\!-\!1)\,(i_{avg})_{int} + 2\,(i_{avg})_{ext}\,] \tag{9}$$

The diffusion equation (Eqn.(5)) associated with the initial and boundary
conditions (Eqn.(5.1) to Eqn.(5.8)) for both multiple and single electrode models are
solved by a numerical method. The Hopscotch finite difference algorithm proposed and
analyzed by Gourlay[34,35] is chosen to accomplish the simulation, because this
algorithm is more accurate than conventional explicit methods, and it also requires less
storage space and is faster than fully implicit methods.

EXPERIMENT

A self–contained three–electrode micro sensor (i.e., working, counter and
reference electrodes all fabricated on the same aluminum substrate) has been
introduced in response to the needs of the plating and surface finishing industry and can
be used here to illustrate the developed model.   The Multiple–Electrode Model
presented earlier characterizes the sensing (working) electrode element (band–typed
electrode array) on the sensor device.   Interactions between element electrodes in the
band–shaped electrode array and their effects on the transient response to a potential
step perturbation were investigated using the reduction reaction of ferric ions.   The
standard electrode potential of the ferric/ferrous redox reaction, $Fe^{3+} + e \rightleftharpoons Fe^{2+}$,
occurs at 0.529 volt positive vs. the calomel reference electrode (SCE) at $25^{O}C$. An

overpotential which is stepped to a value of several hundred millivolts negative from the standard electrode potential of the reaction will produce a net cathodic reaction resulting at a limiting current.

In this section, the cell, solution, and electrical setup for experiments will be given in detail. The experimental results, as well as their comparisons with the Multiple–Electrode Model, will be given later.

Two different band–typed electrode arrays were fabricated on the same substrate by using photographic reduction and thick film metalization techniques: one array consisted of three element electrodes which are spaced with a spacing parameter, r = 0.16 ; the other array also consisted of three element electrodes with a closer spacing giving r = 0.5. Both arrays are of same width, w = 254 μm, and of same length, l = 0.69 cm. Fig.3 shows the physical configuration. The fabrication techniques will be described in the next section.

FABRICATION

Microelectronic fabrication is proven technology for depositing well–controlled, defined metal, semiconductive or insulative materials. The application of this technology for the formation of sensors and microsensors has not, however, been fully exploited. In recent years the fabrication of ion selective field effect transistors, ISFET, as well as microsize physical and electrochemical sensors has demonstrated the viability of using microelectronic fabrication technology. In this discussion, we shall focus on the microelectronic fabrication procedures which are directly relevant to the fabrication of microelectrodes or microsensors, including

1. Lithographic reduction and pattern generation.

2. Thin and thick film and sputtering metallization techniques.

3. Chemical, plasma and other etching techniques.

The details, specificity, and selection of special methods or equipment, can be found exclusively in reference books on microelectronic processes. We plan herein to use our selected system with minor additions to illustrate these processes.

Photolithographic reduction is an established method for transferring a circuit design (in our case, a microelectrode or elements of a microelectrode) into a small scale. Usually the design is first drawn on a large scale either using computer aided design or drafting techniques. A pattern of this design is then produced into a photomask by various techniques. Electron beam and x–ray lithography are advanced techniques providing a high degree of resolution, whereas optical lithography is somewhat limited in its resolution, but less expensive in operation. In our development, optical lithographic techniques are used to produce a photomask which is used for later metallization. If a further reduction and repeat of the pattern is desired, a step–and–repeat camera can be used to accomplish these objectives.

Thick film metallization is basically a silk–screen process. It is relatively simple and straight–forward. By comparison to thin film techniques, thick–film metallization may not provide as high a degree of resolution. Obviously, the choice of which technique to use is based upon the needs. In the thick film process, the desired metal, such as silver, gold or platinum, etc. commercially comes in the form of printing ink which contains the metal and a bonding material. In the thick film printing process, the pattern to be metallized will first be transferred onto a fine mesh stainless screen.

Standard thick film silk–screening equipment involves a holder for the substrate usually employing a vacuum suction mechanism, a mounter for the screen, and a spreader for applying the metallic ink. After the metal is silk–screened onto the

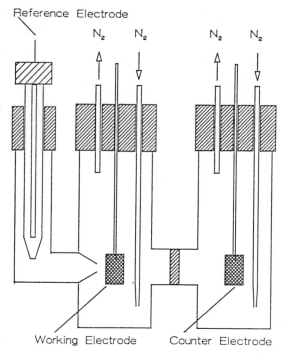

Fig. 4. Three-compartment cell and
electrodes setup.

substrate, the printed substrate is placed in a furnace at a pre–determined temperature to remove the bonding material.

For thin film metallization, the metal deposition is usually done in the chamber of a high vacuum system. The metal is deposited over the total surface area of the substrate. The desired pattern is then developed by etching techniques. Thin film metallization can also be accomplished by sputtering which is also an established microelectronic fabrication process. By comparison, the thin film technique is a little more involved than the thick film process. On the other hand, thin film patterns are usually more refined.

Because in the thick film metallization process the desired microelectrode pattern is directly produced onto the substrate, no etching process is involved. In this film processing, etching is required to produce the pattern. Etching techniques include chemical and plasma etching which, again, are standard microelectronic fabrication techniques.

The experiments reported here employed the thick film process. A one inch square alumina substrate having a thickness of 0.025 inch (Coors Industries) was used for the formation of the sensor. The microelectrode array was platinum and a platinum ink (#A4731, Engelhard Industries) was used in the thick–film metallization. Lead wires were soldered onto the bonding pads and secured with an epoxy containing silver particles for electrical conduction. This working array was then used for experimentation.

TESTING PROCEDURES

The testing procedure described here is quite general for evaluating the performance characteristics of microelectrodes. The reagents employed here, however, is for the reduction of ferric ions. In this study, both working electrode arrays ($r = 0.16$

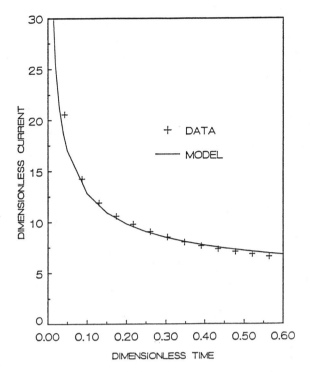

Fig. 5. Model predictions and experimental data at the electrode array of r=016, applied potential: 0.25V, solution composition: $[ZnSO_4]=1M$, $[Fe^{3+}]=45mM$.

and r = 0.5) were tested separately. A three compartment cell as shown in Fig.4 was employed in this experiment. An external platinum counter electrode and a calomel reference electrode(SCE) were used. A potentiostat was used to apply the potential steps, and the resulting current/time transients were recorded on an X–Y recorder. The potential step applied in the measurements was –0.25 volt vs. SCE. The working solution contained ferric ion ranging from 9 mM to 54 mM in a 1 M $ZnSO_4$ solution. In order to suppress the oxygen content in the test cell, nitrogen was purged through the solution prior to any measurement, and the solution was blanketed with nitrogen during the experiment.

MODEL SIMULATION RESULTS AND EXPERIMENTAL VERIFICATION

The validity of the Multiple–Electrode Model presented in the previous section was verified by experimental investigation by using the reduction of ferric ions in a zinc sulfate solution which was deoxygenated. The experimental data are compared with the model predictions.

Since the total number of elements in an electrode array used in the experiments is three, the total current predicted by the Multiple–Electrode Model in comparison according to Eqn.(9), is :

$$I_{sim} = 1\,[\,4\,(i_{avg})_{int} + 2\,(i_{avg})_{ext}\,] \tag{10}$$

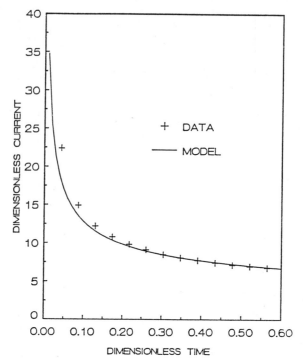

Fig. 6.   Model predictions and experimental
data at the electrode array of
r=0.5, applied potential: -0.25V,
solution composition: [ZnSO$_4$]=1M,
[Fe^{3+}] =45=45mM.

And the total dimensionless theoretical current per unit length is :

$$\bar{i}_{avg} = I_{sim} / nFDC^o l \qquad\qquad (11)$$

This dimensionless quantity will be used to compare with experimental data obtained.

In order to compare with the current transients predicted from the Multiple–Electrode Model, the actual current, I_{exp},(with unit of Amps) measured from experiments is transformed into dimensionless current per unit length, $\bar{i}_{avg}$ by:

$$\bar{i}_{avg} = \frac{N}{nFDC^o l} \times I_{exp} \qquad\qquad (12)$$

In this experiment, n = 1, F(Faraday's constant) = 96500 coul/mole, D (diffusivity of ferric ion) = 6 × 10^{-6} cm^2/sec, C^o(concentration of ferric ion) ranges from 9 to 54 mM, and l(length of the electrode) = 0.69 cm.

The comparisons of results of dimensionless current per unit length with dimensionless time for experimental data(transformed as Eqn.(12)) and model predictions (calculated from Eqn.(11)) are shown in Fig.5 for the electrode array of r = 0.16, and in Fig.6 for the electrode array of r = 0.5. Excellent agreement between experimental data and model predictions are found in the comparisons.

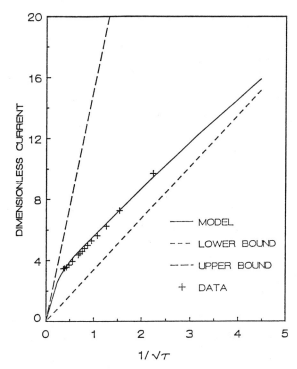

Fig. 7. Dimensionless current of experimental data and model prediction confined by lower and upper bound. Experimental conditions: electrode arrays of r=0.16, solution composition: $[ZnSO_4]=1M$, $[Fe^{3+}]=45mM$.

Two characteristic limiting bounds for electrode arrays are found from the current–time relationship. At very short time period, when the diffusion layer thickness is relatively thin compared to the width of each single electrode, a lower bound is formed since the current response behaves similar to semi–infinite linear diffusion to each electrode. Consequently, the lower bound is described by the Cottrell equation[36] which is rewritten in dimensionless form as:

$$(\bar{i}_{avg})_{low} = \frac{Nw}{\sqrt{\pi Dt}} \quad (13)$$

where N is the number of electrodes having width w.

As time increases, the diffusion layer thickness of each electrode increases gradually, and finally the diffusion layers overlap adjacent electrodes. The array of single electrodes now form an apparent planar electrode. If N (number of element electrodes) is large or r (fraction of conducting area) is small, the apparent planar electrode is large enough to make the edge effect negligible. Therefore, the current behavior approaches an upper bound similar to semi–infinite planar diffusion to an apparent planar electrode:

$$(\bar{i}_{avg})_{up} = \frac{w[(N-1)/r+1]}{\sqrt{\pi D t}} \quad (14)$$

Two characteristic limiting bounds anticipated from model predictions are actually found from experimental results with the platinum electrode arrays of r = 0.16 and N = 3. Fig.7 shows that the current transients at testing electrode arrays are well confined within two limiting bounds :

$$\text{lower bound}: \quad (\bar{i}_{avg})_{low} = \frac{3w}{\sqrt{\pi Dt}} \tag{15}$$

$$\text{upper bound}: \quad (\bar{i}_{avg})_{up} = \frac{40.5w}{\sqrt{\pi D t}} \tag{16}$$

CONCLUDING REMARKS

The unique qualities advantages and limitations of microelectrodes and microsensors have been described. Our study shows the modeling and fabrication of the microelectrode and the experimental results that confirm the modeling. Obviously, the use of microelectronic fabrication techniques to produce microelectrodes and microsensors has many practical applications. Physical, chemical and biochemical sensors in miniature sizes have been produced and have demonstrated the validity of this approach. The incorporation of on–chip or near–the chip signal processing is a logical approach to further extend the usefulness of these sensors. One of the size limitations, however, is the power source required for the operation and signal process of the sensor. This is an important consideration for implantable or in vivo applications. Thus, it is fitting that the development of a microbattery such as discussed in this symposium would have a direct impact on the advancement of microsensors in the years to come.

ACKNOWLEDGEMENT

Support for this research has been provided by American Electroplates and Surface Finishers Society, National Sciences Foundation (CBT – 8696073), Edison Sensor Technology Center of the State of Ohio, and Resource for Biomedical Sensor Technology (NIH).

REFERENCES

1. M. Fleischmann, S. Pons, D. R. Rolison and P. P. Schmidt, Ultramicroelectrodes, (Datatech Systems, Inc., Science Publishers, 1987), and references therein.

2. A.M. Bond, M. Fleischmann and J. Robinson, J Electroanal. Chem., 168 (1984) 299, 172 (1984) 11.

3. M. Fleischmann, F. Lasserre, J. Robinson and D. Swan, J. Electroanal. Chem ., 177(1984) 97.

4. M. Fleischmann, F. Lasserre and J. Robinson, J. Electroanal. Chem., 177 (1984) 115

5. M. A. Dayton, A. G. Ewing and R. M. Wrightman, Anal. Chem., 52 (1980) 2392.

6. M. A. Dayton, A. G. Ewing and R. M. Wrightman, J. Electroanal. Chem., 146 (1983) 189

7. P. M. Kovach, A.G. Ewing, R. L. Wilson and R. M. Wrightman, J. Neuroscience Meth., 10 (1984) 215

8. R. S. Kelly and R. M. Wrightman, <u>Anal. Chem., Acta,</u> <u>187</u> (1986) 79

9. W. G. Kuhr and R. M. Wrightman, <u>Brain Res.</u>, <u>381</u> (1986) 168.

10. H. S. White, G. P. Kittlesen and M. Wrighton, <u>J. Am. Chem. Soc,</u> <u>106</u> (1984) 5375, 7389, <u>107</u> (1985) 7373.

11. J. W. Thackeray, H. S. White and M. Wrighton, <u>J. Phys. Chem.</u>, <u>89</u> (1985) 5133.

12. E. W. Paul, A. J. Ricco and M. Wrighton, <u>J. Phys. Chem.</u>, <u>89</u> (1985) 1441.

13. A. J. Bard, J. A. Crayson, G. P. Kittlesen, T. V. Shea and M. S. Wrighton, <u>Anal. Chem.</u>, <u>58</u> (1986) 2321.

14. N. Sleszynski, J. Osteryoung and M. Carter, <u>Anal. Chem.</u>, <u>56</u> (1984) 130

15. K. Aoki and J. Osteryoung, <u>J. Electroanal.</u>, Chem., <u>125</u> (1981) 315.

16. W. Thormann, P. van der Bosch and A. M. Bond, <u>Anal. Chem.</u>, <u>57</u> (1985) 2764.

17. W. Thormann and A.M. Bond, <u>J. Electroanal. Chem.</u>, <u>218</u> (1987) 187.

18. C. E. Chidsey, B. J. Feldman, C. Lundergren and R. W. Murray, <u>Anal. Chem.</u>, <u>58</u> (1986) 601.

19. B. J. Feldman and R. W. Murray, <u>Anal. Chem</u>, <u>58</u> (1986) 2844.

20. A. G. Ewing, B. Feldman and R. W. Murray, <u>J. Electroanal. Chem.</u>, <u>172</u> (1984) 145.

21. M.A. Dayton, A. G. Ewing and R. M. Wrightman, <u>Anal. Chem.</u>, <u>52</u> (1980) 2392

22. B. Scharifker and G. J. Hills, <u>J. Electroanal. Chem.</u>, <u>130</u> (1981) 81.

23. A. J. Bard and L. R. Faulkner, <u>Electrochemical Methods</u>, (Wiley, New York), 1980, Chapter 4.

24. H. Reller, E. Kirowa–Eisner and E. Gileadi, <u>J. Electroanal. Chem.</u>, <u>138</u> (1982) 65.

25. K. Aoki, K. Honda, K. Tokuda and H. Matsuda, <u>J. Electroanal. Chem.</u>, <u>186</u> (1985) 79.

26. C. A. Amatore, M.R. Deakin and R. M. Wrightman, <u>J. Electroanal. Chem.</u>, <u>206</u> (1986) 23.

27. D. Shoup and A. Szabo, <u>J. Electroanal. Chem.</u>, <u>160</u> (1984) 1.

28. K. Aoki, K. Akimoto, K. Tokuda, H. Matsuda and J. Osteryoung, <u>J. Electroanal. Chem.</u>, <u>171</u> (1984) 219.

29. P. M. Kovach, W. L. Caudill, D. G. Peters and R. M. Wrightman, <u>J. Electroanal. Chem.</u>, <u>185</u> (1985) 285.

30. C. A. Amatore, B. Fosset, M. R. Deakin, R. M. Wrightman, <u>J. Electroanal.Chem.</u>, <u>225</u> (1985) 33.

31. K. B. Oldham, <u>J. Electroanal. Chem.</u>, <u>122</u> (1981) 1.

32. H. Reller, E. Kirowa–Eisner and E. Gileadi, <u>J. Electroanal. Chem.</u>, <u>161</u> (1984) 247.

33. Anderson, J. L. <u>Extended Abstracts</u>, <u>88–1,</u> (The Electrochemical Society, Pennington, NJ, l988) p.657.

34. D. Shoup and A Szabo, <u>J. Electroanal. Chem.</u>, <u>160</u> (1984) 1.

35. A. R. Gourlay, <u>J. Inst. Math Appl.</u>, <u>6</u> (1972) 375.

36. F. G. Cottrell, <u>Z. Phys. Chem.</u>, <u>42</u> (1902) 385.

SOLID STATE BATTERIES

James R. Akridge

Eveready Battery Co., Inc.
Technology Laboratory
P.O. Box 45035
Westlake, Ohio   44145

INTRODUCTION

Significant resources have been devoted to the investigation of solid electrolytes since the early 1960s with the beta-aluminas having had the most monetary resources expended in R&D of any solid electrolyte. The only solid state cell which has achieved commercial success is the Li/I_2 cell and it is probably not truly solid state [1]. Good reviews are available on the state of the art today [2]. This paper will describe advances made in solid state cell systems over a period of several years of development at Eveready Battery Co., Inc.

Solid state cell systems have not progressed much beyond the experimental stages of the laboratory for many reasons. One reason is that they offered no advantages over preexisting cell systems. More importantly, the current rates were so low that it was questionable whether solid state cells could operate existing devices--even CMOS memory backup was questionable over the $-55^{\circ}C$ to $+125^{\circ}C$ operating range of the latest generation of CMOS. Another very critical reason was the question of fabrication of a durable and reliable interface between solid state materials of greatly differing mechanical properties.

The difficulty of identifying advantages over existing cell systems can be addressed by solid state cells' targeting the high temperature end of device performance. If a designer has a circuit for operation up to +125°C or higher, traditional aqueous or nonaqueous systems will not satisfy that temperature range. CMOS backup is an application which should be good for a solid state cell provided the cell has **MORE** current capability than the CMOS it is backing up at all specified operating temperatures. It is important that the cell performance enhance (certainly not reduce) the circuit designer's capability. A solid state cell system must be as stable and long lasting as the device in which it is inserted if the cell is not replaceable. Questions of energy density are not important for state-of-the-art CMOS memory backup. Questions of cell stability over years of in use service must be answered--especially for military applications. The in use service of a cell or battery to maintain memory can exceed 20 years. Solid state cells must be extremely stable to stand the test of time.

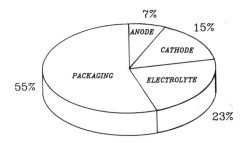

Figure 1. **Typical 2025 Coin Cell Volume Utilization.**

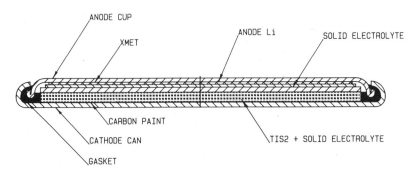

Figure 2. XR Series of Solid State Batteries--Standard Construction.

The other difficulties mentioned--ionic conductivity, interfacial formation, etc. relate to the manufacturability of the solid state cell system. A material of sufficient ionic conductivity exists [3,4]. The problem of interfacial formation (for the chalcogenide glasses) has a satisfactory solution [5].

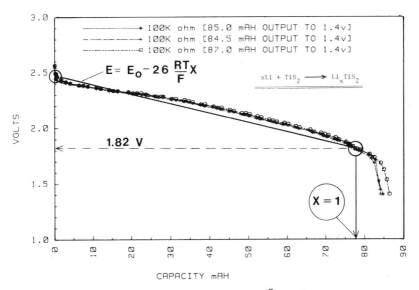

Figure 3. XR2025 Cell Showing 21°C Discharge with Superimposition of Linear Approximation to Discharge Curve.

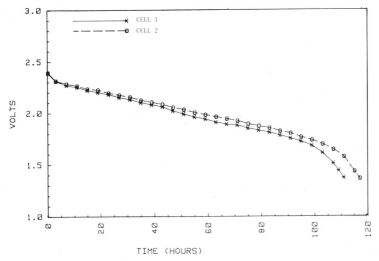

Figure 4. 13K ohm, 0°C Discharge of XR2025 Solid State Cell.

EXPERIMENTAL

The construction of solid state primary cells has been published by Akridge and Vourlis [6] and will not be repeated. The general form of the cell is (refer to Figure 2):

$$\text{Li or Li(Al)}/5\text{LiI} \cdot \text{Li}_8\text{P}_4\text{O}_{0.25}\text{S}_{13.75}/\text{TiS}_2 + \quad \begin{array}{c} \text{Solid} \\ \text{Electrolyte} \end{array}$$

Cells are fabricated in an inert atmosphere dry box and then sealed in standard 2016, 2025, 1620, or 1130 cell packages. The numbers refer to diameter and height in the following manner: The first two numbers are the cell diameter and the second two numbers divided by 10 are the cell height. Dimensions are in millimeters.

BATTERY DESIGN

If one calculates the capacity of lithium metal the following is obtained:

$$\frac{96,487 \text{ Coulombs Hour}}{3600 \text{ Seconds Equiv.}} = 26.80 \text{ Amp Hours/Equivalent}$$

$$\frac{26.80 \text{ Amp Hours}}{6.941 \text{ grams}} = 3.86 \text{ Amp Hours/Equivalent}$$

In units of volume (3.86 Amp Hours/gram) x 0.53 grams = 2.05 Amp Hour/cm^3. If the same calculation is performed using zinc metal--5.85 Amp Hour/cm^3 is obtained. Lithium is a good anode because it produces twice the voltage as zinc with common cathode materials, not because it has a higher capacity density. The theoretical energy density of a Zn/MnO$_2$ cell and a Li/TiS$_2$ cell (solid state incorporating the chalcogenide glasses) is: Zn/MnO$_2$ = 0.96 Watt Hour/cm^3, and Li/TiS$_2$ = 0.95 Watt Hour/cm^3. These numbers have taken into account the volumes of the anode and cathode but not the volumes for the packaging and the liquid or solid electrolyte (concentrated KOH for alkaline cells and vitreous solid electrolyte for solid state cells). When the packaging and all other factors are added, about 20% of the above values are actually realized in the application. Figure 1 shows the relationship between active

346

components, packaging, and electrolyte volumes. Figure 2 shows the cross section of a 20mm diameter primary solid state cell.

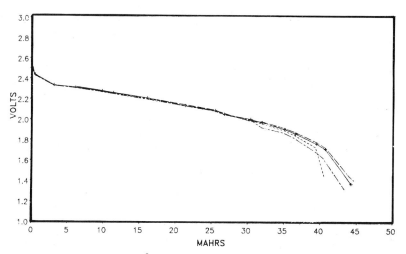

Figure 5. 10K ohm, 120°C Discharge of XR2025 Solid State Cell.

CELL PERFORMANCE

The performance of a solid state cell at ambient temperature is inextricably correlated with the "goodness" of the anode and cathode interfaces with the solid electrolyte. Figure 3 shows a typical discharge curve at room temperature of a $Li/glass/TiS_2$ primary cell. Superimposed upon the discharge curve is a linear approximation of the discharge curve which has been used by Atlung [6]. (See "The Composite Electrode Theory for Solid State Batteries - Attempted Experimental Verification" in this volume.) One can see that the discharge of the solid state system is well represented by the linear approximation. The decreasing voltage of the solid state cell could be used to project remaining capacity by comparison of cell voltage with a 'standard'. The standard could be the linear approximation shown in Figure 3. There are three cells represented in the data of Figure 3. Consistency of discharge is another characteristic of good interfaces.

347

Discharge at subambient temperatures is shown in Figure 4. The discharge load is much greater (approximately $150\mu A/cm^2$) than one expects for a CMOS memory at $0^\circ C$. State-of-the-art CMOS should have current requirements in the nanoampere range at $0^\circ C$. The discharge curve demonstrates that high rates are achievable for solid state systems at low temperature. Discharge curves have been published for temperatures to $-40^\circ C$ [7]. Figure 5 shows typical performance at $120^\circ C$ and Figure 6 at $200^\circ C$. LiI is replaced by LiBr in the solid electrolyte at temperatures above $120^\circ C$. Figure 7 shows the pulse capability of the solid state system at ambient and elevated temperature. The solid state is characterized by high pulse capability (or high current drain) for short times and low continuous discharge currents for long times. As expected, as the temperature is increased, the current capability of the solid state is improved.

Figures 8 and 9 demonstrate another characteristic of solid state energy storage systems. Individual cells were potted in epoxy. Epoxy or other overcoating or overpackaging is required for temperature cycling because differential expansion of the metal vs. plastic gasket will cause some loss of seal integrity. Cells were cycled between $-65^\circ C$ and $150^\circ C$ one hundred times. The results shown in the figures show that the solid state can survive extremes of temperatures. Eight individual cells are

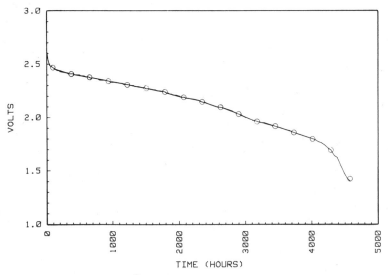

Figure 6. 80K ohm, $200^\circ C$ Discharge of High Temperature Solid State Cell.

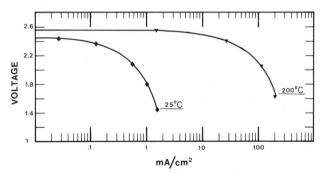

Figure 7. XR2025 Cell Showing 2 Second, 21°C Pulse Performance.

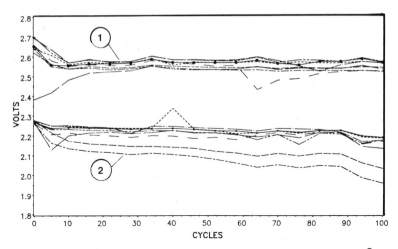

Figure 8. XR2025 Cell Thermal Cycling Between +150°C and -65°C 1) Open Circuit Voltage--21°C After Cycle # Indicated 2) Closed Circuit Voltage--21°C, 2K ohms for 2 Seconds.

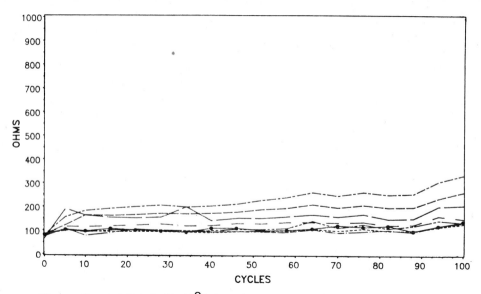

Figure 9. XR2025 Cell 21°C Impedance vs. Thermal Cycle Number.

represented in the figures. The overpackaging for commercial applications is shown in Figure 10.

Shelf life is an important characteristic of any battery system and especially important for solid state systems. Shelf life has been often touted as a strong selling point for solid state cells. Figure 11 shows the performance of fresh, 2-, and 5- year-old solid state cells of the configuration [S.E. = Solid Electrolyte]:

$$Li/5LiI \cdot Li_4P_2S_7/TiS_2 + S.E.$$

The first solid state cells using chalcogenide glasses were fabricated at Eveready Battery Co., Inc. in 1981. The engineering and material improvements to solid state technology since that time resulted in the discovery that isostatic compression at 5.5×10^8 Pascals immediately gave improvements in discharge rates and overall cell integrity. The cell of Figure 11 is showing improved service with age. This can be explained by theorizing that the interface is improving with aging because of diffusion of lithium at the lithium/electrolyte interface into the separator glass layer. Isostatic compression of the cell of Figure 11 was

350

at 2.25 x 10^8 Pascals. This lower pressure does not give optimum service.

The aged cells provided an output of 95+% of their theoretical capacity. This is better than the fresh cell and could be expected due to the before mentioned diffusion effect of improving the Li/glass interface. The 2- and 5-year-old-cells show no difference in output demonstrating that over a 5 year period no degradation of the cell can be detected.

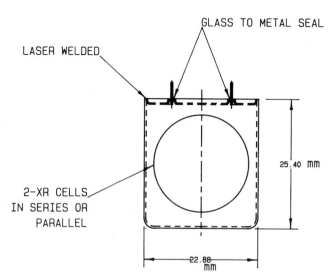

Figure 10. **Packaging Configuration for EVEREADY XP4025 and XP4050 Series of High Temperature Solid State Cells.**

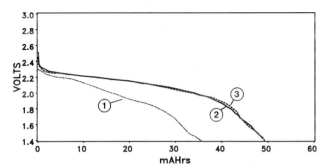

Figure 11. 50K ohm, 21°C Discharge of Solid State Cells (1) Fresh (2) After 2 years ambient shelf (3) 5 years shelf.

CONCLUSION

Solid state energy cells are able to perform over a temperature range which exceeds that of most aqueous and nonaqueous cells. The shelf life data is demonstrating that the solid state cell is stable over extended time periods. With continued research new and improved solid electrolytes will be discovered which will further enhance the capability of solid state cells.

REFERENCES

1) J.B. Phipps, T.G. Hayes, P.M. Skarstad, and D.F. Untereker, Solid State Ionics, 18/19, 1073 (1986).

2) B.B. Owens and M.Z.A. Munshi, Proceedings of the Symposium on "History of Battery Technology", ed. by A.J. Salkind, Proceedings Volume 87-14 of The Electrochemical Society, 1987, pg. 199.

3) J.R. Akridge, U.S. Patent 4,599,284, July 8, 1986.

4) R. Mercier, J.P. Malugani, B. Fahys, and G. Robert, Solid State Ionics, 5, 663 (1981).

5) J.R. Akridge and H. Vourlis, U.S. Patent 4,477,545, October 16, 1984.

6) S. Atlung, Solid State Batteries, pg. 129, ed. by C.A.C. Sequeira and A. Hopper, Martinus Nijhoff Publishers, 1985.

7) J.R. Akridge and H. Vourlis, Solid State Ionics, 18/19, 1082 (1986).

RECHARGEABLE SOLID STATE CELLS

James R. Akridge and Harry Vourlis

Eveready Battery Co., Inc.
Technology Laboratory
P.O. Box 45035
Westlake, Ohio 44145

INTRODUCTION

The **concept** of a solid state energy storage cell has been proven technically feasible by numerous researchers over a period of at least 30 years [1]. The related **concept** of a solid state rechargeable cell has also been proven technically feasible [2,3]. The very important factor of commercial viability for solid state energy storage cells and batteries (solid state is herein meant to be no liquid or gaseous phases throughout the range of cell operating capability) has not been demonstrated [4].

EXPERIMENTAL

The physical form of the cell has been published for the primary construction by Akridge and Vourlis [5]. The secondary cell construction is experimental and designed as a test vehicle to determine chemical formulation of the solid state cell components and engineering parameters required to manufacture a reliable secondary solid state cell. The general form of the cell is [S.E. = $\underline{S}$olid $\underline{E}$lectrolyte]:

$$Li/5LiI \quad Li_8P_4O_{0.25}S_{13.75}/TiS_2 + S.E. + \begin{matrix} Electronic \\ Conductor \end{matrix}$$

Solid State Microbatteries
Edited by J. R. Akridge and M. Balkanski
Plenum Press, New York, 1990

TABLE 1

CATHODE FORMULATION OF RECHARGEABLE CELLS

Components	% By Weight
TiS_2	5 - 51
Solid Electrolyte	42 - 72
Electronic Conductor	0 - 33

This is nearly a twin of the primary construction. Several points require explanation. The anode has been chosen as a pure lithium foil not an alloy of aluminum, silicon, etc. The cathode is TiS_2 admixed with solid electrolyte and metal powders to improve electronic conductivity [6,7]. The solid electrolyte indicated as separator is one of many possible stoichiometries [8].

The anode was selected to be pure lithium metal foil because whether thin film or thick film it has the highest energy density, gives the highest cell potentials, and is the easiest form of lithium to purchase and handle. A few cells were built with Li/Al alloy anodes to compare their performance with pure Li foil anodes. The Li/Al alloy did not enhance cell performance or cycle life and was judged to be no improvement when compared to pure Li metal. Titanium disulfide is chosen because of its well-characterized structure and endurance under repeated intercalation and deintercalation of Li^+. TiS_2 is used as a bench mark cathode material for experimental trials of engineering feasibility of solid state rechargeable cells whether thick or thin film.

The cathode composition is shown in Table 1. The cathode capacity ranges from 1.0mAH to 9.5mAH. The cathode needs no binding agent. The lithium anode was of varying thickness between 150 to 300 microns providing anode to cathode capacity ratios of 5 to 120. The cell packaging was a standard sized 2016 coin cell package. (Coin cell chemistry and size are designated by letter prefixes and number suffixes, e.g., XR2016 is read as XR for solid state and 2016 is read 20mm diameter by 1.6mm height.) The configuration of the package and internal cell components are shown in Figure 1.

354

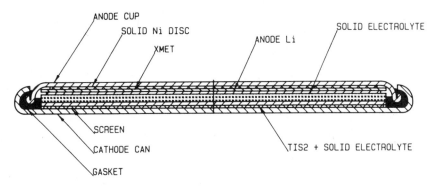

Figure 1. **Cross Section of a Rechargeable Solid State Cell Showing Internal Cell Configuration**

The cell construction is straightforward [9]. The cathode mix is weighed and poured into a hardened steel mold and compressed. The solid electrolyte is then added and compressed. The lithium foil anode of a diameter that will not touch the mold side (which would short the cell) is applied using a high density polypropylene punch. (It is **important** that solid state cells <u>not</u> be shorted.) The entire cell is ejected from the steel mold and encapsulated in a heat sealed polyethylene bag. All steps to this point are executed in an inert atmosphere dry box of about 1ppm H_2O. The encapsulated cell is removed from the dry box and isostatically compressed at 5.5 x 10^8 Pascals in a liquid media (alcohol, water, etc.). The cell is returned to the dry box and closed in the 2016 metal package. All subsequent cell testing is performed in ambient atmosphere.

CELL TESTING REGIME

The packaged cells were subjected to constant current charge/discharge cycling at 21°C to depths of discharge of 12%, 24%, 36%, and 100%. The capacity input/output per cycle was 1.0mAH, 2.0mAH, and 3.0mAH. The cells are built in cathode limited constructions to shorten the engineering development time.

DISCUSSION

The anode has the smallest area and all current densities are referenced to the 1.8 cm^2 of anode surface area. Figure 2 shows the charge/discharge behavior of a cell. The solid state system is clearly

capable of cycling. The gradual voltage fade on discharge with cycle can
be the result of numerous factors. The full understanding of the fading is
not complete but some speculative comments will be offered.

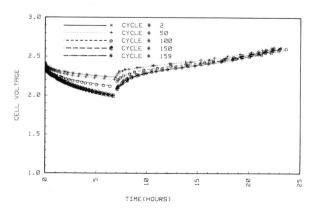

Figure 2. Rechargeable 2016 Cell Showing Consistency of
Charge/Discharge Cycling With Gradual Voltage Fade at
End of Discharge Cycle

The cathode is highly diluted with solid electrolyte which can result
in the electrical isolation of TiS_2 particles in the cathode matrix. The
addition of and the rebalancing of electronic conductivity (both TiS_2 and
added additional electronic conductors) of the cathode has a large effect
on the voltage maintenance of the solid state system (see Figure 3).

Proper stoichiometry of the solid electrolyte is important along with
plasticity under isostatic compression and ability to accept without
permanent deformation the volume changes of the TiS_2 upon intercalation
and deintercalation. The point about plasticity of the cathode matrix has
been made many times by numerous researchers, hence the interest in ionic
conducting polymers. Plasticity is important but it is not the only factor
in the successful design of a solid state energy storage cell. An
extremely important factor is that all materials have kinetic stability in
contact with one another. A second factor is that the interfacial contact
between all components be made. A third factor is that at least some solid
electrolytes, which are clearly not classically plastic in the sense of
metal salt doped PEO, are plastic enough.

One further aspect of any device is its assembly and engineering.
Lithium being a soft metal is easily extruded. The extrusion of Li metal
will occur under uniaxial compressive forces long before sufficient

pressure can be applied to the Li/glass interface to form a good interface. This problem has been solved through the use of isostatic (or hydrostatic) pressure [9]. The difference in performance obtained can be substantial [10].

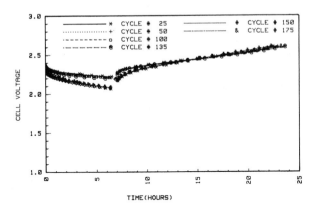

Figure 3. Rechargeable 2016 Cell Showing Improved Voltage Maintenance with Increased Electronic Conductivity of the Cathode

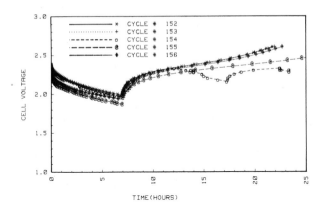

Figure 4. Rechargeable 2016 Cell Displaying a Voltage Instability on Charge

Further analysis of Figure 2 shows that the rechargeability of the solid state cells is without excessive polarization and proceeds nearly 'classically' to other well-established aqueous systems such as Ni/Cd, Pb/acid, etc. Of course as the cathode gradually loses its electronic connectivity, the charge cycle mirrors the polarization of the discharge cycle. The impedance of the cell at $21^{\circ}C$ is between 25 and 100 ohms depending upon where the cell is in the charge/discharge cycle.

Figure 3 shows the performance achievable in a rechargeable solid state cell with proper balance of TiS_2, solid electrolyte, and electronic conductor. Notice that the fade on discharge has been reduced from that shown in Figure 2. The charge curve shows excellent stability and charge acceptance.

The rechargeable solid state system has not yet been developed to a state whereby the cells are consistently reliable. The number of cycles obtainable from any one cell cannot be rigorously predicted. The mechanism of cell failure appears to be consistent but no consistency can yet be given on how many cycles a particular cell construction will deliver.

Figure 4 shows the cell of Figure 2 between cycles 152 and 156. At cycle 154 as shown there is an anomaly in the charge curve where the cell voltage drops away and then recovers towards but not fully to the previous charge curves. Subsequent cycling from cycle 155 to 165 shows normal behavior (Figure 5). At cycle 166 (Figure 5) another anomaly in the charge curve occurs. The subsequent cycles (Figure 6) continue to display the anomaly in the charge curve until at cycle 170 the cell voltage drops sharply. The tendency would be to say that the failure is obviously a Li dendrite. This may or may not be the case. The cell under discussion appears to have suffered a dendrite short. However, doubling, tripling, and quadrupling the thickness of the solid electrolyte layer has no

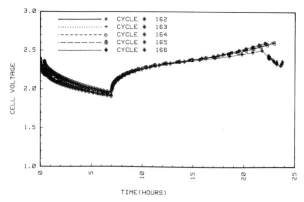

Figure 5. **Rechargeable 2016 Cell Showing Voltage Instability on Charge In Cycles Subsequent to Figure 4**

consistent effect on increasing reliability or on the manner of cell failure. Cell failure is nearly always the same type of failure and can occur at cycle 5 or cycle 200. Dendrite penetration of the separator (solid electrolyte) layer probably does occur and is one possible failure mechanism but it does not appear to be the only or most important mechanism.

Figure 7 is a trace of the charge portion of the cycle of a cell which has displayed an anomaly on charge. The anomaly is not a solitary event but consists of multiple voltage instabilities occurring at random intervals as long as the cell remains on charge. The reduction in voltage of Figure 7 is not below 2.0 volts. Indeed this is the general pattern of failure. The cell voltage almost never drops to zero volts and rarely drops below 2.0 volts. Removing the cell from charge and following the open circuit voltage with time shows the voltage to be stable with time at about 2.05 to 2.15 volts for months. Once the anomaly occurs the cell cannot be fully charged regardless of the length of time the cell remains on charge. In Figure 8 the fine structure of the charge anomaly is shown. The events are both long (many minutes--Figure 7) and short (microseconds--Figure 8). The precise location of the failure (anode/solid electrolyte interface, solid electrolyte/cathode interface, both, etc.) is under investigation using referenced cells and cycling cells with the configurations:

$$Li/solid\ electrolyte/Li$$
and
$$Li_xTiS_2/solid\ electrolyte/TiS_2$$

The correct answer to failure may well be Li dendrites but it is suspected that the cause is more complex than a simple dendrite.

CONCLUSIONS

The phosphorous chalcogenide based vitreous solid electrolytes which were discovered in Besancon, France [11] and expanded upon at Eveready Battery Co. are proving to be a technologically and experimentally interesting class of materials. The glasses are easily prepared and can be readily handled in a dry box. The glasses are 'plastic' so that interfaces between Li metal and transition metal chalcogenides can be prepared using isostatic compression which are extremely well formed and durable. Their high conductivity allows the fabrication of useful energy storage devices with low levels of polarization under load.

A question can be raised at this point as to the usefulness of a rechargeable solid state secondary cell with (so far) low mAH output. Matsushita Electric Industrial Co. has developed a 20mm diameter secondary cell with construction: $LiSn_xBi_y/PC$ + $LiClO_4/C$ (activated carbon) [12] which has a 1mAH rating. More recently Bridgestone and Seiko have jointly developed a Li/polyaniline (AL2016) secondary cell with 3mAH of capacity. Quite recently Matsushita and Japan Synthetic Rubber Co. received attention for a solid state secondary cell based on $RbCu_4Cl_3I_2$ [13]. The rechargeable system discussed in this paper has high temperature operation capability and four times the voltage of the copper-based rechargeable system. The target of the research is the CMOS memory backup market.

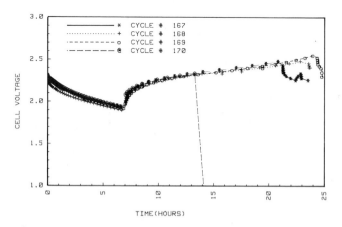

Figure 6. Rechargeable 2016 Cell Showing Catastrophic Failure

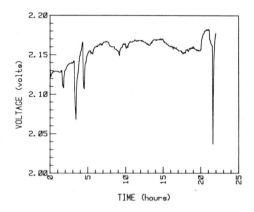

Figure 7. Cell Voltage vs. Time Profile of Charge Portion of
Rechargeable Solid State Cell Showing Anomalous
Behavior Over Long Time Periods

From a fundamental viewpoint little is truly understood about the details of ion transport across metal/solid interfaces at ambient temperature. Primary (XR2016 cells of 40mAH) Li/glass/TiS$_2$ cells consume a thickness of lithium of more than 250 microns during discharge at current densities of greater than $300\mu A/cm^2$ without disruption of the Li/electrolyte nor the electrolyte/cathode interface. The flexing of the cathode particles within the solid electrolyte matrix certainly occurs as does stress beneath the anode. The phosphorous pentasulfide-based lithium ion conducting solid electrolytes are sufficiently deformable to withstand the level of cycling presented in this paper.

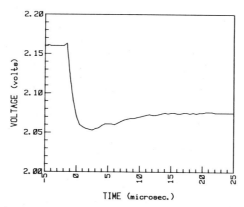

Figure 8. **Cell Voltage vs. Time Profile of Charge Portion of Rechargeable Solid State Cell Showing Anomalous Behavior Over Short Time Periods**

The phosphorous chalcogenide-based electrolytes have additional advantages beyond their 'plasticity' and ease of fabrication and use. The interface formed between lithium metal and the electrolyte compositions discussed in this paper are stable. Stability over periods of years at ambient temperature and stability at elevated temperature for thousands of hours can be demonstrated. The stability of interfaces with vitreous electrolyte network formers SiS$_2$, B$_2$S$_3$, GeS$_2$, etc. is suspect. It would seem that fundamental investigations using the phosphorous chalcogenide-based glasses are warranted.

REFERENCES

1. A. Sator, <u>Publ. Sci. Univ. Alger, Ser. B., Phys.</u> <u>2</u>, 115 (1956).

2. K. Kanehori, <u>Solid State Ionics</u>, <u>9/10</u>, 1445 (1983).

3. A. Levasseur, Materials for Solid State Batteries, pp. 119-138, ed. by B.V.R. Chowdari & S. Radhakrishna, World Scientific Publ. Co., Singapore(1986).

4. J.B. Phipps, et. al, Solid State Ionics, 18/19, 1073 (1986).

5. J.R. Akridge & H. Vourlis, Solid State Ionics, 18/19, 1082 (1986).

6. L.H. Gaines, U.S. Patent 4,091,191, May 23, 1978.

7. S. Manadake, et. al, Japanese Unexamined Patent Application: 57-172661, October 23, 1982.

8. J.R. Akridge, U.S. Patent 4,599,284, July 8, 1986.

9. J.R. Akridge & H. Vourlis, U.S. Patent 4,477,545, October 16, 1984.

10. J.P. Malugani, B. Fahys, R. Mercier, and G. Robert, Solid State Ionics,9/10, 659 (1983) and compare with J.R. Akridge and H. Vourlis, Solid State Ionics, 18/19, 1082 (1986) and the paper 'SOLID STATE BATTERIES' in this volume.

11. R. Mercier, J.P. Malugani, B. Fahys, and G. Robert, Solid State Ionics, 5, 663 (1981).

12. Matsuda, et. al, J. Electrochem. Soc., 131, 104 (1984).

13. T. Sotomura, et. al, Progress in Batteries & Solar Cells, 6, 25 (1987).

THE COMPOSITE ELECTRODE THEORY FOR SOLID STATE BATTERIES

- ATTEMPTED EXPERIMENTAL VERIFICATION

J.R. Akridge, S.D. Jones, and H. Vourlis

EVEREADY BATTERY COMPANY, INC.
Technology Laboratory
P.O. Box 45035
Westlake, Ohio 44145

INTRODUCTION

Composite electrodes consist of a mixture of a solid electrode material and a solid electrolyte. In a composite electrode the electrolyte anion is fixed in the lattice of the solid electrolyte and therefore the electrolyte concentration is constant. This distinguishes them from porous electrodes with a liquid or polymeric electrolyte. The electrode material is an intercalation compound such as TiS_2. The number of voids or isolated particles should be kept at a minimum and the two materials should form contiguous chains that are interwoven to provide a very large contact area between the two materials. This is believed to be the construction of the current solid state cells based on the chalcogenide glass electrolyte.

A theory has been proposed by Atlung(1) that describes the behavior of a composite electrode during discharge. According to the theory, the discharge curve should consist of three distinct regions. The first region should be linear with the square root of time. The second region should be linear with time and the slope of this linear region should be independent of the discharge rate of the battery when plotted against reduced time, T. The third region should be linear with the square root of 1-T.

Earlier attempts to experimentally verify this theory were only moderately successful. The lack of a better agreement with the theory was blamed on the problems associated with fabricating a good electrode for solid state cells. This paper applies the theory to data obtained from commercially available, state-of-the-art, solid state cells to see if it supports the proposed composite electrode theory better than the earlier data.

EXPERIMENTAL

The cells were rechargeable solid state cells that used the chalcogenide glass electrolyte. The construction and characteristics of this type of cell have been described in a previous paper(2). The particular cells used in this work were made with a powdered lithium anode and a cathode capacity of 8.5 mAh. Although the cells were of the rechargeable design, only the initial primary discharge was used for this work. Cells of this construction were chosen primarily for two reasons. First, the cells contain powdered metal in the cathode to enhance electronic conductivity.

Solid State Microbatteries
Edited by J. R. Akridge and M. Balkanski
Plenum Press, New York, 1990

This will simplify the theoretical approach to the composite electrode because electronic conductivity will be much larger than ionic conductivity. Second, it was felt that the polarization of the powdered lithium anode would be smaller than that of a foil anode due to its higher surface area and this would give a more reliable potential for the composite cathode.

The cells were discharged at ambient temperature on a computer controlled data acquisition system. Three different rates of constant current discharge were investigated. The discharge curves are shown in Figure 1.

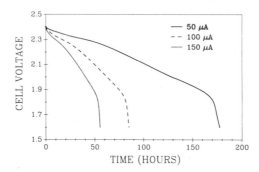

FIGURE 1. Solid State Cells Constant Current Discharge.

DISCUSSION

The equilibrium potential of an intercalation compound electrode is dependent on the degree of insertion. An example of this dependence is the equation proposed by Armand(3) for Li/TiS$_2$:

$$E = E_o - \frac{RT}{F}(\ln[\frac{1 - X}{X}] - f[X - 0.5])\tag{1}$$

The first term is the standard electrode potential (potential at X=.5). The second term is a configurational term that accounts for the distribution of the inserted ions over ideal equivalent sites. The third term accounts for the interaction between the inserted ions. The second and third terms are both functions of the degree of intercalation, X. The interaction parameter f is positive for repulsive interactions and negative for attractive interactions. For Li insertion into TiS$_2$, f is 16.

A simple linear dependence on the degree of intercalation:

$$E = E^* - kX\tag{2}$$

also gives a relatively good approximation of the discharge curve of a Li/TiS$_2$ cell. This is shown in Figure 2 where the data from cells at the three rates of discharge are plotted versus cathode utilization (or X) and then fit to a first order equation:

$$y = -0.563x + 2.428 \text{ for } 50 \text{ μA discharge rate}$$

$$y = -0.575x + 2.418 \text{ for } 100 \text{ μA discharge rate}$$

$$y = -0.584x + 2.410 \text{ for } 150 \text{ } \mu\text{A discharge rate}$$

The average k value for the three linear fits is $0.574 \pm .011$ while the average E* value is $2.419 \pm .009$.

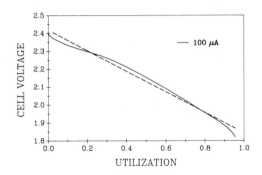

FIGURE 2. Solid State Cell Discharge with Linear Fit.

This linear approximation can be used to analytically solve for the working potential of a composite insertion electrode. The solution also depends on the assumption that once the concentration profile is established it does not change shape during continued discharge. The calculated results show three distinct regions occur during discharge of a composite electrode:

$$T < T_t$$

$$E_c = E^* - k(2[\frac{1 + \beta^2}{(1 + \beta)^2}]\sqrt{\frac{LT}{\pi}} + \frac{\beta L}{(1 + \beta)^2}) \tag{3}$$

$$T_t < T < T_{sat}$$

$$E_c = E^* - k(T + \frac{L}{3}) \tag{4}$$

$$T_{sat} < T < 1$$

$$E_c = E^* - k(1 + \frac{L}{1 + \beta} - \frac{1}{1 + \beta}\sqrt{[3L(1 - T)]}) \tag{5}$$

T is defined as a reduced time and is equal to t/τ_d where τ_d is the total discharge time at a given current assuming 100% utilization. T would also be equivalent to the degree of intercalation, X. E* and k are the intercept and slope values from the linear approximation of the discharge curve. β is the ratio of the ionic conductivity of the composite electrode to the electronic conductivity, κ_i/κ_e. L is termed a "load factor" and is defined as τ_c/τ_d where τ_c is the time constant for the composite electrode which is defined as ℓ^2/D_c. (ℓ is the thickness of the composite electrode and D_c is a pseudo diffusion coefficient for the inserted ion in the composite electrode.) T_t is the transition time between the first two regions of the discharge curve and can be found to be :

$$T_t = [\frac{1 + \beta^2}{(1 + \beta)^2}]\frac{2L}{\pi} \tag{6}$$

T_{sat} is the time at which the surface of the composite electrode becomes saturated with the inserted ions and discharge continues at the inner part of the electrode. This allows 100% utilization of the composite electrode.

At the beginning of discharge, the insertion ions see a semi—infinite phase to diffuse into and the concentration profile is established. This region of the discharge curve should produce a straight line when plotted against $\sqrt{T}$ or $\sqrt{t}$. Figure 3 shows that this is indeed the case.

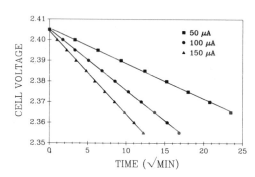

FIGURE 3. Solid State Cell Voltage vs. Time with Linear Fits.

The linear least square fits to the data of Figure 3 are:

$$y = -0.00171x + 2.406 \text{ for } 50 \text{ μA discharge rate}$$

$$y = -0.00295x + 2.405 \text{ for } 100 \text{ μA discharge rate}$$

$$y = -0.00403x + 2.404 \text{ for } 150 \text{ μA discharge rate}$$

From the slope of these lines, values for electrode time constant, τ_c, can be calculated. Assuming $\beta=0$ ($\kappa_e \gg \kappa_i$) and using the average values of E^* and k from the linear fits of the total discharge, equation 3 becomes:

$$E_c = 2.419 - 0.574 \left(2\sqrt{\frac{LT}{\pi}} \right) \tag{7}$$

Because $L=\tau_c/\tau_d$ and $T=t/\tau_d$ the slope of these lines is the following:

$$\text{slope} = m = - \frac{0.574 * 2}{\tau_d} \sqrt{\frac{\tau_c}{\pi}} \tag{8}$$

and therefore τ_c can be found as:

$$\tau_c = \frac{m^2 \, \tau_d^2 \, \pi}{(0.574)^2 (2)^2} \tag{9}$$

Using the above equation the values of τ_c shown in column (a) of Table 1 are obtained. These values are within a factor of 2 of each other. From these values of τ_c, one can now calculate the diffusion coefficient of the composite electrode, D_c, since the time constant, τ_c, was defined as ℓ^2/D_c (ℓ = electrode thickness = 0.053cm). Using this definition the following equation can be used to determine the diffusion coefficient of the compos—

ite electrode:

$$D_c = \frac{\ell^2}{\tau_c} \qquad (10)$$

The diffusion coefficient is defined as $D_c = k\kappa_c/Q_c$ where κ_c is the compound conductivity of both the electronic and ionic networks of the composite electrode and is defined as $\kappa_c = \kappa_e * \kappa_i / \kappa_e + \kappa_i$. However, if κ_e is much larger than κ_i (as assumed for $\beta=0$) then $\kappa_c = \kappa_i$ since $\kappa_e + \kappa_i \simeq \kappa_e$. Q_c is the specific capacity of the composite electrode. (Specific capacity = 446.4 C/cm^3.) One can now calculate the ionic conductivity of the composite electrode from the equation:

$$\kappa_i = \frac{D_c Q_c}{k} \qquad (11)$$

The following values are obtained using the above values for τ_c:

$D_c = 6.45 \times 10^{-8}\ cm^2/s$; $\kappa_i = 5.02 \times 10^{-5}$ S/cm at 50 µA discharge

$D_c = 8.67 \times 10^{-8}\ cm^2/s$; $\kappa_i = 6.74 \times 10^{-5}$ S/cm at 100 µA discharge

$D_c = 1.05 \times 10^{-7}\ cm^2/s$; $\kappa_i = 8.20 \times 10^{-5}$ S/cm at 150 µA discharge

The values for the diffusion coefficient are less than an order of magnitude larger than the literature value of $2 \times 10^{-8}\ cm^2/sec(4)$. The values for the ionic conductivity are about one order of magnitude smaller than the measured conductivity of the electrolyte which is 5×10^{-4} S/cm at 25ºC.

The value of τ_c can also be calculated from the time of the transition from the $\sqrt{t}$ region to the linear with t region, T_t. Using the same definitions for L and T as before and again letting $\beta=0$, equation 6 can be reduced to:

$$t_t = \frac{\tau_c}{\pi} \qquad (12)$$

and therefore:

$$\tau_c = \pi t_t \qquad (13)$$

The values of τ_c shown in column (b) of Table 1 are obtained using the above equation. It can be seen that there is a spread of greater than 2x using the transition region data.

Plotting the data versus the square root of time showed a break in the slope of the line and the appearance of a second linear region. This is shown in Figure 4. This was not expected from the theory and indicates a sudden change in the time constant. The linear least square fits to this second linear region give:

$y = -0.00254x + 2.426$ for 50 µA discharge

$y = -0.00375x + 2.418$ for 100 µA discharge

$y = -0.00467x + 2.413$ for 150 µA discharge

The calculated values of τ_c from the slope of the second linear region are shown in column (c) of Table 1, while the calculated values of τ_c from the transition times are shown in column (d) of Table 1. These values of τ_c also do not show a constant value, particularly those calculated from the transition times.

367

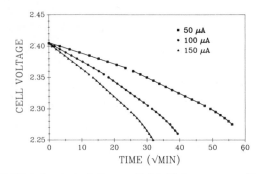

FIGURE 4. Solid State Cell Voltage vs. Time
with Linear Fits.

The second predicted part of the discharge curve arises after the shape of the concentration profile has been established across the entire composite electrode. The concentration of the intercalating ion is increasing at the same rate through the entire electrode. This region of the discharge curve should be linear with time and the slope should be independent of the discharge rate when plotted against the reduced time T. This plot is shown in Figure 5.

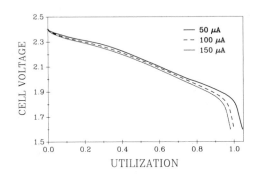

FIGURE 5. Solid State Cell Voltage vs. Reduced
Time.

There does not appear to be a linear region beginning immediately after the square root of time region. From about 20% to 40% utilization there appears to be another transition area that is not explained by the theory. After this unexplained transition region there does seem to be a linear region that is relatively parallel at all three discharge rates. This is shown in Figure 6. The equations of the linear fits are:

$$y = -0.630x + 2.481 \text{ for } 50 \text{ μA discharge}$$

$$y = -0.665x + 2.486 \text{ for } 100 \text{ μA discharge}$$

$$y = -0.669x + 2.473 \text{ for } 150 \text{ μA discharge}$$

It can be seen that the slopes of these equations are relatively constant.

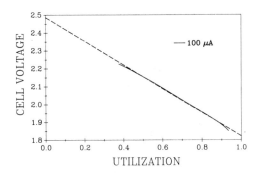

FIGURE 6. Solid State Cell Discharge – Linear
Region with Linear Fits.

According to the theory, it should be possible to calculate a τ_c from the intercept value of this second region that is linear with time. From equation 4 it can be seen that the intercept of this region will be:

$$\text{Intercept} = I = E^* - \frac{kL}{3} \qquad (14)$$

Using the same definition for L as before and transposing the resulting equation it can be seen that the time constant can be defined as:

$$\tau_c = \frac{(E^* - I)3\tau_d}{k} \qquad (15)$$

However, this calculation requires that the intercept is less than the value of E^* (2.419 volts). Therefore this calculation cannot be done. The high intercept values appear to be the result of the increased slope caused by the unexplained transition area from 20% to 40% utilization.

The third region of the discharge curve arises after one or both of the surfaces of the composite electrode become saturated before 100% utilization of the electrode has occurred. This part of the discharge curve should be linear in the square root of 1–T. This region is not seen in the experimental data. This is not unexpected because during the discharge to 100% utilization many of the properties of the composite electrode such as thickness and assumed equilibriums are changing. These changes are not taken into consideration in the theory.

CONCLUSIONS

The experimental data from these solid state cells does not completely verify Atlung's proposed theory for the discharge behavior of a composite electrode. There does appear to be an initial region of the discharge curve that is linear with the square root of time and a second region that is linear with time and whose slope is independent of discharge rate. However there are several characteristics of the discharge curve that are unexplained by the theory. Although the present solid state cells are significantly improved compared to the earlier cells to which this theory was compared, there was no improvement in agreement of the calculated discharge curves and the data. Whether this is still the result of not being able to produce a good composite electrode to study or indicates a need to modify the theory is not clear at this time. However, before a theoretical approach can be used to optimize the solid state battery, better agreement between experimental data and the theory must be obtained.

Table 1. Calculated Values of the Cathode Time Constant, τ_c.

	(a) From First Slope	(b) From First Transition	(c) From Second Slope	(d) From Second Transition
50 µA Discharge Rate	12.1 Hours	28.8 Hours	26.6 Hours	118.6 Hours
100 µA Discharge Rate	9.0 Hours	14.8 Hours	14.5 Hours	47.9 Hours
150 µA Discharge Rate	7.4 Hours	7.9 Hours	10.1 Hours	30.5 Hours

REFERENCES

1. S. Atlung, Solid State Batteries, pp. 129–161, ed. by C.A.C. Sequeria and A. Hooper, Martinus Nijhoff Publishers, Dordrecht, The Netherlands (1985).
2. J.R. Akridge and H. Vourlis, this conference (1988).
3. M. Armand, Thesis, University of Grenoble (1978).
4. D.A. Winn, J.M. Shemilt, and B.C.H. Steele, Mater. Res. Bull., 11, 559 (1976).

MATERIALS FOR MICROBATTERIES

Werner Weppner

Max-Planck-lnstitut
für Festkörperforschung
Heisenbergstr. 1
D-7000 Stuttgart 80, Fed.Rep.Germany

INTRODUCTION

The chemical compatibility of the electrolyte and the electrodes is a stringent requirement for batteries in general, but most critical for microsystems. The composition of the electrodes is changed during discharge; nevertheless, the phases have to remain stable in contact with each other. Kinetically impeded reactions may generally not be taken into consideration since batteries require fast bulk transport and exchange of atomic species across the interfaces. This is most important for the electroactive component, but applies also to the other components.

Stability of the microbattery configuration requires an activity range of the electrolyte which comprises the activities of the mobile component both in the anode and cathode. High energy density batteries require large voltages, i.e., large activity differences and therefore thermodynamically very stable electrolytes. Lithium systems have found much attention in recent years from this point of view and also because of the low atomic weight of Li. Commercially realized however, are only very simple systems so far which involve a small number of components and simple chemical reactions. LiI is employed as electrolyte in pacemaker batteries and is at the same time basically the reaction product of the anode (Li) and cathode (I_2 + electronic conducting charge transfer complex poly-2-vinyl-pyridine) [1]. All phase changes occur along the binary phase diagram which shows the ionic conductor as the only existing binary phase. The battery is very reliable because of the self-healing formation of the electrolyte in case of a short-circuiting contact between the two electrodes. Also, the voltage is high (2.8 V at room temperature) but the power density is low because of the low ionic conductivity. The application of this system is therefore restricted to very special cases. The strategy for improved batteries is therefore to look for

- alternative binary lithium compounds as electrolytes,
- solid ionic conductors for other chemical elements, and
- multinary electrolyte phases if the availability of suitable binary phases is restricted.

Solid State Microbatteries
Edited by J. R. Akridge and M. Balkanski
Plenum Press, New York, 1990

THE THERMODYNAMICS - KINETICS DILEMMA

Investigation of other binary lithium compounds has indicated very high conductivity values for Li_3N [2]. The material is stable in contact with lithium but has a low absolute value of the Gibbs energy of formation and accordingly a small decomposition voltage of 0.4 V at ambient temperature [3]. The material readily reacts with air and moisture and will behave in this way in many cases only as a "support" for a thin lithium oxide or hydroxide film which is then actually the electrolyte.

A look at ionic conductors for other ions reveals a large number of silver and copper ion conductors such as silver and copper halides which show ionic conductivities partially as high as in liquid electrolytes. Again, thermodynamic data show very low decomposition voltages of these fast solid ion conductors, e.g. 0.4 V for α-AgI at 200 °C and 0.5 V for CuBr at 300 °C.

In comparison, thermodynamically sufficiently stable binary compounds show low conductivities. This relation of high ionic conductivity at low thermodynamic stability and vice versa is illustrated in Fig. 1 and appears to be generally true. One may take the activation enthalpy of the conductivity as an indicator for the mobility of the ions. This provides an astonishingly good proportionality to the decomposition voltage for a variety of binary compounds, which is not very sensitively dependent on the specific structure of the material. The relationship is quite plausible in view of the fact that the Gibbs energy of formation is determined by the strength of the bonding of the ions to their crystal lattice position.

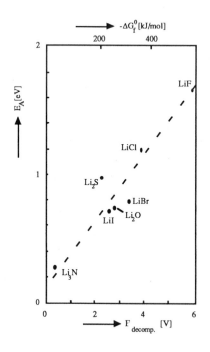

Fig. 1. The thermodynamics-kinetics dilemma. The activation enthalpy for the ionic conductivity of a variety of binary compounds (which is an indication of the ionic resistance) is plotted as a function of the decomposition voltage. High ionic conductivity implies low stability and vice versa. This rule is nearly independent of the crystal structure of the binary compounds.

THE MATERIALS CHEMISTRY CONCEPT

Binary compounds form generally crystal structures of small unit cells with closely interacting anions and cations. This is not necessarily the case for multinary systems. The two or more less mobile ions may form a sublattice which results in larger unit cells and provides a spacing of the atoms as well as a large number of defects for an easy motion of the mobile ions ("structural disorder"). At the same time a high structural stability may occur. The partial lattice of the immobile ions does not collapse upon a large variation of the activity of the mobile electroactive component. The best example is probably the β-alumina family which forms stable spinel blocks with spaceous separating planar voids every 11Å for the mobile ions. Application of voltages higher than 2 V is possible without destroying the spinel substructure in spite of the high ionic mobility. Structural solid state chemistry should be more systematically employed in order to come up with a larger selection of solid ionic conductors.

Even without the possibility to predict reliably suitable crystal structures for fast ion conductors, one may apply materials chemistry to overcome the problems of binary compounds to a certain extent. Additional components allow to modify the thermodynamic stability and in this way also the kinetic parameters, at least within some limits. This provides a very systematically method to improve the energy and power densities of the battery. In order to achieve high ionic conductivity, it may turn out to be advantageous to make the electrolyte less thermodynamically stable (until it just fulfils the stability requirements). This will generally improve the ionic conductivity, but one has to be fortunate to obtain extraordinarily high ionic conductivity by the occurance of a special suitable crystal structure.

In order to develop materials with improved thermodynamic stability and optimized kinetic properties it is necessary to distinguish between conductors for ions which form compounds of lower and higher stability.

Multinary Electrolytes for Ions which Form Less Stable Compounds

In order to improve the stability and conductivity of binary compounds by adding further components one may apply and extend the idea of doping the partial lattice of the mobile ions with another (typically aliovalent) type of ions in order to generate mobile ionic defects in the structure. In the present case the additional component may also be homovalent and it is not necessary to retain the same crystal structure.

The improvement of AgI to come up with a fast silver ion conductor at ambient temperatures by adding RbI and forming Ag_4RbI_5 may serve as an example (Fig. 2). The stability with

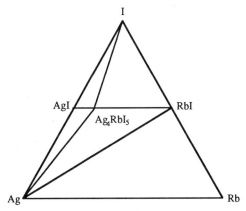

Fig. 2. Phase equilibria of the ternary system Ag-Rb-I. Reaction of AgI with RbI to form the ternary compound Ag_4RbI_5 does not influence the stability with elemental silver. Since RbI is more stable than AgI the presence of Ag does not tend to replace the Rb in RbI and Ag_4RbI_5 to form a more silver rich compound.

elemental silver (which is preferably used as an anode material) remains unchanged by the additional compound, since RbI is more stable than AgI and the solid line of the two-phase regime connects Ag and RbI. Silver compounds are commonly less stable than most other binary salts of the same anions and the approach is generally applicable for most other halides as an additional compound.

The additional more stable salt also improves the decomposition voltage. The lowest silver activity applicable to Ag_4RbI_5 is that of the three-phase equilibrium in the opposite direction to the silver corner of the Gibbs triangle. This is the equilibrium betweeen Ag_4RbI_5, RbI and I_2. The activity a_{Ag} of this regime is lower than that of the equilibrium between AgI and I_2. This may be qualitatively seen from the phase diagram since straight lines through the silver corner may be constructed which first hit the triangle Ag_4RbI_5-AgI-I_2 (which has the same silver activity as for the binary AgI-I_2 equilibrium) before reaching the triangle Ag_4RbI_5-RbI-I_2, keeping in mind that the activity of Ag has to decrease monotonously along each straight line through the silver corner. The decomposition voltage E_{decomp} is increased and may be calculated from the standard Gibbs energies of formation ΔG_f^0 of the equilibrium phases [4]

$$E_{decomp} = -\frac{kT}{q} \ln a_{Ag}(Ag_4RbI_5,RbI,I_2)$$
$$= -\frac{1}{4q} [\Delta G_f^0(Ag_4RbI_5) - \Delta G_f^0(RbI)] \tag{1}$$

(k: Boltzmann's constant, q: elementary charge, T: absolute temperature). This expression is larger than the corresponding value of the Gibbs energy of formation of 4AgI by the Gibbs energy ΔG_r of the reaction between AgI and RbI to form the ternary compound Ag_4RbI_5:

$$E_{decomp} = -\frac{1}{4q}(4\Delta G_f^0(AgI) + \Delta G_r) \tag{2}$$

A similar approach in the case of the comparatively very stable lithium compounds shows quite different results. This will be illustrated by the stability measurements of the ternary phase $LiAlCl_4$. The understanding of the phase relations in this case is a guide to another solution in the case of solid electrolytes for ions which generally form very stable compounds.

Multinary Electrolytes for Ions which Form Stable Compounds

$LiAlCl_4$ is a fast solid and molten ionic conductor [5] which is readily prepared from equimolar ratios of LiCl and $AlCl_3$ since the compound melts at 146 °C. But lithium anodes turn out not to be stable with this electrolyte. Electrochemical investigations of the thermodynamics and phase equilibria [6] show only a stability range for lithium activities between $10^{-28.4}$ and $10^{-73.7}$, corresponding to an electrical potential range between 1.68 and 4.36 V vs. lithium, at 25°C. The material decomposes into $AlCl_3$ and Cl_2 gas when lithium is removed and reacts to LiCl and Al when lithium is added (Fig. 3). Several other Li-Al phases are formed before stability with elemental lithium is reached. The origin of this behavior is the high thermodynamic stability of the lithium compound as compared to the aluminium compound. This is contrary to the situation in the before described case of silver iodide. Any lithium added to $LiAlCl_4$ tends to replace the less strongly bonded aluminium under formation of LiCl and metallic Al.

374

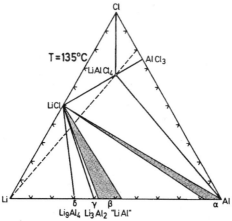

Fig. 3. Phase equilibria of the ternary system Li-Al-Cl$_2$. LiCl is considerably more stable than AlCl$_3$ which results in the tendency of replacement of Al in AlCl$_3$ and LiAlCl$_4$ by Li if this is present as an anode material. LiAlCl$_4$ is only stable within an intermediate Li activity range.

This indicates that AlCl$_3$ might be the wrong choice and a metal chloride with a lower (more negative) Gibbs energy of formation than LiCl should be used. There are, however, hardly any candidates, and those few materials which fulfil the requirement form ternary compounds with very low conductivities. A different strategy is therefore required in the case of lithium ion conductors.

Solid ternary electrolytes which are stable with elemental lithium may be found along the quasibinary tie lines between the two binary compounds on the two legs of a Gibbs triangle adjacent to the lithium corner. Ternary compounds with lower valence of lithium than in the binary compounds are not known which ensures that all compounds formed between the two most lithium rich lithium salts are stable with lithium. Stable ternary lithium ion conductors have therefore lithium as the only cation and two different types of anions.

A large variety of ternary di-anionic systems has been found and investigated in recent years , e.g., LiCl-Li$_3$N [7], LiBr-Li$_3$N [8], LiI-Li$_3$N [9], LiF-LiH, LiF-LiOH, LiF-Li$_2$O, Li$_2$S-Li$_2$O, Li$_2$S-LiCl and Li$_2$S-LiBr [11]. The compositions of the observed ternary compounds are listed in Table 1.

Table 1. Compositions of various ternary or quasi-ternary solid lithium ion conductors

System	Compounds
Li$_3$N - LiCl	Li$_9$N$_2$Cl$_3$, Li$_{11}$N$_3$Cl$_2$
Li$_3$N - LiBr	Li$_6$NBr$_3$, Li$_9$N$_2$Br$_3$, Li$_{13}$N$_4$Br
Li$_3$N - LiI	Li$_5$NI$_2$, Li$_{3x+1}$N$_x$I (1.89 $\geq$ $\leq$ x $\leq$ 2.76)
LiOH - LiCl	Li$_5$(OH)$_3$Cl$_2$, Li$_2$(OH) Cl
LiOH - LiI	Li$_2$(OH)I
LiF - LiH	LiF$_{1-x}$H$_x$ (0 $\leq$ x $\leq$ 1; 550 $\leq$ T[K] $\leq$ 630)

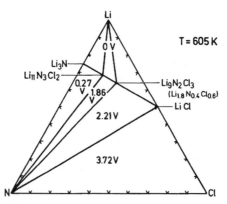

Fig. 4. Phase diagram of the ternary system Li-N-Cl$_2$. The ternary phases along the quasibinary section Li$_3$N-LiCl are all stable in the presence of elemental lithium since both binary compounds are co-existing with Li. The decomposition voltages of the ternary compounds lie between those of the binary phases or are even higher.

In addition to the stability with lithium, the solid electrolyte has to have a sufficiently large decomposition voltage. This voltage corresponds to the lithium activity of the 3-phase equilibrium in the opposite direction to the lithium corner of the Gibbs triangle. As an example, the phase diagram of the ternary system Li-N-Cl is shown in Fig.4. Decomposition of the fast ion conductor Li$_9$N$_2$Cl$_3$ results in this case in the formation of LiCl and N$_2$. As a general law, the binary or ternary compound with the highest decomposition voltage is stable with both gases (assuming that no other phases than those along the quasibinary section exist in the present case). The decomposition voltage decreases monotonously in both directions toward the binary phases. This is the result of the decrease of the activity of the electroactive component along any straight line that originates from the corner of this component in the Gibbs triangle. If this requirement is not fulfilled, different phase equilibria may be constructed which have lower (more negative) Gibbs energies of formation and are thermodanamically more favorable. The decomposition voltages of the ternary phases are always inbetween the values of the two binary phases or higher, but not lower than that of the less stable binary compound.

According to these general thermodynamic rules it is possible to make predictions on the stability of the ternary phases and to select useful materials for batteries. At least one of the employed binary phases should have a decomposition voltage larger than the required battery voltage which is generally 1.5 V or higher. LiF, LiCl, LiBr and LiI have large values of 6.1, 3.98, 3.53 and 2.80 V, respectively, at room temperature. One may therefore readily add a much less stable binary lithium compound such as Li$_3$N with a decomposition voltage of 0.44 V at room temperature. The stabilities of the ternary phases are spread in this case over a wide range and the actual values are structure dependent and have to be determined experimentally [12]. The results are shown in Table 2. By selecting binary compounds with stabilities within a narrower range, it is possible to control the stability of the ternary phase within better precision.

In some cases it is even possible to make a fine tuning of the thermodynamic stability. of the electrolyte. As an example, LiF and LiH have both an NaCl type structure with nearly the same lattice parameter and are completely miscible in the temperature regime from about 550 to 630°C. The thermodynamic stability varies within the solid solution continuously from 6.1 V for pure LiF to 0.7 V for pure LiH at 600°C. The conductivity changes inversely and increases

Table 2. Thermodynamic stability ranges of the ternary lithium nitride halides (except fluorides)

Compound	$\log a_{Li}$		E_{decomp} [V]	$\log p_{Hal_2}$ [atm]	
$Li_9N_2Cl_3$	-34.0	> 0	2.52 (374 K)	-18.6	-6.3 (605 K)
$Li_{11}N_3Cl_2$	-15.5	> 0	1.83 (595 K)	-109	-10.7 (605 K)
Li_6NBr_3	-11.8	> 0	1.34 (573 K)	-16.8	-8.7 (573 K)
$Li_9N_2Br_3$	-9.1	> 0	1.03 (573 K)	-19.3	-11.4 (573 K)
$Li_{13}N_4Br$	-5.8	> 0	0.66 (573 K)	-49.0	-17.4 (573 K)
Li_5NI_2	-26.6	> 0	1.96 (371 K)	-16.6	-4.7 (560 K)
$Li_{6.67}N_{1.89}I$	-7.9	> 0	0.92 (586 K)	-50.0	-12.7 (573 K)

over more than 4 orders of magnitude with increasing LiH content [10]. This phenomenon of decreasing conductivity with increasing stability appears to be a general rule (see also the contribution "Kinetic Aspects of Solid State Micro-Ionic Devices" in the present volume) and one should therefore choose those materials as solid electrolytes with a decomposition voltage which is just sufficiently large for the required stability with the electrodes. Kinetic properties may be optimized in this way by considering thermodynamic data.

Thermodynamic information may be readily obtained by electrochemical methods. Only the application of currents and voltages to the electrolyte, and the reading of emfs is required. This will be described in the next section.

DETERMINATION OF DECOMPOSITION VOLTAGES

The thermodynamic stability and decomposition voltage is related to the Gibbs energies of formation of the electrolyte and the adjacent phases which exist along the straight line through the corner of the electroactive component in both directions relative to the electrolyte phase. The cell voltage of a 3-phase equilibrium of a ternary system with reference to the pure electroactive compound is given by [4]

$$E = \frac{1}{z_k q d} \sum_{i=1}^{3} (-1)^i d_{ik} \, \Delta G_f^0 (A_{\alpha_i} B_{\beta_i} C_{\gamma_i}) \quad (i=1,2,3) \tag{3}$$

z_k, q, d and d_{ik} are the charge number of the electroactive component k, the elementary charge, the determinant formed by the stoichiometric numbers of the three compounds

$$d = \begin{vmatrix} \alpha_1 & \beta_1 & \gamma_1 \\ \alpha_2 & \beta_2 & \gamma_2 \\ \alpha_3 & \beta_3 & \gamma_3 \end{vmatrix} \tag{3a}$$

and the minor formed by eliminating the i-th row and the k-th column of the stoichiometric numbers of the conducting species.

The voltage may be readily measured with the help of a galvanic cell which makes use of the electrolyte under investigation and the adjacent equilibrium phases as an electrode. A convenient technique is to form these phases electrochemically by decomposing the electrolyte. It is only necessary for this purpose to polarize the sample between a reversible electrode and an inert electrode such as Pt or Mo (in the case of a lithium ion conductor) in the same direction as in polarization experiments for measuring the conductivity of the electronic minority charge carriers. The electrical current removes some of the electroactive species from the inert electrode side of the electrolyte. At sufficiently high voltages, the sample starts to decompose at the interface with the inert electrode. The phases of the 3-phase region in the opposite direction to the corner of the electroactive component in the Gibbs triangle are formed. This causes a sharp linear increase in the dc current. The slope should correspond to the ionic conductivity measured by ac impedance techniques. An electromotive force E corresponding to the 3-phase equilibrium, i.e., eq. (3) is observed after switching off the external electrical circuit.

Measurements of the cell voltages which correspond to the activities of the electroactive component of the various 3-phase equilibria allows in reverse to determine the Gibbs energies of formation of all compounds according to eq. (3). With that knowledge it is also possible to determine the activities of all other components. This is especially important in the case of gaseous components in order to control possible reactions or decomposition. Data are included in Table 2 for lithium nitride halides. A graphical representation is given in Fig. 5.

In the case of ZrO_2 solid electrolytes it is not possible to decompose the material because it becomes predominantly electronically conducting before it decomposes. One may use an auxiliary phase, however, with known Gibbs energy of formation, e.g., Pt-Zr alloys, which may decompose the electrolyte. The oxygen activity remains in this case sufficiently large in order to prevent electronic conduction [13].

High energy density of a battery requires also appropriate electrodes of sufficiently large discharge capacity and large differences in the activities of the electroactive component.

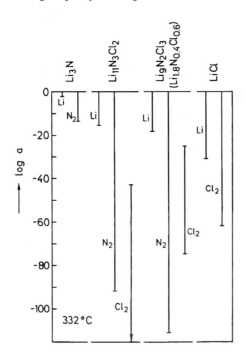

Fig. 5. Thermodynamic activity ranges for all three components of all compounds existing along the quasibinary section Li_3N-LiCl. All phases are in equilibrium with Li. The Li activity range increases monotonously from Li_3N to LiCl. The activities of N_2 and Cl_2 have to be sufficiently low.

STABILITY OF ELECTRODE MATERIALS

Large variations of the composition occur during charge and discharge. Anode and cathode material are in regimes of the phase diagram far away and come closer and closer together. Nevertheless, thermodynamic stability is required at any time.

Constant voltages are wanted for most batteries. This may be easily achieved by consideration of the phase rule. The number of *active* components of the electrolyte and electrodes should be the same as the number of phases at constant total pressure and temperature. In contrast, the voltage changes by using electrode materials in which the stoichiometry changes largely for the active component such as in intercalation compounds. These phases are also unfavorable from a kinetic point of view (see the contribution "Kinetic Aspects of Solid State Micro-Ionic Devices" in this volume). Metallic systems,e.g., alloys, which generally have a wide range of stoichiometry show large concentrations of the electronic species. An enhancement of the motion is therefore impossible [14].

Favorable kinetic properties of electrolytes may require the presence of dopants. This does not have to be a major concern with regard to the stability since it does not influence the thermodynamics and capacity of a system in a major way. On the other hand, kinetics is of less concern in microbatteries.

CONCLUSIONS

Batteries are intrinsically difficult to realize since several requirements are conflicting. It is expected that materials chemistry may overcome most of the problems. Batteries have many important advantages compared to other energy storage systems. The knowledge on materials, especially multinary systems is presently negligibly small and needs to be improved.

REFERENCES

[1]   C.C. Liang, in: Applications of Solid Electrolytes (T. Takahashi, A. Kozawa, Eds.), JEC Press, Cleveland, OH 1980, p. 60

[2]   B.A. Boukamp, R.A. Huggins, Phys. Lett. A58, 231 (1976); U.v. Alpen, A. Rabenau, G.H. Talat, Appl. Phys. Lett. 30, 621 (1977)

[3]   P.A.G. O'Hare, G.K. Johnson, J. Chem. Thermodyn. 7, 13 (1975); R.M. Yonco, E. Valechis, V.A. Maroni, J. Nucl. Mat. 57, 317 (1975)

[4]   W. Weppner, Chen Li-chuan, W. Piekarczyk, Z. Naturforsch. 35a: 381 (1980); H. Rickert, Solid State Electrochemistry; An Introduction, Springer-Verlag, Berlin, Heidelberg, New York, Tokyo (1985)

[5]   W. Weppner, R.A. Huggins, J. Electrochem. Soc. 124: 35 (1977)

[6]   W. Weppner, R.A. Huggins, Solid State Ionics 1: 3 (1980)

[7]   P. Hartwig, W. Weppner, W. Wichelhaus, Mat. Res. Bull 14: 493 (1979)

[8]   P. Hartwig, W. Weppner, W. Wichelhaus, A. Rabenau, Sol. State Commun. 30: 601 (1979)

[9]   P. Hartwig, W. Weppner, W. Wichelhaus, A. Rabenau, Angew. Chem. Int. Ed. Engl. 19: 74 (1980)

[10]   B. Schoch, A. Rabenau, W. Weppner, H. Hahn, Z. anorg. allg. Chem. 518: 137 1984)

[11]   W. Weppner, J. Power Sources 14: 105 (1985)

[12]   P. Hartwig, A. Rabenau, W. Weppner, J. Less. Comm. Metals 80: 81 (1981)

[13]   W. Weppner, J. Electroanal. Chem. Interfac. Electrochem. 84: 339 (1977)

[14]   W. Weppner, in: Transport-Structure Relations in Fast Ion and Mixed Conductors, F.W. Poulsen, N. Hessel Andersen, K. Clausen, S. Skaarup, O. Toft Sørensen, eds., Roskilde, 1985, p. 139.

KINETIC ASPECTS OF SOLID STATE MICRO-IONIC DEVICES

Werner Weppner

Max-Planck-Institut
für Festkörperforschung
Heisenbergstr. 1
D-7000 Stuttgart 80, Fed.Rep.Germany

INTRODUCTION

Ionic transport in solids is known as a phenomenon since the last century, long before the development of the microscopic picture of the crystalline state by X-ray diffraction and the thermodynamic description of solid compounds as ordered mixed phases with an equilibrium
concentration of 0-dimensional defects.

Since that time solid electrolytes are taken into consideration for various practical applications.
The early use of doped ZrO_2 for light sources as a replacement of the then used carbon filament lamps required high temperatures which was just appropriate in view of the enhanced
transport properties in this special case. But, generally, it became desirable to operate solid
state ionic devices at ambient or only slightly increased temperatures. The zirconia based λ–
probe makes use of the hot exhaust gas in automobiles and does not operate properly during
the warm-up period. Various silver and copper compounds were among the earliest discovered fast ion conductors at intermediate temperatures (150 - 500 ºC). Room temperature
solid electrolytes were only reported recently since the past two decades when Ag_4RbI_5, β-
and $\beta"$-Al_2O_3 and "Nasicon" were discovered.

Kinetic requirements differ depending on the type of application which is summarized in Fig.
1. The difficulty generally increases with the required current density. In some cases it is necessary to apply a predominant ionic conductor and in other cases a mixed electronic-ionic
conductor. High performance batteries require both exceptionally high levels of ionic conductivity in the electrolyte and ionic diffusion in the electrodes. Microbatteries have the advantages
of thin electrolytes with lower cell resistances and thin electrodes with shorter diffusion
lengths. This is, however, only a geometrical factor whereas the ionic conductivity and diffusion may differ by orders of magnitude from one material to another. Structure-conductivity
relations are covered to a large extent by other authors [1]. The present contribution will
therefore focus on a variety of other kinetic aspects which play a major role for the selection
and application of solid ionic materials in microbatteries.

Solid State Microbatteries
Edited by J. R. Akridge and M. Balkanski
Plenum Press, New York, 1990

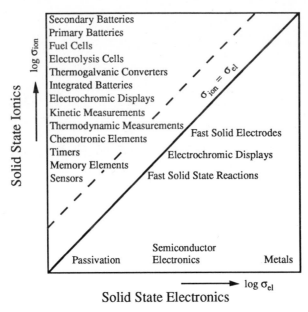

Fig. 1. Summary of practical applications of ionic motion in solids as a function of the required ionic and electronic conductivity.

SOLID ELECTROLYTES AS LOW ELECTRONIC CONDUCTANCE MATERIALS

The electrolyte builds up a voltage between two phases of different activities of the mobile component because of the absence of mobile electronic or other ionic species which could allow a compensation of the charge flux. The different chemical potentials of the neutral components generate a driving force for the ions and electrons which results in the flux densities [2,3]

$$j_n = - \frac{c_n D_n}{kT} \frac{\partial \eta_n}{\partial x},$$ (1)

where n stands for ions and electrons. c_n, D_n, η_n, k and T are the local concentration, diffusivity, electrochemical potential of species n, Boltzmann's constant and absolute temperature, respectively. The electrochemical potential may be split into the chemical potential μ and a term containing the electrostatic potential ϕ, $\eta_n = \mu_n + z_n q \phi$. The diffusivity is related to the electrical mobility u and the general mobility b according to Nernst-Einstein's relation

$$D_n = (kT/ \mid z_n \mid q) \, u_n = kT b_n$$ (2)

Except for an initial period of time (as long as space charges are built up), the fluxes of the various charged species become coupled by the charge neutrality condition

$$\sum_n z_n j_n = 0$$ (3)

Assuming the most common case of one type of ions and electrons to be predominantly mobile, neglecting the fluxes of all other ions and insertion of eq. (l) for the ions i and electrons e into eq. (3) yields the following expression for the local electrostatic potential gradient

$$\frac{\partial \phi}{\partial x} = - (z_i^2 q c_i D_i + q c_e D_e)^{-1} (z_i c_i D_i \frac{\partial \mu_i}{\partial x} - c_e D_e \frac{\partial \mu_e}{\partial x}) \tag{4}$$

Integration between the two electrodes provides an expression for the measurable cell voltage. The chemical potential of electrons in both metallic leads may be considered to be the same (because of the high concentration of electronic species in metallic conductors) and does therefore not contribute. For maximum voltages it is necessary that the product of the electronic concentration and diffusivity is small compared to that value for the ions. Self discharge and consequently loss of capacity are minimized in this way.

The flux of the ions by electronic leakage (under open circuit conditions) may be written in terms of chemical potential gradients as in Fick's first law by eliminating the unknown electrostatic potential gradient with the help of the flux equation for the electrons and eq. (3):

$$j_i = - \frac{c_i D_i c_e D_e}{c_e D_e + z_i^2 c_i D_i} \frac{\partial \mu_i^*}{\partial x} \tag{5}$$

where μ_i^* is the chemical potential of the neutral mobile component i, assuming thermodynamic equilibrium, $\mu_i^* = \mu_i + z_i \mu_e$. The concentration and diffusivity of the ions i should be as large as possible for a fast ion conductor with small ionic leakage currents. This shows from another point of view again the need for a low product $c_e D_e$.

The conductivity $\sigma_e = q^2 c_e D_e / kT$ of the electronic minority charge carriers may be determined from emf measurements according to eq. (4), from transference measurements and from polarization experiments [2]. The first two techniques are based on small voltage and mass changes and are generally inferior to the third method which blocks the ionic current but allows the electrons to be tranferred.

The characteristics of polarization current-voltage curves is a saturation current or an exponential increase of the current depending on the voltage and the type of electronic conduction [4]. This behavior is caused by the breakdown of any potentially formed electrical field because of the high disorder and mobility of the ions. The transport of the electrons is caused exclusively by the influence of Fick's diffusion caused by a concentration gradient which changes exponentially with the applied voltage according to Nernst's law. The plateau current and the intersection of the straight line in the logarithmic representation of the current vs. a linear voltage scale with the current axis provide the information on the electronic conductivities at the activity of the reversible electrode (Fig. 2). It is important to note that the conductivity of the electronic minority charge carriers depends on the chemical potential of the mobile component and may even become dominant whereas the ionic conductivity is frequently independent of the composition because of high ionic disorder.

Low electronic conductivity may be caused by low mobilities or/and low numbers of free electronic species. For the purpose of search for new solid electrolytes the question comes up whether one should look preferentially for materials with low electronic concentrations or low electronic diffusivities. To this purpose we take a look at known fast ionic conductors. It is

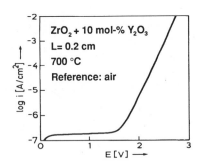

Fig. 2. Polarization current-voltage curve for the minority charge carriers of a fast solid ionic conductor under blocking conditions. The electron and hole minority charge carrier concentrations may be determined from the plateau and the intercept of the straight line (representing the exponential current increase) with the current axis as indicated.

necessary to consider processes which depend on either c_e or D_e. Hall measurements for the determination of the electronic mobility are generally not showing a measurable signal which already turns out that the diffusivities of electronic minority charge carriers in fast ion conductors are very small. Electrochemical relaxation measurement, however, turned out to be useful for the determination of D_e [5]. This technique looks at the rate of equilibration after a disturbance of the local electronic concentration; this process is only dependent on the gradient but independent of the magnitude of the concentration. The concentration of the electrons may be both varied and measured at the interface by the cell voltage using an inert electrode. This is demonstrated for yttria doped ZrO_2.

The voltage relaxation technique may make use of any initial and boundary condition provided Fick's second law may be solved and the solution may be compared to the experiment. An experimentally employed procedure [5] started-up with a linear concentration gradient of the electronic species which was generated by the application of a voltage via an inert Au and a reversible air electrode:

$$(-)\ Au\ |\ ZrO_2\ (+\ Y_2O_3)\ |\ air,\ Pt\ (+)$$

The applied voltage fixes the oxygen activity at the blocking (left) and due to the dissociation equilibrium also the electronic activity. Since the chemical potential of oxygen ions may be considered to be locally independent because of the high disorder and assuming the behavior of an ideal diluted solution for the electrons, we have a Nernst type relationship between the voltage and the electronic concentrations at the blocking ($x=0$) and reversible ($x=L$) interface:

$$E = -\frac{kT}{q}\ \ln\ \frac{c_e(x=0)}{c_e(x=L)} \tag{6}$$

A linear change of the concentration is built up under steady state conditions if the diffusion coefficient of the electrons is insensitive to the concentration. The relaxation process after the applied voltage is switched off is rate determined by the electrons as the second fastest species. The charge flux by the electronic species is readily compensated by the mobile ions so that the transport of the electrons occurs under the only influence of diffusion. The time and local dependence may be obtained from solving Fick's second law, and the concentration of the electrons at the interface with the inert electrode may be measured electrochemically as a function of time by the open cell voltage according to eq. (5). A complication arises from the formation and annihilation of electrons and holes to establish thermodynamic equilibrium, $\mu_e + \mu_h = 0$, if both are present in similar amounts. This equilibrium provides an additional source or sink for the local concentration. Electrons and holes may move into a given area to recombine or to form a new pair of electrons and holes. Fick's second law holds therefore only in the differential form

$$\frac{\partial}{\partial t}(c_h - c_e) = D_h \frac{\partial^2 c_h}{\partial x^2} - D_e \frac{\partial^2 c_e}{\partial x^2} \qquad (7)$$

Using a ratio of 10^{-5} for the conductivities of electrons and holes at the air electrode, the solution is shown in Fig. 3 for various polarization voltages from 1 - 1.8 V under the assumption of equal values for the diffusion coefficients. A characteristic shoulder is observed since the higher voltages generate predominantly n-type conduction at the interface with the blocking electrode whereas the material is p-type at the oxygen partial pressure of the air electrode. A pn-junction is formed within the originally homogeneous material simply by applying a voltage. The relaxation requires a motion of the electrons in the blocking direction. It is first controlled by the diffusion of the electrons at the blocking electrode side. The transport of the holes at this side is negligible by comparison because of their low concentration which only increases slowly due to the equilibrium with the electrons. After a transitional period of time, the holes will diffuse from the side of the air electrode to the blocking electrode which causes a rapid increase of their concentration and accordingly a rapid voltage drop.

The actual relaxation behavior is shown in Fig.4. Compared to Fig. 3 the initial voltage drop is faster and the final decay starts later. This is an indication that the diffusivity of the electrons is faster than that of the holes and the original assumption of equal values does not hold. The diffusivities may be obtained under these circumstances from the short time behavior which is

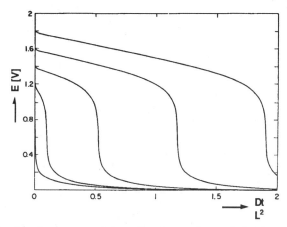

Fig. 3. Theoretical relaxation curves of the cell voltage often polarization voltages in the range from 1 - 1.8 V were applied between an inert and a reversible electrode and steady state conditions were reached. A ratio of $1:10^5$ is assumed for the conductivities of the electrons and holes at the reversible electrode. The diffusivities of both electronic charge carriers are assumed to be the same.

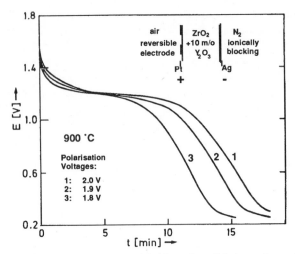

Fig. 4. Experimental relaxation curves for cubic zirconia solid electrolytes. The fast drop at the beginning indicates higher electron than hole mobility. The shoulder indicates the motion of the pn-junction through the electrolyte.

controlled by the electrons and the final decay which is controlled by the holes. After a short initial period of time with a square root of time dependence of the voltage , the voltage changes linearly

$$\frac{dE}{dt} = -\frac{\pi^2 kT}{4qL^2} D_e \tag{8}$$

D_e is proportional to the slope. The delay of the final voltage drop is proportional to the applied polarization voltage E_0. The variation of this voltage with the delay τ is proportional to the diffusivity of the holes D_h:

$$\frac{dE_0}{d\tau} = \frac{\pi^2 kT}{4qL^2} D_h \tag{9}$$

Results for the diffusivities D_e and D_h are plotted in Fig. 5 as a function of temperature. Values of 2×10^{-3} cm²/s for the electrons and 1.5×10^{-5} cm²/s for the holes at 900 °C and the observation of activated processes indicate the presence of trapped electronic charge carriers and the possible application of small polaron models for the electronic species.

The conductivities σ of electrons and holes (as obtained from polarization measurements) and the diffusivities $D_{e,h}$ allow to calculate the concentration of the electrons and holes

$$c_{e,h} = \frac{kT\sigma_{e,h}}{q^2 D_{e,h}} \tag{10}$$

as a function of the oxygen partial pressure. The results are shown in Fig. 6. The intrinsic point of equal electron and hole concentrations is moved to higher oxygen partial pressures

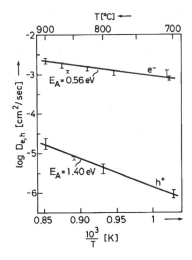

Fig. 5. Diffusivities of the electrons and holes for cubic stabilized zirconia solid electrolytes as a function of temperature.

with increasing temperature. From the temperature dependence of these concentrations one can estimate the electronic band gap to be about 4.1 eV in good agreement with spectroscopic data [5]. Fig. 6 also shows that the electronic concentration varies locally in a battery electrolyte and may be negligible at one side, while the other side may be predominantly electronically conducting.

The results show for the case of yttria stabilized zirconia that the low electronic conductivity is mainly caused by very low mobilities; but concentrations are also small. The small electronic mobility is a general condition for solid electrolytes as we may learn from eq. (5) by considering the requirement for electrolytes to block the electrons effectively. Eq. (5) may be rewritten more similar to Fick's first law:

$$j_i = - \left(\frac{c_e D_e}{c_e D_e + z_i^2 c_i D_i} \frac{\partial \ln a_{i*}}{\partial \ln c_{i*}} \right) D_i \frac{dc_{i*}}{dx} \tag{11}$$

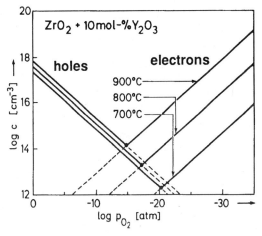

Fig. 6. Concentration of the electrons and holes for cubic stabilized solid electrolytes as a function of oxygen partial pressure for three temperatures.

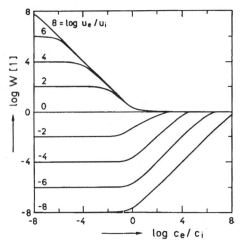

Fig. 7. Wagner factor W as a function of the ratio of the concentrations of the electronic and mobile ionic species and the ratio of the mobilities (parameter).

The additional term in paranthesis needs to become very small compared to 1 in order to stop the motion of the ions in the case of an open electrical circuit. This expression is plotted in Fig. 7 under the assumption $d \ln a_{i*} \approx d \ln c_{i*}$ as a function of the ratio c_e/c_i for various ratios of the diffusivities or mobilities of the electrons and holes [6]. It is seen that the diffusivity of the electrons evidently has to be smaller than the diffusivity of the ions. Otherwise, the motion of the electrons will even enhance the tendency of the ions to move. The electronic concentration should, in addition, not exceed the ionic concentration too much and should also be small.

Small electronic concentrations imply that the phase width (stoichiometric range) of fast ion conductors has to be narrow. In general, any phase has to accommodate a certain chemical potential range of the neutral components which may be considered to be about the same within the order of magnitude for all compounds. This quantity may be split into chemical potential ranges of the ions and of the electrons. Considering preferably materials with high ionic disorder, the chemical potential of the ions is about constant and the chemical potential of the electrons which is roughly proportional to $\log c_{e,h}$, will change predominantly. Small electronic concentrations require much smaller changes of the electronic concentration and therefore the stoichiometry in order to produce the same variation of $\log c_{e,h}$ (i.e., the chemical potential of the neutral component) as compared to large electronic concentrations.

As a result of these considerations, one may select potential candidates for fast ion conduction by looking at the electronic property of low electronic mobility and the narrow phase width. The requirement of low electronic mobility is also important from a practical point of view. If extremely low electronic concentrations would be the reason for low partial electronic conductivities of good solid electrolytes, impurities and variations of the stoichiometry would readily change the electronic concentration to a comparatively large amount and the material would become electronically conducting.

Sufficiently low electronic mobility, however, does not automatically ensure fast ionic transport. It is necessary to have highly mobile ions (or ionic defects) in high concentrations. The mobility is strongly dependent on the binding energy in the crystal lattice (thermodynamics-kinetics dilemma; see the contribution "Materials for Microbatteries" in the present volume).

The concentration of the ionic defects is more readily variable by general principles of materials chemistry.

SOLID ELECTROLYTES AS HIGHLY DISORDERED MATERIALS

The problem in the search for suitable battery electrolytes is to disclose highly disordered compounds which fulfil the thermodynamic stability requirements. Binary compounds have so far been well investigated, but no materials were found which show both high ionic conductivity and high stability. The search has therefore been extended to multinary systems. This approach is quite empirical at the present time and emphasis is therefore given to find some general rules. For this purpose, mostly ternary compounds have been investigated.

In view of the thermodynamic stability requirement, a stable binary salt such as LiCl is used as basic material tò which a less stable salt with much higher ionic conductivity such as Li_3N is added. As a general rule, the thermodynamic decomposition voltage and the ionic conductivity of any of the formed ternary compounds lie inbetween the values for the two binary compounds. Some exceptional phases were found in this way. In the case of the ternary system $LiCl$-Li_3N, a ternary phase with the composition $Li_9N_2Cl_3$ is formed which has an antifluorite type structure. 10% of the lithium sites are vacant ("$Li_{1.8}N_{0.4}Cl_{0.6}$"). In contrast to zirconia which has an analogous fluorite type structure, these vacancies are not generated by bringing in a dopant, but exist as an intrinsic property of the material. In fact, a mixture of LiCl and Li_3N in a ratio of 1:1 with a nominal stoichiometry "$Li_2N_{0.5}Cl_{0.5}$" and the ideal cation : anion ratio of the antifluorite structure forms a two-phase mixture of $Li_9N_2Cl_3$ and $Li_{11}N_3Cl_2$. The vacancies form a structural component which may no longer be considered as a defect in the traditional sense of thermodynamic formation of point defects. Fig. 8 provides a compilation of Li ion conductivities in various mostly multinary and structurally disordered compounds [7].

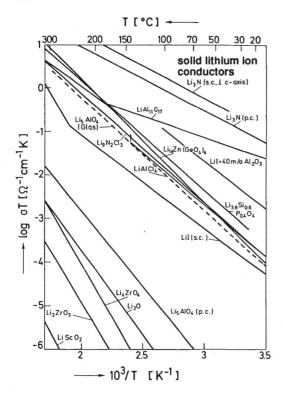

Fig. 8. Compilation of presently available solid lithium ion conductors.

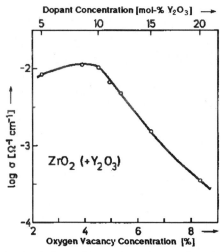

Fig. 9. Ionic conductivity of cubic stabilized zirconia as a function of the oxygen vacancy concentration (as changed by the dopant concentration).

Larger concentrations of defects, however, do not necessarily provide higher ionic conductivities. An example is yttria stabilized cubic zirconia. It is found that the conductivity decreases with increasing vacancy concentration above a certain limit (Fig. 9). This behavior is frequently attributed to the agglomeration of defects, but no experimental proof exists until today. Also, tetragonal zirconia which is structurally very similar to the cubic material and is formed by doping ZrO_2 with only about 1-3 mol-% Y_2O_3 has a lower defect concentration but higher ionic conductivity [8]. Another kinetic aspect of this material is the applicability at much lower temperatures, even if the resistances are higher than that of the cubic material at much higher temperatures. The tetragonal material may be employed in potentiometric devices at temperatures as low as 150°C under reducing conditions which shows that high exchange current densities between the electrolyte and the electrodes is most important for fast kinetics of galvanic cells.

In addition to fast ionic motion in the solid electrolyte, a battery requires fast diffusional ion transport in the mixed conducting electrodes. It is necessary that the ions which have just passed the electrolyte and became discharged at the interface with the electrode do not accumulate at this position and polarize the galvanic cell in this way. By the same reason, the ions which are built into the electrolyte at the opposite site are required to diffuse quickly to the interface. This type of motion in the electrodes is intrinsically different from the ionic transport by electrical fields in the electrolyte.

FAST SOLID MIXED CONDUCTING ELECTRODES

After the ions are discharged at the electrolyte/electrode interface, the driving force for the diffusion into the bulk of the electrode is the chemical potential gradient in this material. This is generated by the deposition of the electroactive component but holds also for the other components because of Gibbs-Duhem's relation

$$\sum N_n d\mu_n = 0 \qquad (12)$$

where N_n is the atomic fraction of the component n. Other components than the electroactive one may be predominantly mobile in the electrode which also results in an equilibration of the electrode, but also possibly to the formation of phases far away from equilibrium in multinary systems. The same consideration holds for delivering the ions to the electrolyte at the opposite electrode.

The transport processes in the electrodes are typical chemical diffusion processes which also occur during corrosion, reduction of ores and solid state reactions in general [2,3]. This chemical diffusion may be considered as an interaction of the fluxes of the two predominantly mobile charge carriers. Since the electrode is a mixed electronic-ionic conductor, we need to consider primarily electrons (or holes) and one type of ions as mobile species which interact for charge neutrality. The fluxes j for any charge carriers are given by eq. (l); the coupling leads to the same result as for the electrolyte, eq. (11). The expression in paranthesis should be large in this case, however, in order to obtain large flux densities.

The interaction of the ions and electrons may be viewed to be caused by an electrical field which acts in such a direction to slow down the faster species and to speed up the slower ones. In the case of electrodes it is desirable to have the electrons speed up the ions. This is the case when the expression in paranthesis in eq. (11) is larger than 1, which is the case when the material is predominantly electronically conducting. In order to have large enhancements it is necessary to choose materials with electronic concentrations and diffusivities within specific regimes. Fig. 10 shows the enhancement as a function of the ratio of the electronic and ionic concentration and the ratio of the electronic and ionic mobility, assuming $d \ln a_{i*} \approx d \ln c_{i*}$ [6].

It is found that the electronic concentration should be small and the diffusivity should be high as compared to the concentration and diffusivity of the mobile ionic defects. The fast electrons generate a high internal electrical field to drag the ions in the same direction as it does the activity gradient. The small number of the electrons is not capable in this case to destroy the electrical field which they have generated by moving backward (as it is the case in metals). The requirements for fast electrodes are ideally fulfilled in semiconductors. Ionically disordered semiconductors should be therefore considered for electrode materials in high power density batteries [9].

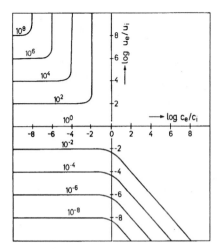

Fig. 10. Enhancement of the ionic motion by the electronic conductivity as a function of the electronic and ionic concentrations and mobilities for a variety of diffusivity ratios.

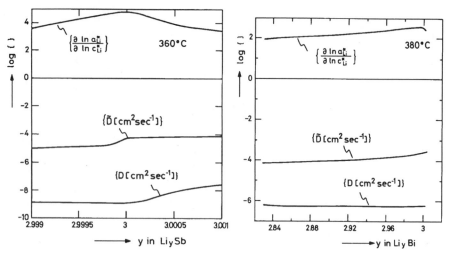

Fig. 11. Enhancement of the lithium ion motion in Li_3Sb and Li_3Bi by internal electrical fields in the presence of concentration gradients of the electronic species.

By the arguments given above for ionic conductors with low electronic concentrations, semiconductors have generally narrow ranges of stoichiometry. One should therefore look for compounds with small stoichiometric widths for enhanced diffusion in electrodes. In contrast, compounds with wide ranges of stoichiometry such as intercalation compounds have been previously considered mostly for electrodes because of the attractive feature of large solubilities for the electroactive component, which does not require phase transitions. The chemical diffusion turned out to be very slow in this case, however. The application of semiconductor electrodes requires phase transitions, but these are fast and also provide a constant cell voltage independent of the discharge state by the equilibrium between several phases which define a fixed activity of the electroactive component. Examples for fast electrodes are Sb and Bi in lithium cells. The discharge products, Li_3Sb and Li_3Bi, show enhancements by factors up to 70 000 at 400 °C as compared to diffusion under the only influence of activity gradients (Fig. 11). The discharge curves are nearly flat until the reaction is completed. The intermediate phases, Li_2Sb and $LiBi$, produce only small voltage drop.

CONCLUSIONS

The engineering of batteries employing solid electrolytes and fast chemical diffusion in mixed conducting electrodes is at a very early stage. A more detailed understanding, especially of multinary compounds is very desirable. It is not to believe that the most suitable materials for batteries have already been discovered. The results of the past have been quite fortuitous. Materials chemistry will certainly provide more insight in the future in order to help to disclose more systematically materials for batteries and other solid state ionic devices.

REFERENCES

[1]   F.W. Poulsen, N. Hessel Andersen, K. Clausen, S. Skaarup, O. Toft Sørensen, eds., Transport-Structure Relations in Fast Ion and Mixed Conductors, Risø Natl. Lab., Roskilde, DK, 1985

[2] H. Rickert, Solid State Electrochemistry - An Introduction, Springer-Verlag, Berlin, Heidelberg, New York, Tokyo, 1985

[3] W. Weppner, R.A. Huggins, J. Electrochem. Soc. 124: 1569 (1977)

[4] C. Wagner, Proc. 7th Meeting CITCE, Lindau, 1955, Butterworth, London 1957, p. 361

[5] W. Weppner, Z. Naturforsch. 31a: 1336 (1976); J. Solid State Chem. 20: 305 (1977); Electrochim Acta 22: 721 (1977)

[6] W. Weppner, in ref. [1], p. 139

[7] W. Weppner, Solid State Ionics 5: 3 (1981)

[8] W. Weppner, Adv. in Ceramics 24 (Science and Technology of Zirconia III): 837 (1988)

[9] W. Weppner, in: Materials for Advanced Batteries, D.W. Murphy, J. Broadhead, B.C.H. Steele, eds., Plenum Press, New York, London, 1980, p. 269; European Patent 81100739.2

THIN FILM SOLID STATE IONIC GAS SENSORS

Werner Weppner

Max-Planck-Institut
für Festkörperforschung
Heisenbergstr. 1
D-7000 Stuttgart 80, Fed.Rep.Germany

INTRODUCTION

Solid state microionic galvanic cells that include gaseous electrodes may be employed as chemical gas sensors. Partial gas pressures are in this case directly transduced into easily and precisely measurable electrical quantities which may be directly processed further for display or process control. Chemical sensors are in high demand for medical applications, environmental protection and industrial automation control ("robotics").

The general principle of electrochemical gas sensors is the direct conversion of chemical energy of the reaction of the mobile electroactive component with the gaseous species into electrical energy or, vice versa, the application of electrical energies to force a chemical reaction to occur. The chemical energy of reaction is dependent on the partial gas pressure which therefore controls voltages and electrical currents.

Commercialized galvanic cell gas sensors are mostly based on liquid electrolytes which need frequent maintenance, require bulky cell arrangements and often show a lack of selectivity. The application of solid ionic conductors readily allows miniaturization and integration into microelectonic circuits. The emf is independent of the size and geometry of the cell. The performance is also generally very tolerant to the processing parameters such as the impurity level, the presence of grain boundaries and surface treatments. Solid electrolytes provide also generally the advantage that one type of ions is by far most mobile which makes these materials selective to this component. This is often not the case for liquid electrolytes which frequently show a dissolution of other gases and the interaction between several mobile ions in the cell reaction.

The electrochemical gas sensor is in the simplest approach a galvanic cencentration cell in which the electrolyte acts as a membrane to separate the gas from a reference electrode (Fig.1). The generated electrical field -grad ϕ compensates the tendency for the ionized gaseous species to diffuse under the influence of the chemical potential gradient grad μ_i:

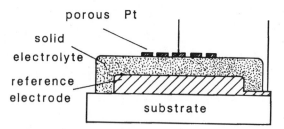

Fig. 1. Schematic drawing of a miniaturized all-solid-state electrochemical gas sensor. An electrolyte which conducts the gaseous species is employed (galvanic cell of type I). As an example, ZrO_2 ($+Y_2O_3$) and a metal/metal oxide 2-phase mixture are employed as oxygen ion conductor and reference electrode, respectively, for oxygen sensing.

$$grad\ \phi = -\frac{1}{z_i q}\ grad\ \mu_i \qquad (1)$$

z_i and q are the charge number of the mobile ions and elementary charge, respectiveley. A logarithmic relationship between the emf E and the partial gas pressure p_i is the result under the assumption of ideally diluted gases ($\mu_i = \mu_i^o + kT\ ln\ p_i$):

$$E = \frac{kT}{z_i q}\ ln\ (p_i^{gas}/p_i^{ref}) \qquad (2)$$

k and T are Boltzmann's constant and the absolute temperature, respectively. The logarithmic dependence is due to the fact that the diffusion is only proportional to the concentration *gradient* (i.e., independent of the absolute value of the concentration) whereas the migration in an electrical field is directly proportional to the number of mobile ions. This relationship (2) allows to cover a wide range of partial gas pressures within the stability window of the electrolyte which generally covers many orders of magnitude. Potentiometric types of solid state ionic gas sensors are commercially successfully employed on the basis of the oxygen ion conducting $ZrO_2(+Y_2O_3)$ to control the air/fuel ratio in combustion engines and heaters, the reduction process of ores and the carburization process for metal hardening [1].

Alternatively, diffusion limited currents are employed which provide a linear relationship between the electrical current I and the diffusional flux J:

$$I = z_i q\ J_i = -\ z_i q\ A\ D_i \frac{p_i}{L} \qquad (3)$$

where D_i, A and L are the diffucion coefficient of the species in the gas phase, the cross-section and the length of the diffusion barriers (Fig. 2). It is assumed that the polarization of the galvanic cell is sufficiently large to neglect the partial pressure at the inner side of the barrier compared to the gas phase [2]. The linear relationship allows to determine more precisely small changes of the partial pressure. E.g., a change of the oxygen content in air by 10% changes the limiting current by the same percentage whereas the cell voltage of a potentiometric device varies only by 2.2 mV at 700 °C.

Thin film potentiometric and amperometric solid ionic gas sensors are presently under development. In addition to oxygen ion conductors, the application of solid electrolytes for

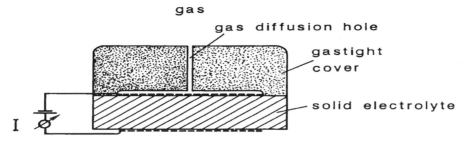

Fig. 2. Limiting current electrochemical gas sensor. A voltage is applied to the solid electrolyte to produce a low partial gas pressure of the electroactive component within the cavity. This results in a diffusional flux through the hole (diffusion barrier) which is transduced into an electrical current.

protons, fluorine and chlorine ions is considered. It is disadvantageous that high operating temperatures are required in most cases in order to achieve sufficiently high conductivities. Even more restrictive is the unavailability of solid electrolytes for other gases. It has to be assumed that it will never be possible to find solid electrolytes for a large variety of practically interesting gases. Among those are the complex gases which are composed of several different atomic species.

A modified type of solid state ionic gas sensors has recently been developed which employs mixed ionic and electronic conducting thin films on solid electrolyte substrates (Fig. 3). This approach allows on the one hand side to make use of the many recently discovered fast solid ionic conductors for a variety of metal ions at ambient or slightly elevated temperature which previously have not been considered for gas sensors, and it allows on the other hand side to determine partial pressures of gases which do not have to be transferred in the electrolyte [3]. The thin film relates the activity of the gas to the activity of the mobile ions and is involved into the virtual cell reaction. The response time is proportional to the square of the film thickness (if bulk transport is rate determining) which provides another advantage of thin films over macroscopic devices.

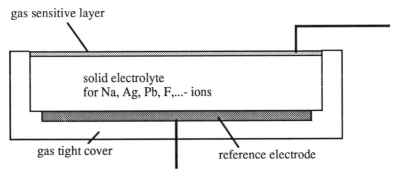

Fig. 3. Schematic drawing of a thin gas sensitive film solid state ionic gas sensor. The thin film controls the galvanic cell reaction in such a way that involves both the gaseous species and the electroactive component.

THEORY

The chemical potentials of the components of the thin film and the electrolyte are dependent on each other according to Duhem-Margules' relation:

$$SdT - Vdp + \sum_i n_i \, d\mu_{i*} = 0 \tag{4}$$

where S, V, p, n_i and μ_{i*} are the entropy, volume, total pressure, number of species i and the chemical potential of the neutral component i, respectively. The first two terms may be neglected if the temperature and total pressure are kept constant. Accordingly, an increased activity of one of the components is reflected by an overall decrease of the activities of the other components.

The absolute values of the chemical potentials are related to the Gibbs energies G according to the Duhem-Gibbs equation

$$G = \sum_i n_i \, \mu_{i*} \tag{5}$$

More conveniently, the chemical potentials may be related to a standard state (e.g., activity a = 1 for any component) which replaces G by the Gibbs energy of formation from all elements in their standard state (both for the thin film gas sensitive layer and the solid electrolyte substrate)

$$\Delta G_f^o = \sum_i n_i \, (\mu_{i*} - \mu_{i*}^o) = kT \sum_i n_i \ln a_{i*} \tag{6}$$

This relation provides a set of two equations which relate the chemical potential of the electroactive species to the partial gas pressure under the assumption that the thin film and the electrolyte belong to the same ternary system which also includes the gaseous component. A quaternary system requires an additional relation of the type of eq. (6). This can be achieved by using a thin film which is composed of two different thermodynamic phases. In general, the thin film has to consist of N-2 different compounds in the case of an N component system. In spite of the fact that the most attractive solid electrolytes are ternary or quaternary compounds, it is in most cases sufficient to employ a single phase because only thermodynamically active components have to be counted, i.e. those which may exchange for equilibration between the various phases. Many components are kinetically inactive because of extremely low mobilities and low partial gas pressures.

Nernst's equation holds for the difference of the chemical potentials of the mobile component A between the thin film and the reference electrode:

$$E = -\frac{1}{z_A q} (\mu_A^{film} - \mu_A^{ref}) = -\frac{kT}{z_A q} \ln (a_A^{film}/a_A^{ref}) \tag{7}$$

The chemical potential or activity of A in the film may be replaced by the activity or partial pressure of the gaseous component B by making use of relation (6). This provides the following equation for the cell voltage as a function of the partial gas pressure p_B:

$$E = \frac{kT(y_1z_2-y_2z_1)}{z_Aq(x_1z_2-x_2z_1)} \ln p_B^{gas} - \frac{z_2\Delta G_{f,1}^{o}-z_1\Delta G_{f,2}^{o}}{z_Aq(x_1z_2-x_2z_1)} + \frac{kT}{z_Aq} \ln a_A^{ref} \qquad (8)$$

for an electrolyte with the composition $A_{x_1}B_{y_1}C_{z_1}$ and a thin film with the composition $A_{x_2}B_{y_2}C_{z_2}$. Eq. (8) is of Nernst type with an off-set as described by the second term. The Gibbs energies of formation ΔG_f^o of both phases are included because the formation or decomposition of these compounds is involved in the virtual cell reaction. If the Gibbs energies are not known, relation (8) may also be obtained experimentally by calibration of the sensor. This method should also be applied if not all stoichiometric numbers are known. In addition, equilibration between the solid phases and the gas phase will establish the correct stoichiometries by itself, i.e., the preparation of the thin film does not necessarily have to be very exact with regard to the composition.

The result of eq. (8) would have also been obtained by the consideration of the law of overall energy conservation . The chemical energy ΔG_r corresponds to the electrical work of the transport of the corresponding number of charges:

$$W_{el} = z_AqE \qquad (9)$$

The virtual cell reaction includes the formation or decomposition of the thin film and the electrolyte. The partial gas pressure B is reflected by the cell voltage because the Gibbs energies of formation depend on this pressure.

PRACTICAL GAS SENSITIVE THIN FILM MICROIONIC GAS SENSOR DEVICES

The presently best characterized and for the new type of sensors applicable solid state ionic conductors are compiled in Fig. 4. The displayed conductivity range corresponds approximately to the range that is practically useful for thin film modified solid state ionic gas

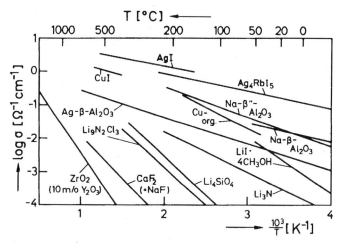

Fig. 4. Graphical presentation of the most suitable presently available solid electrolytes for microionic gas sensors.

sensors. Ag_4RbI_5 and Na- and Ag-ß-aluminas appear to be most suitable at ambient or slightly increased temperatures. It is of interest to compare the conductivities to those of the traditional sensor materials $ZrO_2(+Y_2O_3)$ and $CaF_2 + (NaF)$ which require temperatures around 1000 °C for equal conductivity values.

The gas sensitive thin film is preferably a fast mixed ionic and electronic conductor [4]. The compound should rapidly equilibrate both at the interface and within the bulk in order to communicate variations of the activities in the gas phase to the solid electrolyte phase. The bulk process is controlled by chemical diffusion which may be enhanced by many orders of magnitude as compared to the diffusion under the only influence of a concentration gradient (such as in the case of a tracer diffusion experiment) by internal electrical fields. Effective diffusion coefficients of the same order of magnitude as in liquids and in some cases as in gases may be obtained for the solid state. Large enhancing electrical fields require the presence of a small number of very mobile electronic species. Semiconductors with high ionic disorder may meet the requirements best. Metallic conducting films do not allow the formation of high internal electrical fields and electronic insulators would be blocking for the ionic motion for the reason of charge compensation . In the latter case it is possible, however, to provide the electronic conductivity by a dispersed second phase, e.g., platinum, which is otherwise chemically inert. The interface should be as large as possible since the transport of the mobile component is two-dimensional and occurs by the motion of the ions in one phase and the motion of electrons or holes in the other phase.

Any common thin film preparation technique that is applicable for the compound, e.g., sputtering, evaporation or laser ablation, may be employed. It is not required that the layer is gastight (in contrast to the solid electrolyte); it only has to be involved in the galvanic cell reaction. A most convenient technique is the electrochemical preparation of the gas sensitive film, if necessary combined with conventional techniques. The electrolyte is in this case exposed to the gas and a current of the mobile ions is passed from the reference electrode to a porous inert electrode (e.g., porous Pt or Pt mesh) at the gas side. The compound forms in-situ at the 3-phase interface electrolyte/electronic lead/gas and spreads over the electrolyte surface. If the film requires other elements than the electroactive component and the gaseous species, it is necessary to deposit these at the surface beforehand. The electrochemical preparation technique allows to control the thickness of the layer very sensitively. Thin films may be formed which may equilibrate very readily. It is also advantageous that the gas sensitive compound is predominantly deposited at the location of the transfer from ionic to electronic conductivity in the galvanic cell which generates the galvanic cell voltage.

The auxiliary thin film solid state ionic gas sensor is very generally applicable. The only requirement is that it involves the gas into the actual galvanic cell reaction (which does not have to be the thermodynamically most favorable equilibrium reaction). This allows to measure also very complex gases. The potential danger may be side reactions with other constituents of the gas which may have an influence on the voltage readings. Besides methods of elimination of the interfering gas, one may apply two or more galvanic cell gas sensors in parallel, depending on the number of gaseous species that are involved in the cell reaction. Other components of the gas sensitive thin film have to provide different chemical reactions in order to obtain a set of independent equations between the cell voltage and the various partial gas pressures. It is advantageous from this point of view to employ also divalent ion conductors which generally provide a quite different chemistry than the common monovalent ions.

Ag_2O and $AgCl$ thin films for low temperature oxygen and chlorine gas sensors, respectively, and $NaNO_3$ thin films for NO_2 sensing will be described for illustration. All these sensors are based on β - or β"-alumina ceramic solid electrolytes which appear to be most suitable under most gaseous environments. In the latter case of NO_2 detection an interaction with oxygen does occur.

The β-aluminas may be considered thermodynamically as quasibinary compounds of the mobile Ag- or Na-ions and the remaining immobile Al + O lattice framework since the Al/O ratio is given by the preparation and not changed during the equilibration with the gas. A quasiternary system is formed together with the gas sensitive thin film. Eq. (8) reads in the case of Ag_2O:

$$E = \frac{kT}{4q} \ln p_{O_2} - \frac{1}{2q} \Delta G_f^o (Ag_2O) + \frac{kT}{q} \ln a_{Ag}^{ref} \quad (Ag_2O \text{ thin film}) \tag{10}$$

and for AgCl

$$E = \frac{kT}{2q} \ln p_{Cl_2} - \frac{1}{q} \Delta G_f^o (AgCl) + \frac{kT}{q} \ln a_{Ag}^{ref} \quad (AgCl \text{ thin films}) \tag{11}$$

The standard Gibbs energy of formation of the solid electrolyte does not appear in eqs. (10) and (11) since this phase does not include the gaseous component $(z_1 = 0)$. This is not the case, e.g., when the lithium ion conductor $LiAlCl_4$ is employed together with a thin film of $AlCl_3$.

The experimental curve for Ag_2O at 160 °C is shown in Fig. 5. A straight line that corresponds to the theoretically expected value is observed for high oxygen partial pressures above about 0.1 atm. This demonstrates the feasability of oxygen gas sensors at much lower operating temperatures than previous devices. Ag_2O is not stable at lower oxygen partial pressures which becomes apparent by the bending of the curve.

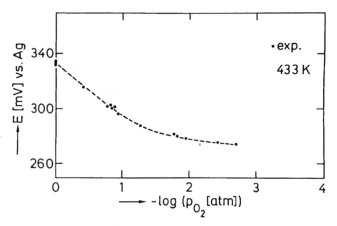

Fig. 5. Emf as a function of the oxygen partial pressure when Ag_2O is employed as the thin gas sensitive film for oxygen. Ag-β"-alumina is used as an ionically conducting ceramic substrate.

The results of Cl_2 partial pressure measurements by AgCl (+ Pt as a dispersed second phase) are plotted in Fig. 6. The measurable range extends down to the sub-ppm regime. The relationship is close to the theoretically expected one. The deviation may be due to uncertainties of the literature values of the thermodynamic data [6] or to solid solubilities which have been neglected in the present treatment.

The cell voltage in the case of an $NaNO_3$ thin film as a function of the NO_2 partial pressure may be derived by taking into account the equilibrium between NO_3, NO_2 and O_2, i.e.,

$$\ln p_{NO_2} + \frac{1}{2}\ln p_{O_2} - \ln p_{NO_3} = -\frac{1}{kT}\Delta G_f^o (NO_2) + \frac{1}{kT}\Delta G_f^o (NO_3) \tag{12}$$

Eq. (8) then reads under the assumption of a quasiternary system of two quasibinary compounds:

$$E = \frac{kT}{q}\ln p_{NO_2} + \frac{kT}{2q}\ln p_{O_2} - \frac{1}{q}(\Delta G_f^o (NaNO_3) - \Delta G_f^o (NO_2)) + \frac{kT}{q}\ln a_{Na}^{ref} \tag{13}$$

Experimental results are shown at 157 °C for variable NO_2 partial pressures while oxygen is the remaining component of the gas up to 1 atm total pressure (Fig. 7). $\log p_{O_2} \approx 0$ holds approximately for the indicated data points. The results prove the underlying theoretical assumptions.

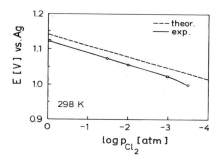

Fig. 6. Emf as a function of the chlorine partial pressure when AgCl is employed as the thin gas sensitive film for chlorine. Ag-β"-alumina is used as substrate material.

LIMITING CURRENT THIN GAS SENSITIVE FILM MICROIONIC SENSORS

Gas sensitive thin films may also be employed in limiting current type gas sensors. In contrast to limiting current devices based on solid electrolytes which transfer the gaseous species [2], a regeneration of the sensor is necessary from time to time since the gas sensitive film is

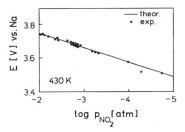

Fig. 7. Emf as a function of the NO_2 partial pressure when $NaNO_3$ is employed as the thin gas sensitve layer for NO_2 detection. The oxygen partial pressure is about 1. Na-β"-alumina is employed as a sodium ion conducting ceramic substrate material.

growing during the operation of the sensor. Very conveniently, this may be done in-situ by applying an electrical current in the reverse direction.

The experimental arrangement of the gas sensitive thin film gas sensor is shown in Fig. 8. The applied voltage supplies the electroactive component to the cavity between the solid electrolyte and the cover which includes the diffusion barrier. There, it reacts with the gaseous species similar to the potentiometric conditions and decreases the partial pressure. The current will then be given by the flux of the gaseous species through the diffusion barrier. Eq. (3) holds as in the conventional type of limiting current gas sensors. The thickness of the thin layer grows proportionally to the partial gas pressure and the response time may slow down. Passing the integrated charge in the opposite direction will rejuvenate the sensor.

The applied voltage may be used to control the chemical reaction and may overcome problems related to selectivity. It may be sensitively tuned to such values which make some reactions thermodynamically and kinetically preferred over the other ones.

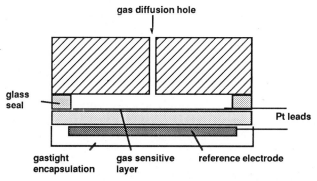

Fig. 8. Limiting current gas sensor employing a thin film surface modification of the solid electrolyte (as compared to Fig. 2) at the side of the cavity. The partial pressure is decreased at this location by passing a current of electroactive species to this side which react with the gas and decreases its partial pressure.

CONCLUSIONS

Gas sensitive thin film microionic devices broaden the applicability of solid electrolytes for gas sensors enormously. This is especially valuable since electrochemical gas sensors have some major advantages compared to other types of gas sensors in general. Gaseous species which never thought to be measurable by solid state electrochemical sensors are now accessible. It is also advantageous that the entire sensor may be miniaturized which even improves the performance.

It is likely that gas sensitive thin films (CaO) have been present - without knowing - in the case of application of CaF_2 for O_2-sensors [5]. Also, it has been reported that the oxygen ion conductor ZrO_2 ($+Y_2O_3$) shows a major loss of selectivity when used in the presence of high fluorine or sulfur activities [6]. This may be understood by the surface reactivity with these elements. Even minor amounts which may not be detected by X-ray examination are sufficient to influence the emf signals.

The problems related to selectivity are lower than in the case of other approaches. Possible side reactions may be overcome by a sensor array as discussed above, but it appears even more attractive to employ kinetic phenomena to select certain reactions. The microionic device readily allows to control the cell reaction by the electrical current, e.g., by the resistance of the voltmeter or by the external application of a (steady state or non-steady state) current. The kinetics of the reaction of the electroactive species with the various gaseous species will differ with the type and activity of the components in the electrolyte and the gas sensitive layer. Calibration then allows to separate between the different gas components.

REFERENCES

[1]   H. Dietz, W. Haecker, H. Jahnke, in: Adv. Electrochem. Electrochem. Eng., Vol. 10, H. Gerischer, C.W. Tobias, eds.), J. Wiley & Sons, New York, NY, 1977, p. 1, Bosch Technische Unterrichtung, L-Jetronic, Robert Bosch GmbH, Unternehmensbereich Kraftfahrzeugausrüstung, Abt. Techn. Druckschriften KH/VDT, Stuttgart

[2]   H. Jahnke, B. Moro, H. Dietz, B. Beyer, Ber. Bunsenges. Phys. Chem. 92:1250 (1988)

[3]   W. Weppner, German Patent DE 2926172 C2 (28 June 1979); US Patent 4.352.068 (28 Sept. 1982); in: Proc. 2nd Int. Meet. Chem. Sensors, J.-L. Aucouturier, J.-S. Cauhapé, M. Destrian, P. Hagenmuller, C. Lucat, F. Ménil, J. Portier, J. Salardienne, eds., Bordeaux, F, 1986, p. 59; Sensors and Actuators 12: 107 (1987); in: Electrochemical Detection Techniques in the Applied Biosciences, G.A. Junter, ed., Ellis Horwood, London, GB, 1988, p. 142
G. Hötzel, W. Weppner, in: Transport-Structure Relations in Fast Ion and Mixed Conductors, F.W. Poulsen, N. Hessel Andersen, K. Clausen, S. Skaarup, O. Toft Sørensen, eds., Risø Natl. Lab., Roskilde, DK, 1985, p. 401; Solid State Ionics 18/19: 1223 (1986); Sensors and Actuators 12: 449 (1987)

[4]   W. Weppner, Solid State Ionics 3/4: 1 (1981); in: Transport-Structure Relations in Fast Ion and Mixed Conductors, F.W. Poulsen, N. Hessel Andersen, K. Clausen, S. Skaarup, O. Toft Sørensen, eds., Risø Natl. Lab., Roskilde, DK, 1985, p. 139

[5]   S.F. Chou, R.A. Rapp, in: High Temp. Metal Halide Chem., D.L. Hildenbrand, D.D. Cubiciotti, eds.,- Princeton, N.J., 1978, p. 392; Abstract, Int. Conf. on Solid Ionic Conductors, St. Andrews, GB, 1978, p. 351

[6]   W. Weppner, Goldschmidt inform. 2/83 (Nr. 59): 16 (1983)

[7]   I. Barin, O. Knacke, Thermochemical properties of inorganic substances, 1973, and I. Barin, O. Knacke, O. Kubaschewski, Supplement, 1977, Springer-Verlag, Berlin, Verlag Stahleisen, Düsseldorf

SOLID STATE ELECTROCHEMICAL IN-SITU TECHNIQUES FOR THE EVALUATION OF MICROBATTERIES

Werner Weppner

Max-Planck-Institut
für Festkörperforschung
Heisenbergstr. 1
D-7000 Stuttgart 80, Fed.Rep.Germany

INTRODUCTION

A profound knowledge of the phases formed during the discharge of a battery, their thermodynamics and their kinetic properties are even more important for the design of microbatteries than for conventional types of galvanic cells for energy storage. The failure of thin film devices is much more sensitive to thermodynamic and kinetic instabilities. The evaluation of the thermodynamic properties, phase equilibria and kinetics of the battery constituents provides the most important cell parameters such as the maximum cell voltage, theoretical energy density and power density. The knowledge of these fundamental properties may also allow to develop strategies to overcome possible limitations.

A common approach is to discharge the battery partly and to take it apart for a microscopic, structural and chemical examination. This procedure destroys the galvanic cell, is very time consuming and is often not sufficiently sensitive to show interfacial processes on a microscopic scale. It will be shown here how electrochemical techniques may be used to analyse the fundamental physico-chemical parameters for the performance of a battery. These are in-situ techniques which can be made using the actual galvanic cell. The method is non-destructive and the data may be determined as a function of the discharge state from a single galvanic cell.

The electrochemical in-situ techniques are based on the fact that galvanic cells transduce thermodynamic and kinetic quantities directly into readily and precisely measurable electrical parameters and chemical reactions may be controlled by charging and discharging processes. The most important technique is called galvanostatic intermittent titration technique ("GITT") [1]. It is different from taking conventional discharge curves in major regards since the battery is allowed to reach equilibrium states inbetween the incremental discharge steps.

The GITT technique combines coulometric titrations over the entire discharge range and electrochemical measurements of the chemical diffusion coefficient by employing galvanostatic processes or other pertubations of the equilibrium state.

Solid State Microbatteries
Edited by J. R. Akridge and M. Balkanski
Plenum Press, New York, 1990

The following general galvanic cell arrangement of the microbattery is considered:

$$A \mid \text{electrolyte for } A^{z+} \text{ ions} \mid A_{y+\delta}B \qquad \text{(I)}$$

$A_{y+\delta}B$ is a binary or multinary cathode material which undergoes a compositional change in A during the discharge of the battery. The voltage E is measured with reference to pure (elemental) A for an electrolyte which conducts A^{z+} ions. The considerations described in the following for the cathode are equally applicable to the anode in case this electrode is of interest. In that case, a cathode of fixed activity is employed.

A current through the galvanic cell changes the composition according to Faraday's law. The time integral of the current $\int I \, dt$ is a precise indication of the variations of the content of A:

$$\Delta\delta = \frac{M}{zmF} \int_0^t I \, dt, \qquad \text{(1)}$$

where M is the molecular weight of the sample, m is the original (starting) mass of the sample and F is Faraday's constant. The resolution of mass changes is extremely high compared to typical balances. Changes of the order of lower than 10^{-10} g may be readily detected.

Considering a virtual cell reaction of passing $d\delta$ A^{z+}-ions into the cathode, provides the following cell reaction

$$A_{y+\delta}B + d\delta \, A = A_{y+\delta+d\delta}B \qquad [\Delta G] \qquad \text{(2)}$$

The corresponding ΔG value is related to the equivalent electrical energy which comes out of the cell or needs to be put in:

$$\Delta G = -z \, d\delta \, q \, E \qquad \text{(3)}$$

The Gibbs energy of reaction (2), ΔG, may be expressed in terms of the difference of the Gibbs energy of formation of $A_{y+\delta+d\delta}B$ and $A_{y+\delta}B$. Therefore, the cell voltage E may be described by the variation of the Gibbs energy of formation of the cathode material with the variation of the stoichiometry of the electroactive component in a differential manner [3]:

$$E = -\frac{1}{zq} \frac{d[\Delta G_f^o(A_{y+\delta}B)]}{d\delta} \qquad \text{(4)}$$

Integration of this relation allows in an easy way the determination of he standard Gibbs energy of formation of $A_{y+\delta}B$ for any composition:

$$\Delta G_f^o (A_{y+\delta}B) = -zq \int_0^\delta E \, d\delta = -zq \int_{\delta_0}^\delta E d\delta + \Delta G_f^o (A_{y+\delta_0}B) \qquad \text{(5)}$$

It should be emphasized that E is the equilibrium (open circuit) cell voltage. Energy losses by various polarizations have not been taken into account for the energy conservation law (eq. (3)). It is therefore necessary to switch off the electrical current from time to time and wait for equilibration to read the emf of the battery.

The difference to the integral of the actual cell voltage under current provides information on the energy loss due to the various polarizations which may not be transformed into electrical energy. The standard Gibbs energy as indicated in eq. (5) is the theoretical limit for the convertible chemical energy of the battery.

Further information on the thermodynamics of the cathode material may be obtained by taking into consideration the variation of the equilibrium cell voltage with the temperature which is experimentally obtained by taking either the coulometric titration curve at different temperatures or by variation of the temperature after each increment of current flux. The standard entropy of formation of the cathode material is obtained from the integral of the variation of the equilibrium open cell voltage with the temperature:

$$\Delta S_f^o (A_{y+\delta}B) = - \frac{\partial \Delta G_f^o(A_{y+\delta}B)}{\partial T} = zq \int_0^\delta \frac{\partial}{\partial T} E(\delta, T) \, d\delta \tag{6}$$

With the knowledge of both the Gibbs energy and the entropy of formation, the enthalpy of formation of the cathode material may be calculated:

$$\Delta H_f^o (A_{y+\delta}B) = \Delta G_f^o (A_{y+\delta}B) + T\Delta S_f^o (A_{y+\delta}B)$$

$$= - zq \left[\int_0^\delta E(\delta, T) d\delta - T \int_0^\delta \frac{\partial}{\partial T} E(\delta, T) \, d\delta \right] \tag{7}$$

With this fundamental thermodynamic information, a large variety of further thermodynamic quantities may be derived and an impressively comprehensive knowledge of the thermodynamics of the battery is obtained without taking the battery apart as a function of the discharge state.

Because of stability requirements it is of special interest to get knowledge of the activities of the components of the cathode. These have to be compatible with the stability window of the electrolyte [2]. The activity of the electroactive component A may be readily determined from the equilibrium open circuit cell voltage as a function of δ by considering Nernst's law:

$$a_A(\delta) = \exp\left(- \frac{zq \, E(\delta)}{kT}\right) \tag{8}$$

The activity of the component B in a binary system follows Duhem-Gibbs' relation and is expressed in the case of $A_{y+\delta}B$ by [2]:

$$a_B(\delta) = \exp\left[\frac{(y+\delta) zq \, E(\delta) + \Delta G_f^o (A_{y+\delta}B)}{kT} \right] \quad \text{or} \tag{9a}$$

409

$$a_B(\delta) = \exp\left[\frac{zq}{kT}\left((y+\delta)\ E(\delta) - \int_0^\delta E\ d\delta\right)\right] \qquad (9b)$$

The phase diagram for the cathode material may be readily determined from the slope of the coulometric titration curve. A cathode of N components shows composition-independent activities (of any component) as long as the maximum number of N phases are in equilibrium. Relative changes of the amounts of the different phases do not change the activities of the components and keeps the cell voltage constant. A voltage plateau is therefore observed for any region of the equilibrium of the maximum number of phases. It should be taken into consideration that not all components are always chemically active. This is especially the case if the ratio of two or more components in any of the phases remains unchanged by thermodynamic or kinetic reasons upon variation of the concentration of the electroactive component by an electrical current through the battery.

An example of equilibrium open circuit cell voltages as a function of the composition is shown in Fig. 1 for discharging Sb and Bi in a lithium battery. In the Sb system, the first plateau corresponds to the equilibrium of the two phases Sb and Li_2Sb. When all Sb is converted into Li_2Sb, a drop in the open circuit cell voltage E occurs within the region of the only existence of Li_2Sb at the cathode side. When Li_2Sb has reached its maximum lithium content, Li_3Sb will be formed in the case of further supply of lithium and another plateau occurs for the equilibrium of the two binary phases Li_2Sb and Li_3Sb before the voltage drops rapidly to 0 within the range of the only existence of Li_3Sb at the cathode side.

The curve for a Bi cathode in a lithium battery looks similar, except that a very large solubility of lithium in elemental (molten) bismuth is observed before the 2-phase-regimes $Bi(+Li)/LiBi$ and $LiBi/Li_3Bi$ occur which are followed by a wide range of non-stoichiometry of Li_3Bi. It was also found by coulometric titration that different intermediate phases, Li_2Sb and $LiBi$, are formed in spite of the similar chemistry of Bi and Sb. The small differences in

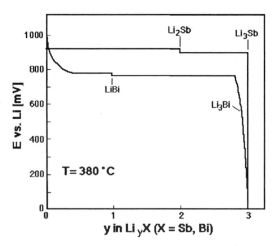

Fig. 1. Examples of coulometric titrations of lithium into Sb and Bi. After passing a current for a certain period of time, the external electrical circuit is interrupted and the equilibrium open circuit cell voltage is measured and plotted as a function of the lithium content. Equilibria of two phases of the binary system are seen as voltage plateaus, whereas the voltage drops within single-phase regions.

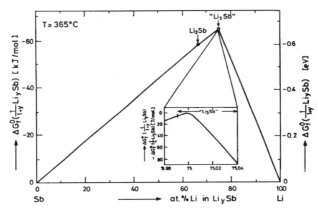

Fig. 2. Standard Gibbs energy of formation of $Li_{3+\delta}Sb$ as a function of stoichiometry. The data are obtained by integration of the coulometric tiration curve over the single-phase regime. The Gibbs energy of formation of stoichiometric Li_3Sb is taken for reference. A resolution better than 1 J/mol is readily obtained.

the two plateaus in both cases are due to the fact that the Gibbs energies of the formation of Li_3Sb and Li_3Bi by the reaction of Sb and Bi with lithium are not very much different from those of the formation of Li_2Sb or LiBi relative to the same number of Li atoms. The maximum energy densities of these types of lithium batteries are determined from the integral of the coulometric titration curves according to eq. (5). Integration of these curves within the regimes of the occurance of the single phases allows to determine the Gibbs energy of formation of the compounds as a function of composition. This is shown for the case of Li-Sb in Fig. 2. The relative resolution is extremely high, i.e., of the order of 1 J/mol. Because of the narrow ranges of stoichiometry of Li_2Sb and Li_3Sb, it is only necessary in this case to determine the temperature dependence of the plateau values in order to determine the standard entropy and enthalpy of the compound Li_3Sb. In the case of Li-Bi, also the temperature dependence of the voltage drops for the solubility of Li in Bi and the non-stoichiometry of Li_3Bi have to be considered, but this has a minor effect on the entropy and enthalpy values of Li_3Bi.

Due to the high resolution in controlling the stoichiometry, even very narrow ranges of stoichiometry may be readily resolved, including so-called line phases. This is important because the slope of the coulometric titration curve in these regimes needs to be known for the evaluation of the kinetic data.

Also, the phases that are formed in the course of discharge of a multinary cathode material may be readily detected from reading the equilibrium cell voltage. An example is shown in Fig. 3 for the determination of the quite complex phase diagram of the system Li-In-Sb in the case of application of a lithium ion conducting electrolyte [4]. In the ternary case, plateaus are observed in the presence of 3-phase equilibria. In order to obtain the complete phase diagram it is necessary to vary the content of the electroactive component for several ratios of the other two components (e.g., In/Sb in the case of Li-In-Sb). Some thermodynamic considerations may be employed to minimize the number of sample preparations. The phase diagram has to be constructed in such a way that the voltage has to increase monotonously along the path of composition with increasing distance from the corner of the electroactive component. In addition, regions of equal voltages for different ratios of the other components have to belong to the same 3-phase equilibrium. Compared to

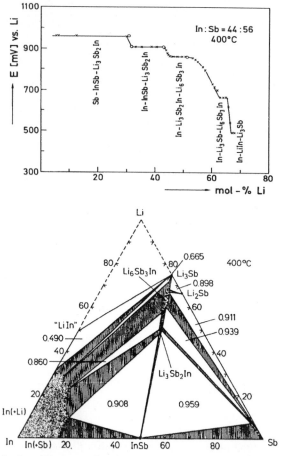

Fig. 3. Coulometric titration curve for the variation of the lithium content in the ternary system Li-In-Sb for a given In/Sb ratio. In this case, 3-phase equilibria are indicated by voltage plateaus, whereas 2- and 1-phase regimes show a voltage drop. The resulting ternary phase diagram is shown below.

conventional techniques of phase equilibrium investigations, the electrochemical in-situ technique requires only the number of sample preparations of a system with one lower degree of freedom.

In addition to the thermodynamic information, the electrical current may be analysed for a very comprehensive information on the kinetic parameters of the battery.

RELAXATION MEASUREMENTS FOR THE DETERMINATION OF KINETIC PARAMETERS

The current I which is driven through the galvanic cell by an external current or voltage source determines how many electroactive species are added to (or taken away from) the cathode and become discharged at the interface electrolyte/cathode. A chemical diffusion process occurs within the electrode and the current corresponds to the motion of mobile ionic species within the cathode just inside the phase boundary with the electrolyte (at x = 0)

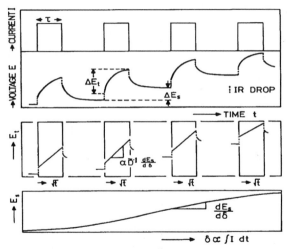

Fig. 4. Principles of the GITT technique for the evaluation of thermodynamic and kinetic data of battery materials. A constant current I_0 is applied and interrupted after certain time intervals τ until an equilibrium cell voltage is reached. The combined analysis of the relaxation process and the variation of the steady state voltage results in a comprehensive picture of fundamental materials properties.

$$I = -S\, z_i\, q\, \tilde{D}\, \frac{\partial c_i}{\partial x}(x = 0) \tag{10}$$

S is the area of the sample-electroyte interface.

The principles of the procedure are illustrated in Fig. 4. Starting with a homogeneous composition throughout the cathode, corresponding to the cell voltage E_0, a constant current I_0 is applied to the cell at $t = 0$ with the help of a galvanostat. According to eq.(10), this produces a constant concentration gradient $\partial c_i/\partial x$ of the mobile ions i at the phase boundary which is in contact with the electrolyte. The applied cell voltage increases or decreases (depending on the direction of the current) with time in order to maintain the constant concentration gradient. A voltage drop due to the polarizations of the cell is superimposed, but the IR drop due to the current flux through the electrolyte and the interfaces remains constant with time and does not change the shape of the time dependence of the cell voltage. This behaviour makes galvanostatic processes advantageous compared to other relaxation techniques which often involve large changes of the cell current, e.g., in the case of a potential step which requires (theoretically) an initially infinitely large current which finally becomes zero. After a constant current is applied for a certain time interval τ, the current flux is interrupted whereupon the composition has changed according to eq.(1). During the following equilibration process, the voltage reads the interfacial composition of the cathode material and approaches a new steady state value E_1. After the electrode has again reached an equilibrium, the procedure may be repeated, making use of E_1 as a new starting voltage. The process is continued until a phase change occurs in the electrode.

In order to relate the time dependence of the voltage E to the ionic transport in the cathode, Fick's second law has to be solved for the concentration of the mobile ions at the interface $x = 0$ which is experimentally observed by reading the cell voltage. With the appropriate initial and boundary conditions of a homogeneous concentration throughout the cathode at $t = 0$, a constant concentration gradient at $x = 0$ at any time (as given by eq. (10)) and a zero concentration gradient at the opposite surface of the cathode (because of an assumed impermeability of the ions at this location), the following solution for the concentration of the mobile species at the interface is derived as a function of time is [1,5]:

$$c_i(x=0,t) = c_0 + \frac{2I_0\sqrt{t}}{Sz_i q\sqrt{\tilde{D}}} \sum_{n=0}^{\infty} \left[\text{ierfc}\left(\frac{nL}{\sqrt{\tilde{D}}\,t}\right) + \text{ierfc}\left(\frac{(n+1)\,L}{\sqrt{\tilde{D}}\,t}\right)\right], \tag{11}$$

where $\text{ierfc}(\lambda) = [\pi^{-1/2}\exp(-\lambda^2)] - \lambda + [\lambda\,\text{erf}(\lambda)]$ is the error function.

For short times, $t \ll L^2/\tilde{D}$, the infinite sum can be approximated by the first term. In this case, the concentration c_i should change with the square root of time:

$$\frac{dc_i(x=0,t)}{d\sqrt{t}} = \frac{2I_0}{S\,z_i\,q\,\sqrt{\tilde{D}\,\pi}} \qquad (t \ll L^2/\tilde{D}) \tag{12}$$

The problem is now: The voltage provides an information on the activity according to Nernst's law whereas eq. (12) indicates the time dependence of the concentration. This may be overcome by expanding eq. (12) by dE and by considering the relation between changes in the concentration and the stoichiometry according to $dc_i^{\cdot} = (N_A/V_M)d\delta$, where N_A is Avogadro's number and V_M is the molar volume of the cathode material:

$$\frac{dE}{d\sqrt{t}} = \frac{2V_M\,I_0}{SF\,z_i\,\sqrt{\tilde{D}\,\pi}}\,\frac{dE}{d\delta} \qquad (t \ll L^2/\tilde{D}) \tag{13}$$

$dE/d\delta$ is the slope of the coulometric titration curve. With this information, the chemical diffusion coefficient may be calculated from the time dependence of the cell voltage durring the application of the constant current..

If the time over which the current is passed through the cell is short as compared to $L^2/\tilde{D}$, (i.e., eq. (13) is valid for the entire time of the current flux) and the change of the voltage is sufficiently small to consider $dE/d\delta$ to be constant, $dE/d\delta$ may be replaced by the variation of the equilibrium steady state voltage ΔE_s over the variation of the stoichiometry $\Delta\delta$ that corresponds to the current flux I_0 for the period of time τ. In addition, if one makes sure that the voltage E versus $\sqrt{t}$ shows the expected straight line behaviour over the entire time period of the current flux, $dE/d\sqrt{t}$ may be replaced by the variation of the transient voltage ΔE_t (disregarding the IR drop) over $\sqrt{\tau}$. This provides the following simple expression under the indicated assumptions for $\tilde{D}$ as a function of the cathode composition:

$$\tilde{D} = \frac{4}{\pi\tau}\left(\frac{m_B\,V_M}{M_B S}\right)^2\left(\frac{\Delta E_s}{\Delta E_t}\right)^2 \qquad (t \ll L^2/\tilde{D}) \tag{14}$$

A variety of other kinetic quantities may be readily derived from this information and the coulometric titration curve. Just to give a few examples, the following relationships are listed:

The relation between the chemical diffusion coefficient and diffusivity (sometimes also called component diffusion coefficient) is given by the Wagner Factor (which is also known in metallurgy as the thermodynamic factor) $W = \partial \ln a_A / \partial \ln c_A$. W may be readily derived from the slope of the coulometric titration curve since the activity of the electroactive component A is related to the cell voltage E (Nernst's law) and the concentration is proportional to the stoichiometry of the cathode material:

$$W = \frac{\partial \ln a_A}{\partial \ln c_A} = -\frac{z_A \, q \, c_A \, V_M}{kT \, N_A} \frac{dE}{d\delta} \tag{15}$$

In view of this relationship, the diffusivity D_A is given by

$$D_A = \left(\frac{\partial \ln a_A}{\partial \ln c_A}\right)^{-1} \tilde{D} = -\frac{4kT \, m_B \, V_M \, I_o}{\pi \, c_A \, z_A^2 \, q^2 \, M_B \, S^2} \frac{\Delta E_s}{(\Delta E_t)^2} \quad (t \ll L^2/\tilde{D}) \tag{16}$$

The diffusivity indicates the capability of the ions to move within the solid by random motion without the influence of a concentration gradient. This motion is independent of the motion of any other species (e.g., electrons or holes) and is not influenced by internal electrical fields such as in the case of chemical diffusion processes which require the simultaneous motion of electronic or other ionic species. D is related to the general and electrical mobility according to Nernst-Einstein's relation [6]. The partial ionic conductivity of the mixed ionic and (predominantly) electronic conducting cathode is given by the product of the concentration and the diffusivity and may be related to the variations of the steady state and transient voltage [1,7]:

$$\sigma_A = \left(\frac{\partial \ln a_A}{\partial \ln c_A}\right)^{-1} \frac{(z_A \, q)^2 \, c_A \, \tilde{D}}{t_e \, kT} = -\frac{4}{\pi} \frac{m_B \, V_M \, I_o \, \Delta E_s}{t_e \, M_B \, S^2 (\Delta E_t)^2} \quad (t \ll L^2/\tilde{D}) \tag{17}$$

where t_e is the transference number of the electrons. This method of determining partial ionic conductivities is especially valuable in the case of minority charge carriers such as the ions in predominantly electronically conducting electrodes since there is a lack of other techniques. Often, data are available for the tracer diffusion coefficient which is related to the diffusivity by the correlation factor f_A, $D_{Tr} = f_A \, D_A$. f_A is structure dependent and takes commonly values between 0.3 and 1.

In addition, the parabolic rate constant of the growth of new phases at the cathode/ electrolyte interface may be determined from the previously described knowledge of the stoichiometric dependence of the chemical diffusion coefficient $\tilde{D}$ or the diffusivity D_A:

$$k_t = \frac{1}{y+\delta} \int_o^L \tilde{D} \, d\delta = \frac{q \, z_A}{kT} \int_o^L D_A \, dE \tag{18}$$

The integration runs over the entire thickness L of the phase. This quantity provides a good indication for the current that may be passed through the battery if the cell voltage or current are fixed.

415

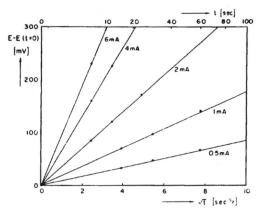

Fig. 5. Time t dependence of the cell voltage E as a function of time after the application of a constant current for Li₃Sb as cathode material. The √t law is in agreement with the assumption of bulk diffusion as the rate determining process.

As an example, the transient voltage during a constant current flow in a lithium battery using an Li_3Sb cathode is shown in Fig. 5. In GITT this process is repeated over the entire range of existence of Li_3Sb. This results in the compositional variation of the chemical diffusion coefficient as shown in Fig. 6. A sharp increase is observed at the ideal stoichiometry which indicates most likely a transition between different types of mobile defects (lithium vacancies and interstitials). Taking into account the Wagner Factor W as determined from the slope of the coulometric titration curve (which shows a maximum value of 70 000), the diffusivity as a function of composition is obtained as shown in Fig. 7 which still reflects the increase but in a more smooth way.

It is of interest to point out the very high chemical diffusion coefficient of Li_3Sb. This is a result of the simultaneous motion of electronic and ionic species. A concentration gradient for both lithium ions and electrons or holes exists in the case of compositional inhomogeneities. The faster electrons drag the ions behind by generating high internal electrical fields. This phenomenon may only be observed in semiconductors which are therefore more suitable for battery electrodes for high power density applications (see the contribution "Kinetic Aspects of Solid State Micro-Ionic Devices" in the present volume).

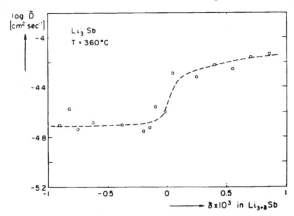

Fig. 6. Chemical diffusion coefficient as a function of the stoichiometry for $Li_{3+\delta}Sb$. The kink at nearly ideal stoichiometry is most likely due to a change in the diffusion mechanism from a vacancy to an interstitial process.

416

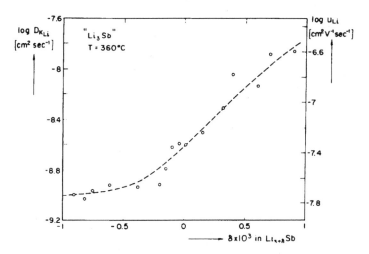

Fig. 7. Diffusivity (or component diffusion coefficient) as a function of the stoichiometry of $Li_{3+\delta}Sb$.

The partial lithium ion conductivity as determined by eq. (17) is plotted in Fig. 8. The increase is due to the larger number of more mobile ions, i.e., most likely lithium interstitials.

The parabolic reaction rate constant k_t is plotted in Fig. 9 for $Li_{3+\delta}Sb$ for different boundary conditions. The curves (a), (b) and (c) show k_t as a function of the lithium activity for electrodes which are kept at the opposite side at a lithium activity of the equi-librium value of Li_2Sb/Li_3Sb, at an intermediate activity of $a_{Li} = 10^{-4}$ and and at $a_{Li} = 1$, respectively. The growth or shrinkage of the phases is the highest if high lithium activities are provided at which the lithium ions with the highest mobility (lithium interstitials) are present. Increasing the lithium activity difference over ranges of lower activity does not provide a major increase in the reaction rate anymore since only lithium defects of lower mobilities are introduced additionally.

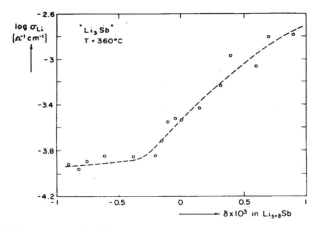

Fig. 8. Partial lithium ion conductivity of the predominantly electronically conducting compound $Li_{3+\delta}Sb$ as a function of stoichiometry. The variation in the conductivity is due to changes of the transport mechanism.

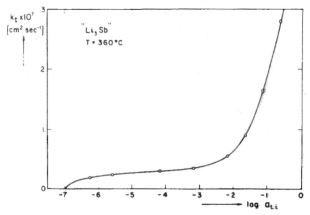

Fig. 9. Parabolic reaction rate constant for the formation of Li_3Sb as a function of the lithium activity for Li_2Sb as substrate material. The most suitable discharge conditions with the lowest polarization loss/power ratio may be derived from this curve.

In conclusion, the combination of thermodynamic measurements over single and multiphase regimes and kinetic measurements within single phase regions allows to obtain a comprehensive overall picture of the battery performance. The obtained thermodynamic and kinetic data may be converted into the rate constants that describe the growth of new phases in the electrode in the course of discharge.

REFERENCES

[1]   W. Weppner, R.A. Huggins, J. Electrochem. Soc. 124: 1569 (1977), J. Solid State Chem. 22: 297 (1977)

[2]   R.A. Huggins, in: Fast Ion Transport in Solids; Electrodes and Electrolytes (P. Vashishta, J.N. Mundy, G.K. Shenoy, eds.) North-Holland, New York, 1979, p. 53

[3]   W. Weppner, R.A. Huggins, J. Electrochem. Soc. 125: 7 (1978)

[4]   W. Sitte, W. Weppner, Z. Naturforsch. 42a: 1 (1987)

[5]   H.S. Carslaw, J.C. Jaeger, Conduction of Heat in Solids, Oxford University Press, 1959; J. Crank, The Mathematics of Diffusion, Oxford University Press, 1956

[6]   H. Rickert, Solid State Electrochemistry, An Introduction, Springer-Verlag, Heidelberg, 1983

[7]   W. Weppner, R.A. Huggins, Ann. Rev. Mater. Sci. 8: 269 (1978)

[8]   W. Weppner, R.A. Huggins, Z. Physikal. Chem. N.F. (Frankfurt) 108; 105 (1977)

A Warning for the Wagner Polarization Cell

Michael P. Setter

Center for Solid State Science
Arizona State University
Tempe, Arizona 85287-1704

In the study of electrolytes it is crucial to determine the relative tranport numbers of ionic species and electronic species.   One elegant method used to separate the elctronic components is through the use of a DC potential and a blocking electrode.   This technique and the corresponding analysis was described by Carl Wagner in his classic 1957 paper[1]. The particular cell arrangement used in this technique involves a mixed conductor electrolyte, MX, and two electrodes, only one of which, M, is reversible with respect to MX,

$$M \quad | \quad MX \quad | \quad \text{electronic conductor.}$$

It is the asymmetry of the cell and of the applied DC potential that stops a continuous ionic current and hence enables a determination of the total electronic conductivity of the electrolyte.

Unfortunately, there is a problem in Wagner's analysis of the electronic conductivity regarding its separation into electron and hole components.   In keeping with Wagner's original paper, the present author will maintain his assumption that the ionic conduction of MX is primarily by cation defects, M^{z1}.   Furthermore, Wagner's use of symbols will be followed and those equations taken directly from his 1957 paper will keep his numbering system (with an appended W to distiguish from the present author's work).   Also, the analysis will be restricted to the steady state case of a DC potential applied between the two electrodes, with the blocking electrode being at the higher potential.

The basis of the analysis is the two reversible reactions established,

$$M^{z1} + z_1 \text{electrons} = M \qquad (8\text{-W})$$

$$X + |z_2| \text{electrons} = X^{|z2|}, \qquad (9\text{-W})$$

where: z_1 = charge of cation defect,
z_2 = charge of anion defect, $X^{|z2|}$.

From these reactions the electochemical and chemical potentials can be related by

$$\eta_1 + z_1 \, \eta_3 = \mu_M \qquad (10\text{-W})$$

$$\mu_X + |z_2| \, \eta_3 = \eta_2, \qquad (11\text{-W})$$

Solid State Microbatteries
Edited by J. R. Akridge and M. Balkanski
Plenum Press, New York, 1990

or in differential form,

$$d\eta_1 + z_1 d\eta_3 = d\mu_M \qquad (1)$$

$$d\mu_X + |z_2| d\eta_3 = d\eta_2, \qquad (2)$$

where: $\eta_i = \mu_i + z_i F\Phi$, electrochemical potential of species i,
Φ = electrostatic potential,
μ_i = chemical potential of species i,
F = Faraday's constant, and
i=1 for cation defects, 2 for anion defects, 3 for
electrons, M for metal, X for nonmetal.

Equations (1) and (2) may also be written exculsively in terms of the
chemical potentials,

$$d\mu_1 + z_1 d\mu_3 = d\mu_M \qquad (3)$$

$$d\mu_X + |z_2| d\mu_3 = d\mu_2. \qquad (4)$$

Implicit in these equations though not actually stated by Wagner is the
assumption that all species are in equilibrium, even when a current exists
in the electrolyte. In other words, the net amounts of the various
species at a given point are not changed by the reactions (8-W) and (9-W).
There is some confusion regarding the species involved in reactions (8-W)
and (9-W). Wagner refers to M and X as metal and nonmetal, respectively,
even within MX. M^{z1} and $X^{|z2|}$ are likewise referred to as cation and
anion. No mention is made of defects. However within the analysis
itself, M^{z1} is treated as the charge carrier with M being immobile. In
this paper, the author has interpretted M and X to be the immobile,
majority of metal and nonmetal species present, in their normal
crystallographic locations in MX. M^{z1} and $X^{|z2|}$ refer to the mobile
defects present in MX.
Wagner obtains the expression,

$$\frac{\delta\mu_M}{\delta x} = z_1 \frac{\delta\eta_3}{\delta x} , \qquad (22\text{-}W)$$

by using the boundary condition that the ionic current at the blocking
electrode interface is zero. This is true thoughout the electrolyte only
if equilibrium between all species is established. Wagner uses equation
(22-W), throughout the elctrolyte, to obtain

$$\mu_M(x) = \mu^\circ_M - z_1 F\Delta E(x), \qquad (23\text{-}W)$$

where: $\mu_M(x)$ = the chemical potential of M a distance x from the
metal electode,
μ°_M = the chemical potential of M at the metal electrode,
$\Delta E(x)$ = the electrostatic potential difference at x compared
to the metal electrode.

Wagner uses reaction (8-W) and the large inherent defect concentrations in
the materials of interest to write

$$d\mu_3 = d\mu_M/z_1, \qquad (27\text{-}W)$$

and finally,

$$\sigma_3 = \sigma^\circ_- \exp[(\mu_M(x)-\mu_M^\circ)/z_1 RT] + \sigma^\circ_+ \exp[-(\mu_M(x)-\mu_M^\circ)/z_1 RT], \qquad (31\text{-}W)$$

where: σ_i = the conductivity of i, $^\circ$ is at the metal electrode,
i = + for holes, − for electrons,

R = the ideal gas constant,
T = the absolute temperature.

By combining equations (23-W) and (31-W), the final result of Wagner's analysis appears;

$$|J| = \{\sigma^o_-[1-\exp(-EF/RT)] + \sigma^o_+[\exp(EF/RT)-1]\}(RT/FL), \quad (32-W)$$

where: $|J|$ = the current density through the electrolyte,
E = the electrostatic potential difference across the electrolyte of thickness L.

Now it is possible to write equation (22-W) as

$$d\mu_M = z_1 \, d\mu_3 - z_1 Fd\Phi \qquad (5)$$

and
$$d\mu_3 = d\mu_M/z_1 + Fd\Phi. \qquad (6)$$

When equation (6) is compared to equation (27-W) it is easily seen that $d\Phi$ must be zero in order to satisfy the conditions for which equation (32-W) is valid. This translates into the requirement that no electrostatic potential exist within the electrolyte, a trivial situation, in order for Wagner's analysis to be applied.

Wagner's error comes in writing equation (27-W). Since there is no ionic current in the electrolyte (from the requirement to obtain equation (23-W)), the only change in the amount of cation defects must be due to the reaction (8-W). Similarly, since steady state is assumed, the only change in the amount of metal present must be due to this same reaction, rather than any diffusion processes. From the stoichiometry involved it can be written that

$$|dC_1| = |dC_M|, \qquad (7)$$

where C_i = the concentration of species i at a given point.

Now surely there are normal more metal atoms present than cation defects, even in highly defective materials. This means that

$$C_M > C_1 \qquad (8)$$

and
$$1/C_M < 1/C_1. \qquad (9)$$

This leads to
$$|RTdC_M/C_M| < |RTdC_1/C_1|. \qquad (10)$$

To which the definition of chemical potential can be applied to yield

$$|d\mu_M| < |d\mu_1|. \qquad (11)$$

This means that the change in chemical potential of the cation defects is greater in magnitude than that of the metal atoms, if there is no change in their respective activity coefficients. Since equation (3) can be written as

$$d\mu_3 = (d\mu_M - d\mu_1)/z_1, \qquad (12)$$

when Wagner obtained equation (27-W) he neglected the larger term of equation (12). It should also be noted that if equilibrium exists, as Wagner assumes, then dC_1 is zero. this again leads to a trivial situation.

In summary, the expression developed by Wagner to separate hole and electron contributions to the electronic conductivity of a mixed conductor is not strictly valid. The assignment of hole or electron conductivity based solely on the shape of the current-potential profile of such a cell

421

is not conclusive.   The present author is preparing an analysis of the
Wagner polarization cell that avoids the theoretical pitfall of Wagner's
expression[2].   This warning however places no restriction on the use of
this polarization technique to determine total electronic conductivity.
As originally observed by Hebb[3], the ionic current is suppressed at the
blocking electrode interface, leaving only the electronic current to be
measured.

REFERENCES

1.   C. Wagner, Proc. 7th CITCE, Lindau (1957).

2.   M. Setter, to be publiished.

3.   M. H. Hebb, J. Chem. Phys., 20, 185 (1952).

MICROBATTERY OPPORTUNITIES IN MILITARY ELECTRONICS

John F. Bruder

Quadri Corporation
300 N. McKemy Ave.
Chandler, AZ   85226-2618   USA

The United States military industrial complex has many opportunities for microbattery development and applications. Funds for research are frequently available for a new technology, especially if it solves a problem which cannot be solved practically by any other means. It is also true that many times the military will pay a high price for a relatively small amount of product during development. These factors can assist the progress of a new product by enabling it to generate a profit early during its product life.

Conversely, the deployment of a product can be slowed by the necessity of proving that the new technology can operate in the extremes of military temperature ranges and mechanical stress. Qualification procedures and the accompanying red tape can also slow development and be frustrating at times.

The least stressful applications are found where electronic equipment is used by the military in a nonmilitary type of environment such as a supply depot, accounting office, or in other nonbattlefield surroundings. Shock, vibration, temperature cycling, and temperature extremes are not of concern here, but a battery located inside an integrated circuit could be exposed to a constant 90°C temperature.

Because equipment requirements in this environment are usually satisfied by commercially available hardware, development funds are rarely

Solid State Microbatteries
Edited by J. R. Akridge and M. Balkanski
Plenum Press, New York, 1990

available for technologies which would only be applicable to this near commercial work place.

Shipboard-sheltered environments tend to be more benign than most battlefield situations because of the low vibration and moderate temperatures. Equipment located outside on the decks will see wider temperature ranges, and in the case of gun platform mounted units, high vibration levels. The executive document for shipboard equipment design is MIL-E-16400. It separates the temperature requirements into the following ranges:

Class 1	-	-54°C to +65°C
Class 2	-	-28°C to +65°C
Class 3	-	-40°C to +50°C
Class 4	-	0°C to +50°C

The Navy equipment generally is not space limited and may have cooling air or water available. As a result the temperature rise inside the equipment is not excessive, and in some cases, it may be nearly ambient. Temperature transitions tend to be slow, so temperature cycle stresses are not extreme.

Except for missiles, the harshest environment is in military aircraft. Space is limited, vibration is severe and temperature changes are wide, rapid and frequent.

The general specification for airborne electronic equipment is MIL-E-5400. It separates thermal environment into the following classes:

Class	Continuous Operating Temperature	Intermittent Operating Temperature
1	-54°C to +55°C	30 Min. at +71°C
1A	-54°C to +55°C	30 Min. at +71°C
1B	-40°C to +55°C	30 Min. at +70°C
2	-54°C to +71°C	30 Min. at +95°C
3	-54°C to +95°C	30 Min. at +125°C
4	-54°C to +125°C	30 Min. at +150°C

Temperature changes similar to these ranges may occur thousands of times during the lifetime of the equipment. This implies that the coefficients of expansion between a microbattery and the package in which it is mounted must be matched or mechanically isolated in order to preserve mechanical integrity.

The different levels of mechanical stress are specified by a set of linear 20 G vibration curves extending from 90 Hz to 2000 Hz and these forces are only applied externally to the equipment. The components comprising the equipment may experience levels higher than this if the unit is not sufficiently rigid or is improperly dampened. Shock can go as high as 30 G with a duration of 11 milliseconds.

Missile requirements are application specific and can be exceptionally severe. Mechanical shock value may be in the range of thousands of G's, and operating lifetimes are so short that repeated temperature cycling is not as important.

Equipment specifications will usually include mean time between failure (MTBF).

As an example, a typical airborne computer could specify a MTBF of 3000 hours. This means that the component parts would each have a predicted MTBF in excess of one million hours.

MTBF is not the same as service life. The service life of a piece of equipment is many times the MTBF of the same unit. Ideally, it could be said that the service life of all the components in a piece of equipment could be exactly the same as the equipments' service life, but the MTBF of each component part would have to exceed that of the equipment by many times.

The service life of airborne electronic equipment tends to be relatively short. Ten years is a typical time span before it is either replaced or updated. Shipboard equipment lasts much longer, however. As an example, the USQ-20B Univac computer was manufactured between 1963 and 1973. Recently they have all been returned to the manufacturer for refurbishing and returned to the fleet - possibly for another 20 years of service. A microbattery that would last for 20 or 40 years would have been ideal in this application.

The testing and specifications for military microcircuits is controlled by two documents - MIL-STD-883 and MIL-M-38510. MIL-STD-883 defines the test methods and procedures to be used in manufacturing and MIL-M-38510 contains the specifications for each device.

A sample of some of the integrated circuit tests called for in MIL-STD-883 are as follows:

1. BAKE - 1000 HOURS AT 125°C
2. TEMPERATURE CYCLE - -54°C to +125°C   10 CYCLES IN AIR
3. TEMPERATURE SHOCK - -55°C to +124°C   15 CYCLES IN LIQUID
4. BURN-IN - 125°C POWER APPLIED FOR 160 HOURS
5. CONSTANT ACCELERATION - 5000 TO 25000 G 3 AXIS
6. MECHANICAL SHOCK - 3000 G FOR .3 MILLISECONDS 3 AXIS
7. VIBRATION FATIGUE - 50 G, 60 Hz, 96 HOURS 3 AXIS
8. VIBRATION ELECTRICAL NOISE - 90 G, 20 to 20000 Hz

These levels are about average for military microcircuits. The severity of the tests depends upon the reliability and quality levels specified for the devices. For instance, "S" (SPACE) level parts will require more stringent testing and inspection than "B" level parts used in a shipboard application.

The basic MIL-M-38510 document defines the quality sampling plans, package outlines, manufacturing procedures, inspection criteria, personnel training, and many other requirements too numerous to mention here. In addition to this, there are 38510/XXXX documents which are the complete performance specifications for each approved part. These "Slash Sheets" vary in size from one or two pages to more than 150 pages depending on the complexity of the device being specified. Presently, there are approximately 600 slash sheets published that cover about 1100 devices.

Conformance to the two military specifications just described is not necessary for devices to be used in military equipment. If there is no device specified by MIL-M-38510 that will suffice, a nonstandard part waiver can be obtained which will allow the unapproved part to be used for a specified length of time. A good example of this would be a CMOS Memory I.C. with an integral 20 year microbattery. No such device is now specified in a 38510 slash sheet, but it would readily be incorporated into equipment because of its obvious advantages.

A number of U.S. Government Agencies have been commissioned to assist in incorporating new technologies into military hardware. Below is a partial list.

DARPA   - Defense Advanced Research Projects Agency. Established to accelerate early development of promising new technologies - especially radiation hard approaches.

DNA   - Defense Nuclear Agency. Almost anything to be used in a nuclear environment.

RADC   - Rome Air Development Center. Interested in signal processing hardware.

CECOM   - Communications and Electronics Command. The U.S. Army agency for new electronic hardware development.

NAVAIR - Naval Air Weapons Center. U.S. Navy electronic hardware development center.

Another method for getting new technology accepted is to approach some of the prime government contractors with a proposal to incorporate the technology in their equipment. For this to be successful, however, the technology must be developed sufficiently to make the contractor very confident of its merits and reliability.

Assuming that a new and promising technology is incorporated into the design of a piece of military hardware, several years may pass before any production volume is needed. The equipment qualification tests almost always uncover mechanical, thermal or performance weaknesses which necessitate redesign and new fabrication of various equipment elements. Three to six year delays are common. The military customer wants to be absolutely certain that there are no serious flaws.

In summary, it can be said that approaching the military marketplace with a promising new technology, like microbatteries, can have definite financial advantages during the early development stages; but the military's fear of failure has generated numerous specifications and tests that tend to delay higher production volumes.

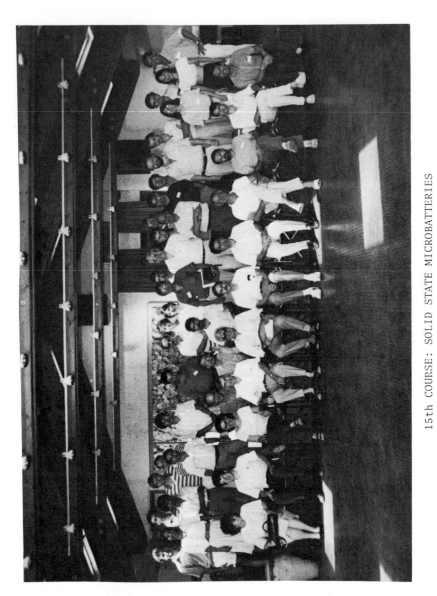

15th COURSE: SOLID STATE MICROBATTERIES

JULY 3-15, 1988 ERICE-TRAPANI-SICILY, ITALY

Ms. T.M. ABRANTES
TECHNICAL UNIVERSITY OF LISBON
DEPARTMENTO ENGENHARIA QUIMICA
LABORATORIO DE QUIMICA-FISICA
AV. ROVISCO PAIS
1096 LISBOA CEDEX
PORTUGAL

DR. J.R. AKRIDGE
EVEREADY BATTERY CO., INC.
P.O. BOX 45035
WESTLAKE, OHIO   44145
USA

MR. P. ALDEBERT
GEN-GRENOBLE
DRF/SPh/PHYSICO CHIMIE MOLECULAIRE
85X
38401 GRENOBLE CEDEX
FRANCE

DR. M. BALKANSKI
LABORATOIRE DE PHYSIQUE DES SOLIDES
DE L'UNIVERSITE PIERRE ET MARIE CURIE
4 PLACE JUSSIEU - TOUR 13 - 2^e ETAGE
75252 PARIS CEDEX 05
FRANCE

DR. J.M.B.F. DINIZ
ESCOLA SUPERIOR DE TECNOLOGIA DE TOMAR
2300 TOMAR, PORTUGAL

MR. R. DORMOY
LABORATOIRE DE CHIMIE DU SOLID DU CNRS
351 COURS DE LA LIBERATION,
33405 TALENCE CEDEX
FRANCE

MR. P. DZWONKOWSKI
LABORATOIRE DE PHYSIQUE DES SOLIDES
DE L'UNIVERSITE PIERRE ET MARIE CURIE
4 PLACE JUSSIEU - TOUR 13 - 2^e ETAGE
75252 PARIS CEDEX 05
FRANCE

MR. M. EDDRIEF
LABORATOIRE DE PHYSIQUE DES SOLIDES
DE L'UNIVERSITE PIERRE ET MARIE CURIE
4 PLACE JUSSIEU - TOUR 13 - 2^e ETAGE
75252 PARIS CEDEX 05
FRANCE

DR. I. ESME
MARMARA UNIVERSITESI
ATATURK EGITIMI FAKULTESI
FIZIK BOLUMU 81040
KADIKOY-ISTANBUL
TURKEY

DR. C.A.D.S. FARIA
FABRICA TUDOR
CASTANHEIRA DO RIBATEJO
2600 VILA FRANCA DE XIRA,
PORTUGAL

MR. J. FOREMAN
DEPARTMENT OF MATERIAL SCIENCE
UNIVERSITY OF PENNSYLVANIA
3231 WALNUT STREET
PHILADELPHIA, PENNSYLVANIA 19104
USA

DR. J. GARBARCZYK
WARSAW UNIVERSITY OF TECHNOLOGY
INSTITUTE OF PHYSICS
SOLID STATE IONICS LABORATORY
CHODKIEWICZA 8, WARSAW
POLAND

DR. J.B. GOODENOUGH
CENTER FOR MATERIALS SCIENCE AND ENGINEERING
UNIVERSITY OF TEXAS AT AUSTIN
ETC 5.160
AUSTIN, TEXAS 78712
USA

MR. E. HARO
LABORATOIRE DE PHYSIQUE DES SOLIDES
DE L'UNIVERSITE PIERRE ET MARIE CURIE
4 PLACE JUSSIEU - TOUR 13 - 2^e ETAGE
75252 PARIS CEDEX 05
FRANCE

DR. C. JULIEN
LABORATOIRE DE PHYSIQUE DES SOLIDES
DE L'UNIVERSITE PIERRE ET MARIE CURIE
4 PLACE JUSSIEU - TOUR 13 - 2^e ETAGE
75252 PARIS CEDEX 05
FRANCE

DR. P.M. JULIAN
ENERGY RESEARCH LABROATORY
NIELS BOHR ALLE 25
DK-5230 ODENSE M
DENMARK

DR. R. LATHAM
LEICESTER POLYTECHNIC
P.O. BOX 143
LEICESTER LE1 9BH
UNITED KINGDOM

MISS N. LEBRUN
LABORATOIRE D'IONIQUE ET D'ELECTROCHIMIE DU SOLIDE
E.N.S.E.E.G.
38402 SAINT MARTIN D'HERES CEDEX
FRANCE

DR. A. LEVASSEUR
LABORATOIRE DE CHIMIE DU SOLIDE DU CNRS
351 COURS DE LA LIBERATION
33405 TALENCE CEDEX
FRANCE

MR. M. LEVY
LABORATOIRE D'IONIQUE ET D'ELECTROCHIMIE DU SOLIDE
E.N.S.E.E.G.
B.P. 75
38402 SAINT MARTIN D'HERES CEDEX
FRANCE

DR. C.C. LIU
DIRECTOR ELECTRONICS DESIGN CENTER
BINGHAM BUILDING
CASE WESTERN RESERVE UNIVERSITY
CLEVELAND, OHIO 44106
USA

MRS. M.A.G. MARTINS
RUA GENERAL HUMBERTO DELGADO NO. 15
1 ESQ. - FORTE DE CASA
2625 - POVOA DE STA. IRIA
PORTUGAL

MISS A. MENNE
MAX PLANCK INSTITUTE FUR FESTKORPERFORSCHUNG,
HEISENBERGSTR. 1
POSTFACH 80 06 65
D 7000 STUTTGART 80
WEST GERMANY

MR. G. MEUNIER
LABORATOIRE DE CHIMIE DU SOLIDE DU CNRS
UNIVERSITE DE BORDEAUX I
351 COURS DE LA LIBERATION
33405 TALENCE CEDEX
FRANCE

MRS. A. MEUNIER
 LABORATOIRE DE CHIMIE DU SOLIDE DU CNRS
UNIVERSITE DE BORDEAUX I
351 COURS DE LA LIBERATION
33405 TALENCE CEDEX
FRANCE

MR. D.B. MORSE
BOX 29-1
DEPARTMENT OF CHEMISTRY
UNIVERSITY OF ILLINOIS
505 S. MATHEWS AVENUE
URBANA, ILLINOIS 61801

DR. R.J. NEAT
MATERIALS DEVELOPMENT DIVISION
B552 HARWELL LABORATORY
UNITED KINGDOM ATOMIC ENERGY AUTHORITY
OXFORDSHIRE, OX11 ORA
UNITED KINGDOM

MISS S. ODABAS
PHYSICS DEPARTMENT
MIDDLE EAST TECHNICAL UNIVERSITY
06531 ANKARA
TURKEY

DR. S. OH
MAX PLANCK INSTITUTE FUR FESTKORPERFORSCHUNG,
HEISENBERGSTR. 1
POSTFACH 80 06 65
D 7000 STUTTGART 80
WEST GERMANY

MR. M. OUESLATI
LABORATOIRE DE PHYSIQUE DES SOLIDES
DE L'UNIVERSITE PIERRE ET MARIE CURIE
4 PLACE JUSSIEU - TOUR 13 - 2^e ETAGE
75252 PARIS CEDEX 05
FRANCE

DR. M. RIBES
LABORATOIRE DE PHYSICOCHIMIE DES MATERIAUX SOLIDES
U.A. 407
U.S.T.L.
PLACE E. BATAILLON
34060 MONTPELLIER CEDEX
FRANCE

MR. I. SAIKH
LABORATOIRE DE PHYSIQUE DES SOLIDES
DE L'UNIVERSITE PIERRE ET MARIE CURIE
4 PLACE JUSSIEU - TOUR 13 - 2^e ETAGE
75252 PARIS CEDEX 05
FRANCE

MR. M.I. SAMARAS
LABORATOIRE DE PHYSIQUE DES SOLIDES
DE L'UNIVERSITE PIERRE ET MARIE CURIE
4 PLACE JUSSIEU - TOUR 13 - 2^e ETAGE
75252 PARIS CEDEX 05
FRANCE

DR. J. SANDAHL
DEPARTMENT OF PHYSICS
CHALMERS UNIVERSITY OF TECHNOLOGY
UNIVERSITY OF GOTHENBURG
S 412 96
SWEDEN

DR. J. SARRADIN
LABORATOIRE DE PHYSICOCHIMIE DES MATERIAUX SOLIDES
U.A. 407
U.S.T.L.
PLACE E. BATAILLON
34060 MONTPELLIER CEDEX
FRANCE

MRS. W.S. SCHLINDWEIN
LEICESTER POLYTECHNIC
P.O. BOX 143
LEICESTER LE1 9BH
UNITED KINGDOM

DR. B. SCROSATI
DEPARTMENT OF CHEMISTRY
UNIVERSITY OF ROME
'LA SAPIENZA',
PIAZZALE A. MORO 5
00185 ROME, ITALY

MR. M. SETTER
CENTER FOR SOLID STATE SCIENCE
ARIZONA STATE UNIVERSITY
TEMPE, ARIZONA 85287-1704
USA

DR. J.L. SOUQUET
LABORATOIRE D'IONIQUE ET D'ELECTROCHIMIE DU SOLIDE
E.N.S.E.E.G.
38402 SAINT MARTIN D'HERES CEDEX
FRANCE

MISS B. STASIAK
LEHISTUHL FUR PHYSIKALISCHE CHEMIE I
UNIVERSITY OF DORTMUND
POSTF. 500 500
4600 DORTMUND 50
WEST GERMANY

MR. M.C. TAPLAMACIOGLU
GAZI UNIVERSITESI
MUHENDISLIK MIMARLIK FAKULTESI
ELEKTRIK ve ELEKTRONIK MUHENDISLIGI BOLUMU
ARAZTIRMA GOREVLISI (RESEARCH ASSISTANT)
MALTEPE-ANKARA
TURKEY

DR. F. TRUMBORE
'AT&T BELL LABORATORIES'
30 GLEN OAKS AVENUE
SUMMIT, NEW JERSEY   07901
USA

DR. M. TSAKIRI
LABORATOIRE DE PHYSIQUE DES SOLIDES
DE L'UNIVERSITE PIERRE ET MARIE CURIE
4 PLACE JUSSIEU - TOUR 13 - 2^e ETAGE
75252 PARIS CEDEX 05
FRANCE

DR. H.L. TULLER
DEPARTMENT OF MATERIALS SCIENCE AND ENGINEERING
ROOM 13-3126
MIT
CAMBRIDGE, MA 02139
USA

DR. W. WEPPNER
MAX PLANCK INSTITUTE FUR FESTKORPERFORSCHUNG,
HEISENBERGSTR. 1
POSTFACH 80 06 65
D 7000 STUTTGART 80
WEST GERMANY

DR. P. WESOLOWSKI
WARSAW UNIVERSITY OF TECHNOLOGY
INSTITUTE OF PHYSICS
SOLID STATE IONICS LABORATORY
CHODKIEWICZA 8
WARSAW, POLAND

MR. A. WEST
DEPARTMENT OF CHEMICAL ENGINEERING
UNIVERSITY OF CALIFORNIA
BERKELEY, CALIFORNIA 94720
USA

DR. C. WIBBELMANN
UNIVERSITY OF ABERDEEN
DEPARTMENT OF CHEMISTRY
OLD ABERDEEN AB9 2UE
SCOTLAND
UNITED KINGDOM

DR. W. WIECZOREK
INSTITUTE OF INORGANIC TECHNOLOGY
WARSAW UNIVERSITY OF TECHNOLOGY
UL. NOAKOWSKIEGO 3, 00-664
WARSAW, POLAND

DR. C.N. WIJAYASEKERA
DEPARTMENT OF PHYSICS,5
 CHALMERS
UNIVERSITY OF TECHNOLOGY,
S-412 96 GOTEBORG
SWEDEN

DR. A. VAN ZYL
MAX PLANCK INSTITUTE FUR FESTKORPERFORSCHUNG,
HEISENBERGSTR. 1
POSTFACH 80 06 65
D 7000 STUTTGART 80
WEST GERMANY